MW00451328

Fabless Semiconductor Implementation

About the Author

Rakesh Kumar, Ph.D., Fellow IEEE, is a semiconductor industry veteran with extensive experience in managing IC implementation in both fabless and fab environments. He is the Founding Partner and President of TCX, Inc., a company that has assisted many fabless companies (http://www.tcxinc.com). He has held technical and executive positions at Cadence Design Services, Unisys, and Motorola. Dr. Kumar serves on the executive committee of the IEEE Solid-State Circuits Society and the Board of Governors of the IEEE Technology Management Council.

Fabless Semiconductor Implementation

Rakesh Kumar, Ph.D., Fellow IEEE

New York Chicago San Francisco
Lisbon London Madrid Mexico City
Milan New Delhi San Juan
Seoul Singapore Sydney Toronto

The McGraw-Hill Companies

3 4 5 6 7 8 9 0 EPAC/EPAC 1 9 8 7 6 5 4 3 2 1

ISBN 978-0-07-150266-5
MHID 0-07-150266-1

Sponsoring Editor: Stephen S. Chapman
Production Supervisor: Pamela A. Pelton
Editing Supervisor: Stephen M. Smith
Project Manager: Jeremy Toynbee, Keyword Group Ltd.
Copy Editor: Nic Walker
Proofreader: John Bremer
Indexer: Martin Hargreaves
Art Director, Cover: Jeff Weeks
Composition: Keyword Group Ltd.

Printed and bound by Epac Technologies.

Dedicated
to my international family
for their encouragement and support throughout my eduction and career
my wife Julie and sons Matthew and Nicholas,
my parents Shakuntala and (Late) Mahavir Prasad,
and my siblings Ajaya (Sadhana) Kumar, Sushma (Dr. Birendra) Prasada
and Prabhat (and Ranjana) Kumar.

Contents

Preface

Since the invention of the first transistor in 1947 and the first IC in 1958, the semiconductor industry has grown to be worth nearly $300B annually. Fabrication of semiconductor ICs using silicon wafers is a cornerstone for the trillion dollar electronics industry. Electronic products using the latest semiconductor technology are omnipresent. Products that were considered "hi-tech" and used only by the "techies" 20 years ago have given way to widespread use of sophisticated products such as laptop computers, high definition TVs, gaming devices, and cellular telephones worldwide. Some of the ICs used in these systems have very powerful capabilities and are developed by fabless IC and system companies.

The development and sourcing of ICs has seen much change in the last 20 years. Before that, companies that shipped semiconductors had to have their own wafer fabrication facility ("wafer fab"). "Real men must have fabs" is a famous saying attributed to Jerry Sanders, an industry pioneer and the founder of Advanced Micro Devices (AMD).

Early investment in the semiconductor infrastructure was made by system companies and semiconductor companies in the U.S., Europe, and Japan. These vertically integrated companies invested in the infrastructure required to design, develop, and fabricate the silicon and also in the packaging, assembly and testing required to manufacture ICs. Semiconductor companies became known as Integrated Device Manufacturers (IDMs).

Outsourcing of assembly processing to the Far East started in the 1970s. The first independent silicon foundry was founded in 1987. The availability of independent sources of silicon wafer processing offered fertile ground for creative entrepreneurs who launched "fab-less" IC companies in the early 1990s. A symbiotic relationship developed between the foundries and the fabless companies as they pushed each other to higher levels of competency. Today there is a phenomenal supply chain and ecosystem available for the design, development, and manufacture of semiconductors. There are over a thousand independent fabless IC companies using a dis-aggregated semiconductor supply chain for the design, development and manufacture of ICs. Fabless companies have many suppliers to choose from and have the opportunity to select "best-in-class" suppliers best suited to meet their technical requirements and business goals. The recent growth rate of the fabless segment has been higher than that of the whole semiconductor industry. In 2006, nearly 20% of annual, worldwide semiconductor revenue came from the fabless segment.

Fabless companies leverage the investment of capital equipment, facilities and R&D (Research and Development) made by supply-chain partners. They focus on their own core competencies, such as chip/system architecture, hardware and software development, chip design, verification, and testing. The technical capabilities, quality and cost structure provided by the independent supply chain enables fabless companies to develop competitive IC products. While many fabless companies thrive, there have also been many that have not been successful. This book opens with five reasons why many fabless start-up companies have failed. This book provides many pointers for positioning a fabless company for success. As a background, Chap. 1 also provides the reader with a historical perspective of the semiconductor industry and the emerging fabless industry.

Chapter 2 provides a perspective of the new value chain, the necessity of addressing customer needs, and gives examples of typical implementation schedules of recent products.

Chapter 3 discusses the lifecycle of a fabless company including strategic decisions about product positioning, specification, the business plan, funding process, partitioning of the development cycle, and possible exit strategies.

In Chap. 4, the focus shifts to possible implementation approaches, sourcing models, and design infrastructure.

Chapter 5 outlines technology selection options in the areas of process, design, and packaging. The material in Chap. 6 details IC implementation by the typical emerging fabless company. Chapter 7 is devoted to a discussion of unit cost and development cost, while Chap. 8 covers quality and reliability considerations important for the company. Program management and the importance of managing relationships with suppliers is emphasized in Chap. 9.

Chapter 10 discusses the threats to the industry and its responses to them. The continued escalation of the cost of semiconductor fabrication facilities ("wafer fabs"), semiconductor process development, and the cost of designing ICs has cast a shadow over the future of the semiconductor industry. Many IDM companies are reducing their capital investments in wafer fabs and are being classified as either "fab-lite" or "asset-lite". For the past 20 years there have been continual doubts about the viability of fabricating devices with minimum dimensions in the sub-one-micron (early 1990s), in the deep sub micron (late 1990s), and, more recently, in the nanometer dimension ranges. However, technologists in the industry have overcome many hurdles over the years and new technologies are continuing to be introduced at a predictable, yet feverish, pace. This resilient industry has defied predictions of slowing technology introductions consistently. Some believed that business considerations would slow down the introduction of new process technologies. Even this has not happened, although there appears to be a stratification of companies that can use leading edge technologies. There is also a trend towards a virtual "re-integration" of the distributed supply chain, especially in the context of leading-edge ICs.

Although this book assumes a fabless start-up, most of the information can be applied to all fabless companies. Large fabless companies, for example, can follow similar principles in launching and executing the development of a new IC product. The book has been written to provide an executive level and broad overview for managers and technical professionals. As will become obvious to the reader, this is indeed a complex technology with a breadth of issues and considerations. It is my intention to provide the reader with a high level view of the important issues and considerations involved in the launching and managing of a successful fabless IC company. It is also my intention to provide cross-disciplinary exposure to specialists and practitioners who may come from other areas of expertise. It is hoped that the content will provide a resource of interactive issues, a background in the various disciplines involved, and references for further research. The advanced technology information included will serve as a reference for companies not yet designing at the leading edge. An extensive Bibliography is included and more detailed information is provided in the Appendices. A Glossary of terms is included, together with a schedule of Acronyms in alphabetical order to assist the reader in following the text.

For the sake of completeness, I have assumed that the reader has limited experience in operating a fabless IC company. Of course, the first piece of advice, besides reading this book, is to surround yourself with some experienced people!

Although many have succeeded in this field, fabless semiconductor implementation is not for the faint-hearted. There are many technical, business, and supplier management related issues that must be dealt with. In the execution phase, companies usually find that the "devil is in the details." While the fabless model offers many choices and allows for the selection of best-in-class supply chain partners, there are also risks and "gaps" that must be bridged. In the past, vertically integrated companies were the "glue" that bridged such gaps internally. Now these gaps must be resolved across company boundaries. While some of the gaps have been reduced as individual suppliers have expanded their individual roles to provide more complete solutions, they are still a problem. Sometimes the gaps become more obvious if the particular IC is pushing limits of mainstream capability at the time.

While use of the fabless model generally results in lower unit and development costs, executives must realize that some internal infrastructure investments are still required at the fabless IC company. This is contrary to common perception. Examples of such investments are discussed in Chaps 6 and 7.

Much of the material in this book applies to semiconductor implementation in general, whether you have a fab or not. I emphasize issues as they apply to fabless semiconductor implementation.

And yes, unfortunately, I am not able to solve the *LUCK* factor! It is indeed true that in spite of the start-up team doing all the right things, luck does play a factor in a company's success. Being aware and flexible can help, but ...

Why This Book?

While the fabless semiconductor industry has seen very rapid growth, only a small percentage of the many companies launched every year are successful. Success is defined here as becoming profitable, getting acquired, or going public. Five common reasons why the majority of fabless start-ups are not successful are:

1. No customer engagement until it is too late.
2. Not understanding customer expectations for a complete and "turn-key" solution. Customers now expect solutions that include a reference design, test suites, software, and application know how. Having an IC with an outstanding feature is no longer sufficient.
3. Overly aggressive product specification, technology selection, and cost targets cause disappointments in schedule failings, and product and development cost increases. This results in unhappy customers and investors.
4. The "kitchen sink syndrome"—too many re-starts and do-overs in the design development phase, due to requests for new features. This leads to delayed product availability and can cause a missed design-in window at the customer.
5. Lack of experience managing the distributed supply chain causes a myriad of problems. To name a few:
   - poor due diligence in partner selection, commitment and engagement;
   - manufacturing capacity unavailable when needed;
   - poor due diligence in selecting the IP imlellectual property block—the block had never been implemented and validated in the process technology and the foundry of choice;
   - overly optimistic build ahead of inventory has been known to cause serious financial burden.

The primary motivation of this book has been to create a resource to help avoid some of the pitfalls and gaps in the development of complex ICs. While the availability of a distributed supply chain has made it possible for many entrepreneurs to develop ICs and bring them to market, there is no one reference covering the breadth of issues involved. This book will provide a basis for making knowledgeable decisions and trade-offs and will get you started on the path to success in the fabless market and avoid the pitfalls common to most failures. It will also help the reader understand the breadth of issues affecting the successful formation and implementation of a fabless IC company. Leads are provided for the reader to follow up. The reader should note, however, that execution details require him or her to build on the foundation provided here. In many areas, the information here provides the appetizer with the main course yet to come.

This book is a sharing of the wealth of experience and learning I have enjoyed during my 35 years in the semiconductor industry. It has been supplemented with publicly available information and with expert opinions and validations in many areas. It is my hope that readers will enjoy the book and be successful in their endeavors as a result.

Key Points

- The growing fabless segment of the semiconductor industry implements new ideas into ICs without large capital investments in fabrication facilities via the use of best in class suppliers in a distributed supply chain.

- Successful implementation of ICs using the fabless model requires product differentiation, proper technology, methodology and partner selection, and the bridging of gaps that can result from the use of a distributed supply chain.

- There are many reasons why the success rate of startups is relatively low. This book provides exposure to the breadth of issues that must be addresssed in order to be successful.

Acknowledgments

I would like to acknowledge the contributions of many mentors, customers, supervisors, colleagues, subordinates, and teachers throughout my career. Learning from them has been crucial to my successes.

There are many colleagues who reviewed and made invaluable suggestions in the preparation of this book: Behrooz Abdi, Bill Adamec, Bill Bidermann, Professor Mark Bocho, Henry Chang, Jim Clifford, Ron Collett, Gilbert Declerck, H. K. Desai, Jim Fiebiger, Tom Gregorich, Michael Han, Jack Harding, Bryan Harding, Russ Harris, Brian Henderson, Professor Dave Hodges, Merrill Hunt, Charlie Kahle, Ken Kundert, John Luke, Kevin Meyer, Matt Nowak, Rich Palys, Glen Possley, Wally Rhines, Rich Rice, Naveed Sherwani, Professor Charles Sodini, Kurt Stoll, Fifin Sweeney, and Gyan Tiwary.

Permission from IEEE to use the following figures and related information is acknowledged: Figs 2.10, 2.11, 4.5, 4.7, 5.1, 5.3, 5.4, 5.8, and 5.11 and Tables 2.2 and 9.1.

Reference material derived from the FSA (Fabless Semiconductor Association) is acknowledged. As of December 2007, this organization is known as the GSA (Global Semiconductor Alliance).

Fabless Semiconductor Implementation

CHAPTER 1
Industry Perspectives

1.1 Semiconductor Industry

1.1.1 History and Background

The semiconductor industry was born over 50 years ago. A time line of major milestones since the invention of the transistor 60 years ago is shown in Table 1.1. Pioneering research was performed at Bell Laboratories, Shockley Laboratories, Fairchild, Texas Instruments, Motorola, IBM, and Sony. The first monolithic IC was demonstrated in 1958 at Fairchild Semiconductor and Texas Instruments. More details of the early semiconductor history can be found in the articles in references [1.1–1.9].

1.1.1.1 Chip Complexity

The 4004 microprocessor, announced by Intel in 1971, was the first highly integrated semiconductor IC, incorporating 2300 transistors. Since then the complexity of ICs has increased by over five orders of magnitude at the rate of roughly doubling every 12 months initially, then every 18 months (1975 to mid 1990s), and now every 24 months. Prediction of such a complexity increase was made by Gordon Moore, one of Intel's founders, and has become known as "Moore's Law." His earliest prediction was that the number of components per IC was going to increase by 2^{16} in 16 years, from 1959 to 1975—a twofold increase every year [1.10]. Refinements to Moore's Law have been made since then [1.11–1.13]. In the 1970s and the early 1980s, a new generation of technology was introduced every three years [1.13]. Including the effect of increasing die sizes, the transistor density increased approximately fourfold every 3 years, or twice every 18 months, in that time frame. Since the mid-1990s, a new technology generation has been introduced every two years and the transistor density has increased approximately twice every two years.

Year	Milestone
1947	Transistor invented at Bell labs by Bardeen and Brattain; William Shockley was the research leader
1948	First transistor unveiled by Bell labs
1954	First commercial transistor announced by Texas Instruments
1955	Shockley Semiconductor founded. Motorola introduced first commercial power transistor. Beginning of Silicon Valley
1957	Fairchild Semiconductor founded by Robert Noyce and 7 others that left Shockley labs
1958	First monolithic IC invented at Fairchild by Robert Noyce and Jean Hoerni, and at TI by Jack Kilby
1967	First calculator announced by TI
1968	Intel founded by Gordon Moore and Robert Noyce
1969	AMD founded by Jerry Sanders
1971	First 4-bit microprocessor (4004) announced by Intel

Fabless Semiconductor Association (GSA) founded with 40 founding members in 1994

Table 1.1 Major Milestones in the Early Semiconductor Industry

The increase in the maximum number of transistors per leading-edge microprocessor chip over the years is shown in Fig. 1.1 [1.12–15]. Overall, the CAGR (Compound Annual Growth Rate) of chip complexity has been nearly 40% every year between 1971 and 2006. It is interesting to note that in the early years, leading edge process technology was applied to and driven by the fabrication of high density memories at large semiconductor companies in the U.S., Europe, Japan, and Korea. The standalone memory market is still served by companies with internal fabrication facilities optimized for their applications, and is not addressed in this book.

Intel's announcements in 2007 and 2008 of a microprocessor product with nearly 300 million and 2 billion transistors per piece of silicon ("die" or "chip") indicates that the industry has come a long way [1.25, 1.26]. Since the introduction of the 4004, the complexity has increased by over five orders of magnitude in transistor count.

The industry coined "scales of integration" terms that were related to the number of transistors or "gates" per chip. A 2-input NAND gate with four transistors became the de facto standard to measure logic complexity per chip. While there are no standards defining the terms, the following list illustrates approximate gate and

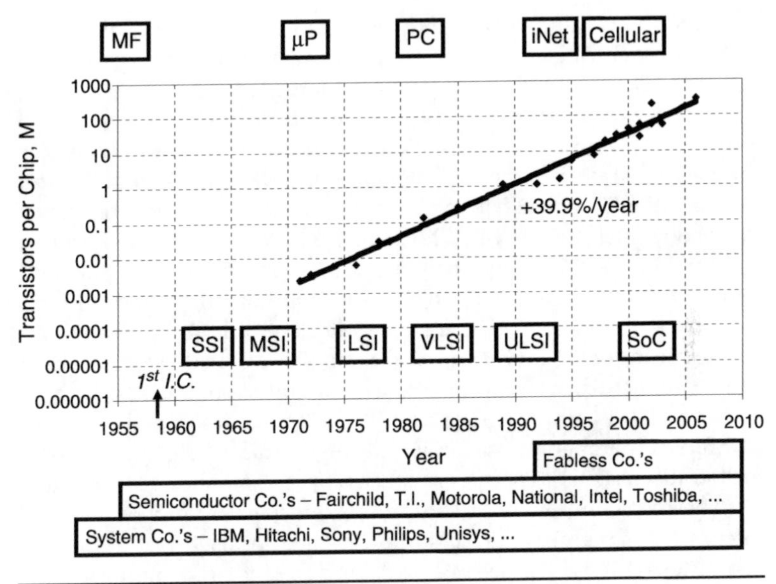

FIGURE 1.1 Transistor complexity trend for leading microprocessors.

transistor complexities associated with each term. These terms are shown in Fig. 1.1, along with the approximate time frames when the various classes of ICs were in use. While SSI, MSI, LSI, and VLSI were commonly used, the term ULSI never became very popular. As discussed later, the SoC term has become commonly used.

SSI (Small Scale Integration)	< 10 gates (or 40 transistors)
MSI (Medium Scale Integration)	< 100 gates (or 400 transistors)
LSI (Large Scale Integration)	< 10 K gates (or 40 K transistors)
VLSI (Very Large Scale Integration)	< 100 K gates (or 400 K transistors)
ULSI (Ultra Scale Integration)	> 100 K gates (or 400 K transistors)
SoC (System on a Chip)	≥ 10 M gates (or 40 M transistors)

As indicated in Fig. 1.1, the major system product driver in the early years was the Mainframe ("MF") computer developed at leading manufacturers such as IBM, Burroughs, and Sperry in the U.S. and Hitachi and NEC in Japan. Consumer application of early ICs was in products such as televisions and hand-held calculators.

The computer system companies invested heavily in development and manufacturing capabilities to gain strategic advantage for their products. The use of ICs with increased integration levels resulted in significant improvement in the cost, performance, and size of the computers. Their investments included the setting up of semiconductor wafer fabrication facilities ("wafer fabs"), packaging, assembly, and test facilities in-house. In the meantime, merchant semiconductor companies such as Fairchild, Texas Instruments, National, Motorola, and Intel in the U.S., Sony, Toshiba, Hitachi, and others in Japan, and Philips in Europe focused on the development of semiconductor memory and logic products for computer and military applications. The semiconductor companies also set up "Integrated" Development and Manufacturing environments for integrated circuits and became known as IDMs. Approximate time lines for system companies, IDMs, and fabless companies are shown at the bottom of Fig. 1.1. There is further discussion of the integrated environment later in this chapter.

The chip complexity "treadmill" was driven by higher levels of integration in memory and microprocessor chips used in the personal computer (PC), which became a major driver for IC volumes in the 1980s. While the first PCs were introduced in the late 1970s, they became a commercial success in the early 1980s. Internet related products and cellular phones have become the high volume product drivers for semiconductor products since the mid-1990s.

Performance and capability were the primary drivers for IC products in the 1980s and 1990s. Since the late 1990s, however, low cost, low power usage, and schedule are the primary drivers. These factors are increasingly important in today's consumer driven marketplace.

1.1.1.2 Scaling

In 1974 Robert Dennard and his IBM colleagues published a paper [1.16] that articulated a theory for scaling MOSFET's (Metal Oxide Semiconductor Field Effect Transistor). Dennard postulated that in order to design a new transistor with smaller dimensions, one must properly transform three variables—dimension (L), voltage (V), and doping in the semiconductor. Table 1.2 indicates the ratios proposed by Dennard for a few of the key variables. Here the scaling factor λ is assumed to be greater than 1 and is typically 1.4. There is further discussion of scaling and the resulting cost advantages in Chaps 5 and 7.

1.1.1.3 Minimum Feature Size

Moore's complexity trend prediction and Dennard's scaling theory formed the basis of, and a driven for, a continued reduction of the minimum dimension printed on a silicon wafer. The historical trend of the minimum dimension ("feature size") for leading microprocessors is shown in Fig. 1.2. This trend shows an average reduction of approximately 12% per year. Based on the scale factor $\lambda = 1.4$, the

Scale factor	λ
New device dimension, T_{ox}, L, W	$1/\lambda$
Voltage, V	$1/\lambda$
Doping concentration, N	λ
Circuit delay	$1/\lambda$

TABLE 1.2 Basic Elements of Dennard's Scaling Theory

industry has introduced a new process generation or "node" with a minimum feature size approximately 70% (= $1/\lambda$) of the previous node every two to three years. While the minimum feature size used on the earliest ICs was around 10 micrometers (μm), the latest process technology node in production in 2008 has minimum dimensions of 45 nanometers (nm). This is a ratio of roughly 220:1. Table 1.3 is a listing of the industry process technology nodes used during the last 20 years in both nanometers (nm) and micrometers (μm). Twelve technology nodes have been introduced in the last 22 years, an average of approximately one every other year. It is also interesting to note that semiconductor production still exists on the oldest of these nodes! Data shown in Chap. 5 indicates that the 0.5 μm or larger process nodes contributed approximately 10% of foundry revenues in 2006.

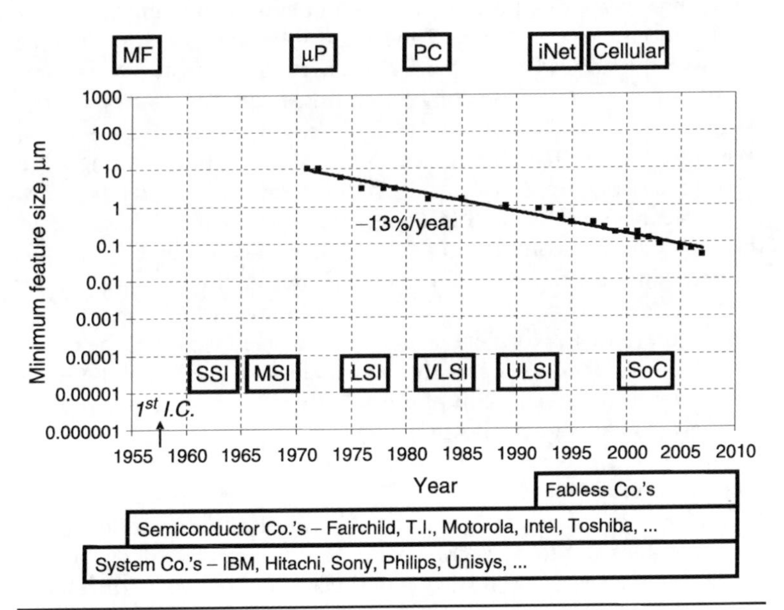

FIGURE 1.2 Minimum feature size trend for microprocessors.

Technology node (μm)	Technology node (nm)
1.5	1500
1	1000
0.8	800
0.6	600
0.5	500
0.35	350
0.25	250
0.18	**180**
0.13	**130**
0.09	**90**
0.065	**65**
0.045	**45**
0.032	**32**

TABLE 1.3 Process Technology Nodes in Nanometers as Well as Micrometers. Bold Lettering in the Grey Regions Indicates Commonly Used Terminology

1.1.1.4 Cost per Transistor

While there is a more thorough discussion of the economic aspects of scaling and unit cost in Chaps 5 and 7, it is interesting to point out that one consequence of the six orders of magnitude increase in packing density on a single silicon chip is that transistors are relatively inexpensive now. This is shown in Fig. 1.3, based on calculations using 2006 information. The average cost of a transistor on a leading edge microprocessor chip is now around 6 micro-cents! This matches data presented recently by Intel [1.17]. Also shown in Fig. 1.3 is a trend that the cost of silicon per square millimeter has continued to go up because of more sophisticated wafer fabrication equipment, processing, and facilities. As is discussed in Chap. 5, semiconductor economics focuses on reducing the cost per function (CPF) in spite of the rising cost of silicon. This concept is fundamental to the success of the semiconductor industry especially as the industry solves manufacturability and design challenges of leading edge ICs.

1.1.1.5 Performance

In addition to density and cost, the chip and transistor performance are important parameters for users. The circuit time delay for an inverter and a basic NAND gate have been used to track performance improvement from one process node to the next. Dennard's scaling

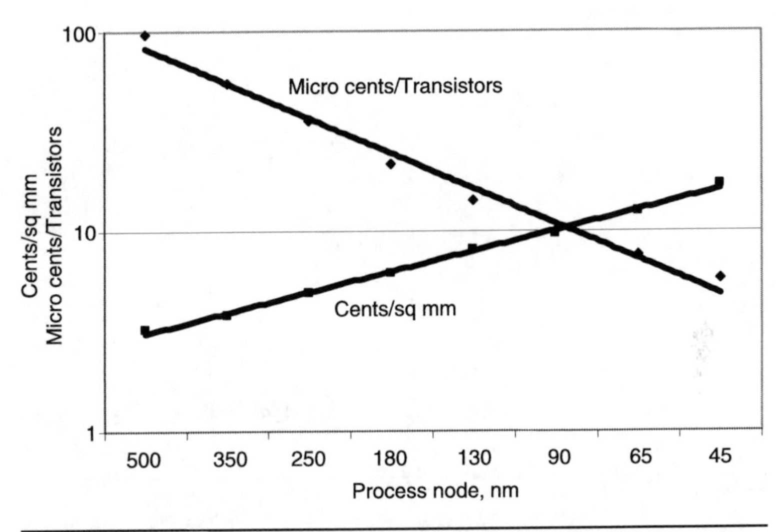

FIGURE 1.3 Transistor cost continues to decrease while the cost per square millimeter of silicon goes up from one process technology node to the next.

theory would predict circuit delay improvement of 30% per node, if $\lambda = 1.4$. In actual practice, the improvement in gate delay has only been 15–30%. One key reason for this is that the supply voltage has not scaled down in accordance with Dennard's theory. The industry stayed with 5 volts for a long time. The transitions to 3.3 volts, and then to 2.5 volts, 1.5 volts, 1.2 volts, 1.0 volt, and most recently 0.9 volts have been difficult and slow in coming. Discussion in Chap. 5 identifies other consequences to device characteristics as a result. Another way to study the improvement in IC performance is the clock frequency trend of leading edge microprocessors, as shown in Fig. 1.4. [1.15]. This data shows an annual growth rate of 38%. However, note that not all of this performance improvement comes from technology scaling. Circuit logic and physical, design, and architectural enhancements play key roles in this robust increase in performance.

1.1.1.6 Integrated Development Environment

As mentioned earlier, system companies and semiconductor companies (IDMs) made significant investments in the design, development, and manufacturing of ICs. A simplified, pictorial representation of the "vertically integrated" environment at these companies is shown in Fig. 1.5.

The entire set of activities required to develop and manufacture ICs was internal to the company. These included chip architecture, specification, design, the process technology research and development, fabrication, packaging, assembly, and testing. System companies had core competencies at the system level in hardware and

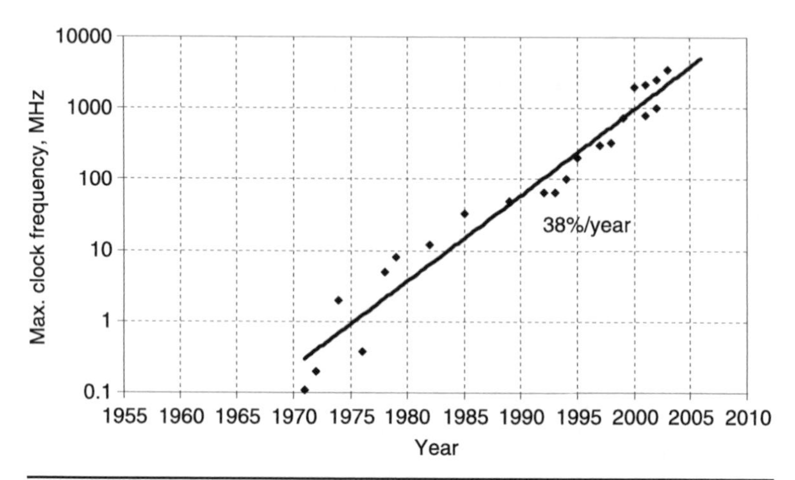

FIGURE 1.4 Clock performance improvement for leading edge microprocessors shows a 38% annual growth rate. (*Courtesy of IC Knowledge [1.15].*)

software, as they had been developing computers using electromechanical devices and vacuum tubes since the early 1900s. These companies also made investments in semiconductor development and fabrication. Conversely, semiconductor companies generally focused on the chip design, technology development, and the chip implementation blocks in Fig. 1.5. Many semiconductor companies have been successful over the years—Intel, Freescale (Motorola), Texas Instruments, Samsung, Toshiba, Hynix, and ST Microelectronics are among the top 10 revenue producers. These companies manufacture standard ICs, some ICs targeted at specific markets and applications, and in some cases also offer services to manufacture Application Specific ICs ("ASICs").

1.1.2 IC Types Defined

There has been much confusion in the industry about the term ASIC. The term became popular in the mid-1980s as gate array devices were introduced by companies such as LSI Logic. While this class of ICs was popular at the time, their use has declined significantly, as discussed in Chap. 4. Sometimes, the decline of gate arrays is equated with a decline of ASICs. Keeping a more global definition in mind, however, it is my view that ASICs will be around for a long time. As long as there are innovative engineers and entrepreneurs, there will be ASICs. This is because fabless companies, as well as others, that design and implement ICs targeted at specific applications usually refer to their IC as an ASIC. However, some analysts and the media continue to use the term ASIC in the context of traditional gate arrays.

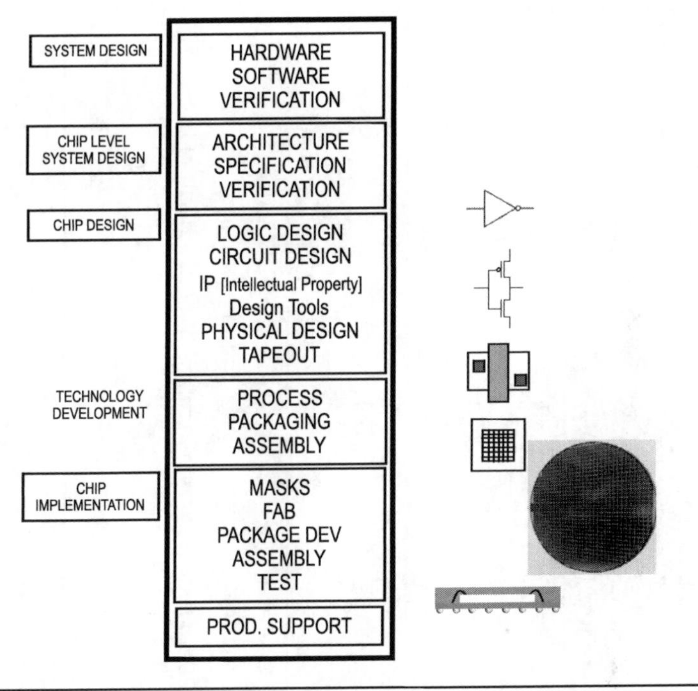

FIGURE 1.5 Integrated development environment.

Overall, integrated circuits can be either "standard" or application specific. ASICs are partitioned into two categories—ASSPs (Application Specific Standard Products) and ASICs. The following Table 1.4 clarifies the distinctions between standard ICs, ASSPs, and ASICs, although the treatment is simplistic and some of the boundaries can be "fuzzy." The criteria used to distinguish these categories are:

- Can the IC be found in an industry or a company catalog?
- How many users/applications?
- Who does the development and manufacturing?

Standard ICs are generally developed, manufactured, and sold by semiconductor companies and are usually based on an industry standard, whether formalized or not. ASSPs are generally driven by a system company or a standard. The chip development could be done by either a semiconductor company or a fabless system or IC company; they could be manufactured by an IDM or by a foundry on behalf of a fabless semiconductor company. On the other hand, ASICs are usually developed by a system company or a fabless IC company. They can be manufactured using a variety of sourcing options, as is discussed in Chap. 4.

	Integrated circuits		
		Application specific	
	Standard	"ASSP"	"ASIC"
Industry catalog	Yes	Sometimes	No
Semi co. catalog	Yes	Sometimes	No
Fabless co. catalog	Yes	Yes	Yes
# Users	Masses	Numerous (> 3)	Few (1–3)
System spec	Industry standard	System Co. or Standard	System Co. or Standard
Chip developer	IDM	IDM or Fabless Co.	Fabless Co.
Manufacturer	IDM	IDM or Distributed	IDM or Distributed
Examples	Logic 74xxx	GSM Baseband	CDMA Modem
	μProcessors	e-net Controller	Sun μP
	DRAM Memory	MPEG4 Decoder IC	LCD Display IC
	SRAM Memory	802.x Router IC	
	NV Memory	MEMS Accelerometer IC	

Table 1.4 Distinction between Standard ICs, ASSPs, and ASICs

A fabless start-up designing a single IC for a specific application, for example an 802.11 baseband with a unique feature, develops the IC as on an ASIC. If the chip is then accepted by and sold to more than a few customers, it may be reclassified as an ASSP. The boundary is somewhat "fuzzy." However, Sun Microsystems' microprocessors, designed specifically for their servers, remain ASICs. Qualcomm's mobile station modem ICs developed for the CDMA applications are ASICs, but could be reclassified as ASSPs as they are sold to many cell phone OEMs (Original Equipment Manufacturers). LCD display driver IC chips could start their life cycle as ASICs and become ASSPs as they are accepted in the marketplace. An ethernet controller, an MPEG4 decoder, an 802.x router and a MEMS accelerometer, are all examples of ICs designed for use by many system companies and are categorized as ASSPs. It is important to understand this classification because there are no standard definitions. Different industry reporting agencies use their own classifications and so interpreting their numbers for metrics such as revenue and number of design starts becomes difficult.

1.1.3 ITRS Roadmap [1.14]

1.1.3.1 Background

In 1992 the Semiconductor Industry Association (SIA, http://www.sia.org) sponsored a workshop attended by nearly 200 leading technologists to create a long term vision for the semiconductor industry. The outcome was a 15 year roadmap of semiconductor technologies. As the industry became global, the cooperative effort was extended to Europe, Japan, Korea, and Taiwan in 1998. Since then there has been an annual publication of either full revisions or updates of the International Technology Roadmap for Semiconductors (ITRS, http://www.itrs.net).

The roadmap provides an executive summary and a comprehensive roadmap in various areas of semiconductor development and manufacturing. Working groups include international experts in the following areas:

- chip size/technology node study group;
- system drivers and design;
- test and test equipment;
- process integration, devices and structures;
- radio frequency and analog/mixed-signal technologies for wireless;
- emerging research devices and materials;
- front-end processes;
- lithography;

- interconnect;
- factory integration;
- assembly and packaging;
- environmental safety and health;
- yield enhancement;
- metrology;
- modeling and simulation

The reason for including these topics here is to provide the reader with a glimpse of the disciplines and areas of consideration in this complex and dynamic industry. The ITRS is also a good resource for exploring next level details in the various disciplines.

One of the prime values of the roadmap has been a unified benchmarking tool. There are targets for benchmarks in the various disciplines involved. It is interesting, though, that the industry has continually outdone the predictions of minimum feature size reduction. For example, the 65 nm node was expected to be in production in 2010. It has actually been in production since 2006 at the leading fabs.

1.1.3.2 "Moore" and "More than Moore"

As the industry drove to maintain the doubling of transistors per chip every two years in accordance with Gordon Moore's prediction, chips with 10 million gates (40 million transistors) became a reality in the late 1990s. The term SoC became commonplace. While there are no real systems completely contained on a single piece of silicon, the use of the SoC term is appropriate for leading edge chips that include multiple functional blocks [1.18]. A typical SoC might include functions such as:

- Logic;
- embedded memory—static, dynamic, non-volatile, read only, etc.;
- embedded microprocessor cores such as those from ARM™ and MIPS™;
- mixed signal blocks;
- re-usable IP (Intellectual Property) blocks;
- real world interfaces such as USB, xDSL, IEEE1394, Ethernet, etc.

Over the last five to ten years, discussion of approaches to achieve single chip SoCs for systems such as cellular phones has led to the conclusion that proper partitioning of the system is very important. Requirements such as RF and high power devices just do not integrate well on the same piece of silicon with low power logic and analog functions. There is more discussion of this topic in Chaps 5 and 6. What has emerged is a very powerful concept that enables integration of multiple chips in the same package. This leads to SiPs (System in Package). Multi-die stacked in a single package have been shipping for many years now. A big advantage from a system perspective is a reduced footprint on the board. ITRS and the industry is using the term "More than Moore" to articulate possibilities for advancing semiconductor technology faster and cost effectively. One such possibility is to integrate non-digital content in the same SiP package, as depicted in Fig. 1.6. Such an approach can be important for the fabless start-up. Depending on the class of innovative idea and differentiation, SiPs could be an interesting alternative to consider.

This is the industry's way to look for alternative ways to capture and apply cost, performance, and size advantages beyond leading edge silicon process capability scaling. The industry is continuing on the scaling treadmill. However, as many predictions have been made of a slowing down of Moore's law, the industry has embarked on a "more than Moore" treadmill. This is represented in Fig. 1.6 from the ITRS 2005 [1.14]

There are some interesting ways for a fabless IC company to leverage best available solutions for their product, e.g., in multi-chip packaging. The "SiP" is a popular version of multi-chip packaging.

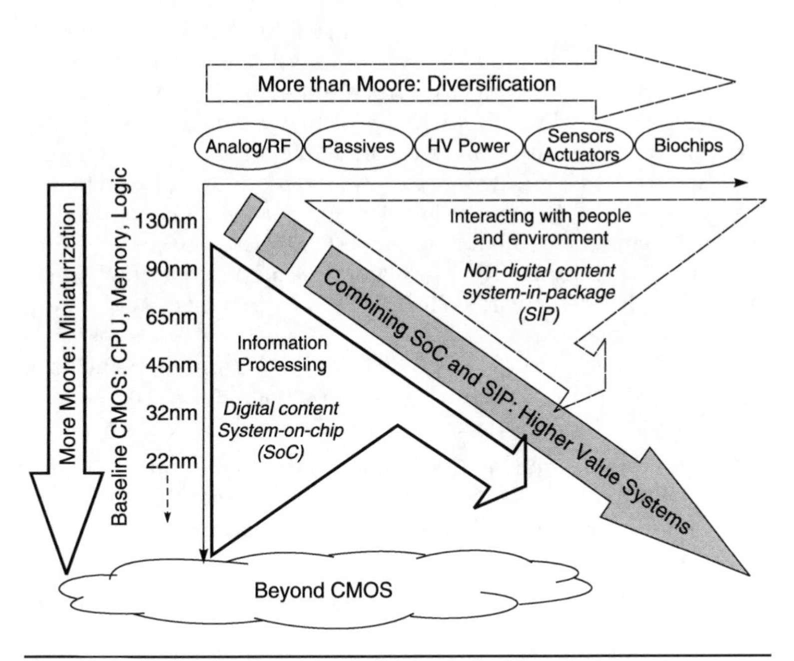

FIGURE 1.6 High value system optimization using Moore's law and a 'More than Moore' approach. (*Courtesy of ITRS [1.14].*)

The fabless company must judiciously select the right functions to include in one or more silicon chips such that their ASIC offers and the best silicon and packaging solution in their customer's application. As is the case with SoCs, SiPs do not really create a full system in one package either. SiP technology does allow the designer to integrate multiple chips with different types of technologies, e.g., sensors, MEMS, and optoelectronics to be assembled in the same package. For example, one could package a nonvolatile, flash memory chip inside the same package that has a leading edge, 65 nm SoC built using a digital process. The Flash chip could be built on a 90 nm or 130 nm process that was optimized for nonvolatile memories. A traditional "more Moore" approach would require modifying the baseline digital process to include the flash within the digital chip. This approach could be expensive because of increased wafer process steps, as well as a potentially lower yield risk. In such an example, SiP packaging usually offers a superior optimization of cost and functionality.

For continuity I will maintain a silicon-scaling-centric theme throughout this book. There will, however be a discussion of some of these "More than Moore" alternatives in the context of packaging in Chap. 5.

1.1.4 Financial Trends

Since the introduction of the first commercial IC in 1961, the semiconductor industry has experienced a healthy growth of approximately 15% CAGR [1.15]. In the meantime, semiconductor sales have grown more rapidly than worldwide electronics sales and the worldwide GDP. Currently, semiconductor sales are roughly 20% of the worldwide electronics sales and about 2% of the worldwide GDP [1.15]. Semiconductor sales in 2006 were nearly $250B (Fig. 1.7), whereas the worldwide electronics sales are expected to be around $1.4T. Fueling this growth has been increasing demand for components used in personal computers, automotives, mobile wireless devices, and consumer products. Although the growth rate is predicted to slow down, the industry has demonstrated much resilience in combating technical and business challenges.

Since 2000, the growth rate has slowed to 6% CAGR. Some have characterized this as a maturing of the dynamic and competitive industry. Some significant trends are:

- Re-composition of the industry characterized by Mergers and Acquisition (M and A) activities and the formation of Joint Ventures (JVs) leading to privatization of some large IDMs.
- Migration of IDMs towards an "asset-lite" or "fab-lite" mode [1.19].
- Downward ASP (Average Selling Price) pressures as the IC unit volumes increase faster than revenue growth.

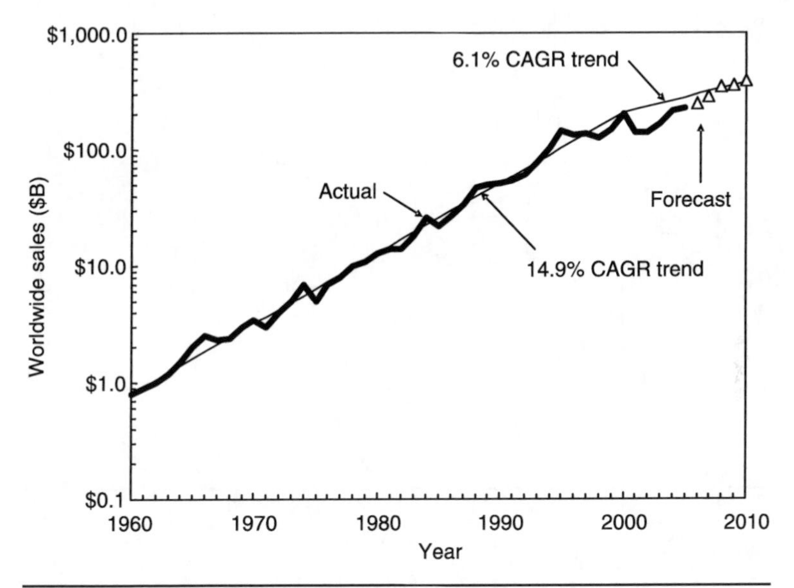

FIGURE 1.7 World-wide semiconductor revenue growth. (*Courtesy of IC Knowledge.*)

1.1.4.1 Wafer Fab, Equipment and Process Development Cost

The persistent quest for scaling down minimum feature size and packing more transistors per chip has required increasing investment in wafer fabs, process development, process equipment, design tools, and design methodologies. There is more detailed discussion of this in later chapters. Figure 1.8 illustrates the trend for wafer fab cost increase [1.15]. A "mega-fab" for the processing of 300 mm diameter silicon wafers costs around $2–5B. If we assume that such a mega-fab has a capacity to manufacture 30,000 wafers per month and we assume a 5-year straight-line depreciation method, there will be a $1400 addition to wafer cost for capital depreciation. Increased process development and equipment investment is required every 2 years for new process nodes. A major investment is required when the wafer diameter is increased, since this requires a new wafer fab. As discussed in Chap. 5 and as shown in Fig. 5.8, the last two transitions occurred 8 and 10 years after the previous step. It will also be shown that the cost per IC (or unit) is lower when using a larger diameter wafer in spite of the higher wafer fab cost.

As will be pointed out throughout the book, there are many processes and design challenges in the nanometer technologies currently in use. The good news is that semiconductor process development engineers have risen to the challenge at every node to formulate and implement solutions. However, solving the issues is becoming increasingly expensive. In addition, it is believed that the sub-45 nm

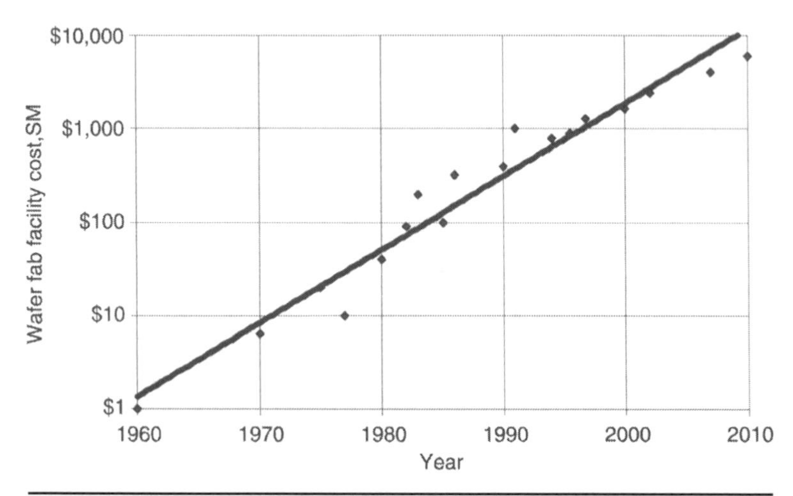

FIGURE 1.8 Wafer-fab cost trend. (*Courtesy of IC knowledge.*)

nodes will require close cooperation between process engineers and designers to overcome some of the new challenges [1.20]. The result is an escalation of process development and design costs. Design challenges and costs are discussed later in the book. Figure 1.9 shows the trend of process research and development costs. For a leading process node, such as 45 nm, the expected R&D costs are likely to exceed $2B [1.21].

The cost of process R&D escalates exponentially if you are the first developer of a new technology node, as was pointed out to me

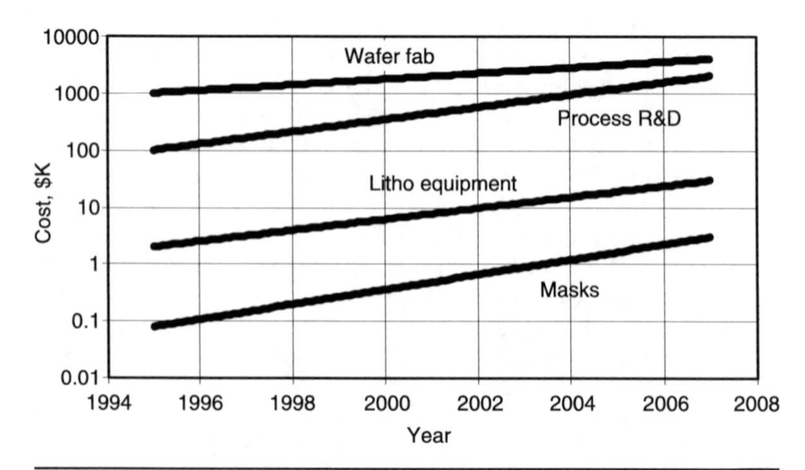

FIGURE 1.9 Escalating cost of wafer fabs, process R&D, lithography equipment and masks [1.21].

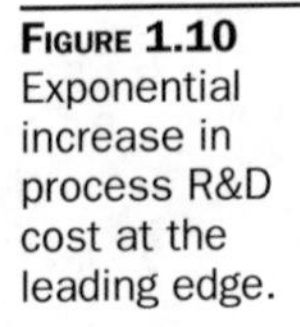
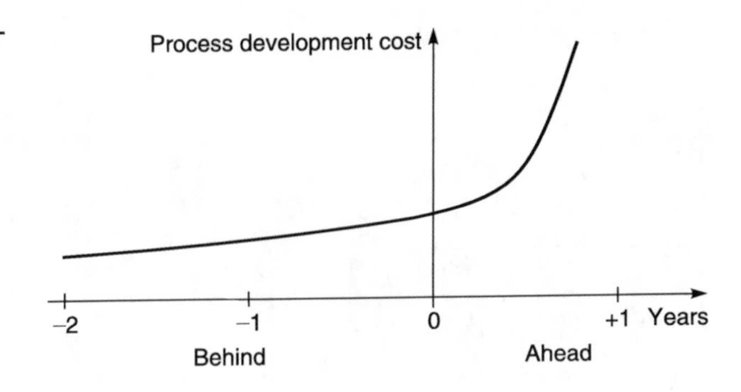

FIGURE 1.10 Exponential increase in process R&D cost at the leading edge.

many years ago by Gerry Parker, former Senior Vice President of Technology and Manufacturing at Intel. Conceptually, this is captured in Fig. 1.10. Costs can be significantly lower if one chooses to be a "follower" and be a year or more behind in implementing new technologies. This makes it difficult for the IDM to compete in leading edge products and negates some of the benefits of owning the fab. This phenomenon is one of the key factors in decisions being made at IDMs these days to go asset-lite or fab-lite. Recently it has been postulated that only a handful of companies, with a revenue stream greater than $7–8B, will be able to justify an internal leading edge wafer fab [1.22].

1.1.4.2 Lithography and Mask Cost

Lithography equipment and masks for the patterning of dimensions smaller than the wavelength of light have escalated exponentially. The cost of immersion and EUV (extreme ultra violet) lithography tools is around $30M and $50M, respectively. Overcoming the many challenges in mask making to print nanometer dimensions on the wafers has caused an escalation in the cost of a set of masks. A set of masks for a 45 nm design is currently around $2M. The good news is that this cost will decrease significantly in the next couple of years due to the normal manufacturing learning curve and other factors such as competition and supply and demand.

It is important to point out that the mask costs for process technologies already in mature production are significantly lower, as shown in Fig. 1.11. This information is derived from quarterly pricing surveys conducted by the GSA (Fabless Semiconductor Association) [1.23]. As discussed in Chap. 7, the mask set cost for a new technology starts off high. The starting cost for each new nanometer node is higher than the previous. However, the cost does drop rapidly in the first few years, due to the manufacturing learning curve.

A judicious choice of the right process technology node is absolutely a must for an emerging fabless company. Designing at the very

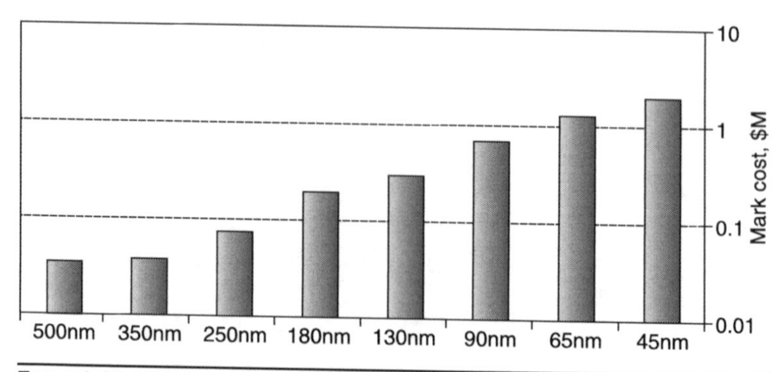

Figure 1.11 Mask set costs in mature technologies can be a lot lower.

leading edge is rapidly becoming restricted only to the large fabless players, which service applications with very high volumes. It is expected that there will be a slowing down of the total number of designs that will be committed early in the life cycle of very leading-edge process nodes. A stratification of the industry is occurring, as is discussed in Chap. 10. Read on for some practical hints and rules of thumb to allow navigation through a complex maze of fabless semiconductor implementation.

1.2 Fabless Industry

1.2.1 Trends

1.2.1.1 Outsourcing trend

Let us now discuss some of the key trends that led to the growth of the fabless industry. In the late 1970s an outsourcing trend emerged as vertically integrated companies set up assembly and packaging plants in the Far East to increase their profit margins. The Far East facilities required lower capital investment and generally offered lower operating cost. They served as "second source" operations for facilities in the U.S., Europe, and Japan. Based on local investments there, the region soon became the hub of high volume package and assembly operations. In the late 1980s the Taiwan government launched an initiative together with Philips Semiconductor, which led to the formation of the first silicon foundry, TSMC (Taiwan Semiconductor Manufacturing Company). Other foundries emerged in following years. The foundries were initially a second source for low cost manufacturing on process technologies lagging by one or more process generations. As is discussed in later chapters, it is now possible to get access to leading edge process technologies from some of the foundries. Approximate time lines for outsourcing are represented in a simplified manner in Fig. 1.12.

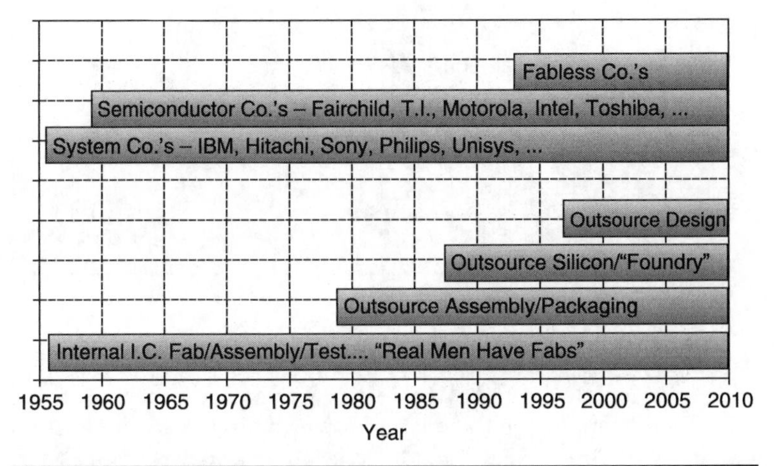

Figure 1.12 Evolution of outsourcing and the fabless industry.

In the 1970s, when vertically integrated semiconductor companies dominated the semiconductor industry, the idea that you could separate design from manufacturing was considered absurd. Who would be willing to spend millions of dollars of capital on manufacturing equipment, absorb major losses during recessionary periods when the fab was under utilized and then forego the opportunity for high component prices and profits when the recovery came? Some major semiconductor companies offered foundry services to fill available fab capacity during down turns. However, there was no guarantee that capacity would be available when the peak of the semiconductor cycle returned. If a wafer fab could manufacture designs from a variety of diverse sources, then maybe the ups and downs of the semiconductor cycle would not be so severe and manufacturing loading could be more uniform. Chips and Technologies, under Gordie Campbell, was one of the early companies to base a large business on foundry sources in the mid-1980s. Seiko Epson also built a large business on this hypothesis. With the creation of TSMC, Morris Chang provided a model that had no conflict with the company's business, i.e., all the customers were fabless companies so wafer shortages during the peak of the semiconductor cycle would be fairly shared. The economies of scale and uniform loading of capacity worked well to harness the creativity of designers with new ideas and capabilities for chip design. While profitability has been an issue at many foundries, TSMC has achieved a high degree of profitability from the

(Continued)

economies of scale associated with volume, yields, and amortization of R&D costs. The fabless industry has successfully demonstrated the ability to design ICs and manufacture them using a distributed supply chain.

Walden Rhines, Ph.D.
Chairman and CEO, Mentor Graphics

In the meantime, system companies such as Unisys created alternate ways to develop ICs leveraging leading edge process technology and design tools to get the best competitive advantage for their systems. With the continuing push to scale down minimum feature size to keep up with Moore's Law and the escalating cost of wafer fabs and process development, they became leaders in reaching out to the high volume semiconductor companies to establish partnerships. As early as 1981, Unisys entered into a partnership with Intel and later with Motorola to leverage their process technologies. It was realized that internal capabilities could be focused on the specific needs of the system companies. One such need was to have a very fast turnaround time when making engineering changes to designs. Another factor was the need to have a large number of design types with a relatively low volume demand per design. System companies also realized the value of packaging technology to improve cost/performance of the computers. Advanced packaging technologies such as flip-chip bumping and multi-chip modules were in use in computers in the 1980s. The importance of Computer Aided Design ("CAD") tools was also recognized early on. As new process technologies were developed and more transistors became available, designers became challenged to be able to find ways to take advantage of the capabilities. The treadmill was under way! In order to capture benefits from the improved process capability and increased transistors on the chip, there was a need for:

- more automated ways to design for improved designer productivity;
- better ways to verify functionality before committing design to silicon;
- better ways to test the product;
- higher pin count packaging;
- reliability verification.

By 1990, Unisys had decided to shut down its wafer fab and to leverage external sources for the fabrication of its ICs. This was an example of the first major system manufacturer going fabless.

1.2.1.2 Gate Array ASICs

With the introduction of personal computers came a significant surge in the demand for microprocessors, memories, and other semiconductor products. The ASIC industry was born as a way to consolidate many SSI and MSI functions on to a single chip. These ASIC companies provided the design tools and methodologies to convert a logic description of ideas into ICs thru "gate arrays". The ASIC company provided packaged and tested finished IC units. While the leading edge microprocessor companies such as Intel set the bar for the highest level of chip integration levels using the best available semiconductor process and chip design capabilities (Fig. 1.1), other companies leveraged the ASIC development capabilities of suppliers such as LSI Logic.

1.2.1.3 EDA Industry is Born

With the escalation of the number of available transistors per chip, two major challenges were recognized in the design world.

- It was becoming increasingly more difficult to connect the transistors and to verify the connections.
- Highly automated design tools were required to help boost productivity of the designer.

Initially, system companies developed their own CAD "place and route" tools and "libraries" of gates and functions that could be implemented on the chip in a customized fashion. As the design challenges escalated further with the increasing transistor packing capabilities, the EDA (Electronic Design Automation) industry was born in the late 1980s, with the formation of Solomon Design Associates, later renamed Cadence Design Systems.

1.2.1.4 Fabless Companies

Leveraging the availability, albeit limited, of the tools and methodologies for implementation of custom designs into silicon, the first independent fabless companies were launched in the early 1990s. Entrepreneurs realized a way to fabricate their logic ideas into silicon without having a semiconductor wafer fab of their own. Those with an idea, a business plan, a core team, and venture funding launched many new companies. Their intent was to get their ideas into specialty ICs, generally ASICs. Examples of the first fabless companies are Chips and Technologies, a manufacturer of PC chip sets and Brooktree Corporation.

The first fabless companies had to cross many hurdles to design their ICs, get the masks built, fabricate the silicon, package, test, and qualify their ICs. These companies were able to successfully demonstrate the feasibility of a distributed supply chain for the design and manufacture of ICs using the COT (Customer Owned Tooling) approach.

The term "tooling" here refers to the glass/quartz masks that are used in implementing a design onto the silicon wafer.

As competitive silicon process, packaging, and assembly capabilities became available, the 1990s saw the beginning of the fabless IC industry. The combined elements of the supporting design infrastructure industries are now being called the "design ecosystem." The design ecosystem includes the EDA and the Intellectual Property (IP) providers. Over the last 5–10 years we have also seen the establishment of many IC design services companies. Licensed or purchased IP is a means to obtain pre-defined library building blocks of functions. The availability of design blocks for re-use from one technology to another and from one design to another can be a major asset in reducing design time, cost and risk.

Introduction of internet related products and cellular handsets in the 1990s has challenged the IC industry with shorter time to market, shorter product life cycles and more complex ICs at a lower cost. Many fabless IC companies have been created to make available ICs with unique features targeted at specialty applications. With heavy investments, especially in the Far East, the distributed supply chain has become a norm in the industry.

Success in the fabless world requires a transformation of management methodologies used in the development and implementation of ICs. In Chap. 9 there is a discussion of the main management challenges associated with the coordination of the fabless, distributed supply chain.

In 2007, the GSA reported the existence of 1400 fabless companies around the world; the trend is, as shown in Fig. 1.13. It is also reported that approximately 50 new fabless companies were started between 2004 and 2007: a listing of the top 10 fabless IC companies is shown in Appendix A. In recent years there has been a significant growth in the number of fabless companies in Taiwan, China and India.

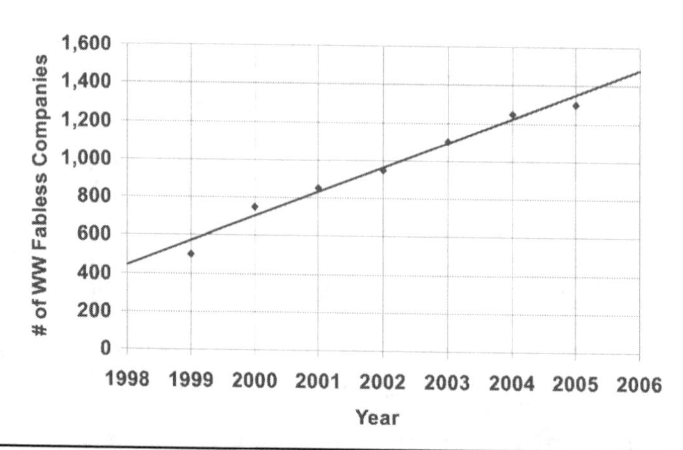

FIGURE 1.13 Number of fabless companies worldwide. (*Courtesy of GSA.*)

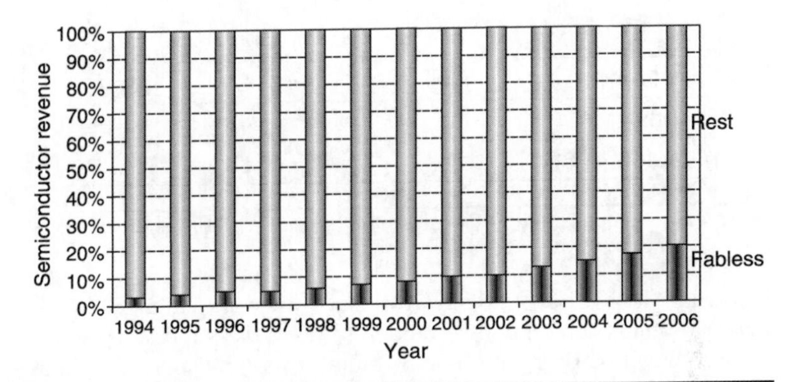

FIGURE 1.14 Fabless company revenue as a fraction of worldwide semiconductor sales. (*Courtesy of GSA.*)

Revenue from fabless companies is approximately 20% of the worldwide semiconductor revenue, as shown in Fig. 1.14 [1.23, 1.24]. It is predicted by GSA and Morgan Stanley that, by 2010, over 35% of the worldwide semiconductor revenue will be from outsourced operations. It is also important to note that the growth rate of the fabless segment over the past five years has been consistently higher than the semiconductor industry, as shown in Fig. 1.15. The 2005 to 2006 growth of the fabless segment was 27% versus 9% for the semiconductor industry. As shown in Table 1.5, fabless companies have aggregated a a 26% CAGR since 1994, versus 6% for IDM's.

Over the last 10–20 years the semiconductor foundries have made enormous capital and resource investments to create leading edge wafer fabs with leading edge process technologies. They provide high quality silicon at competitive prices based on the customer's design input. Since the early 1990s, many system companies have

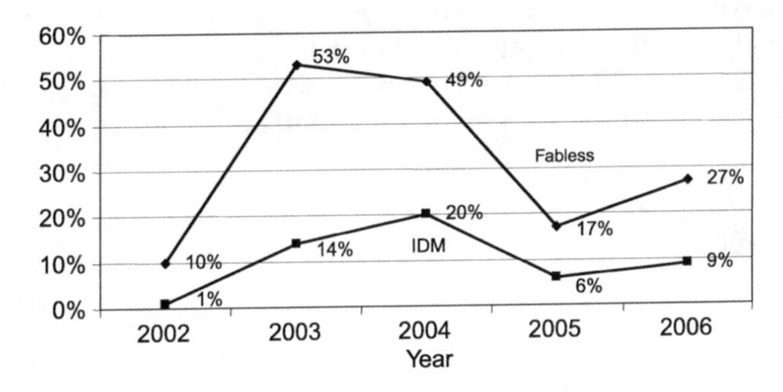

FIGURE 1.15 Fabless and IDM company revenue growth rate, YoY relative to the previous year. (*Courtesy of GSA.*)

	IDM	Fabless
CY 2006 Revenue	$198B	$49.7B
YoY Growth	5%	27%
13 Yr CAGR (94-06)	6%	26%

TABLE 1.5 Comparison of Fabless vs IDM Companies (*Courtesy of GSA.*)

made decisions to not build new wafer fabs and have actually adopted either a fabless or a "fab-lite" model. In the fabless model they outsource silicon fabrication to one or more of the semiconductor foundries. In the "fab-lite" model the company maintains an internal fab for specialty technologies or to maintain some special features in their products. However, the company outsources some or all process development and manufacturing at a foundry partner. More recently, an increasing number of semiconductor companies have adopted expanded use of the "fab-lite" model.

Compared to IDM companies, fabless companies have higher ROE (Return On Equity) and ROA (Return On Asset) ratios, as shown in Table 1.6.

1.2.2 The Past, Present and Future Challenges

In this section there is a discussion of how the fabless world has evolved since the early 1990s. I discuss four, five-year periods: early 1990s, late 1990s, 2000–04 and 2005 onwards.

Fabless companies founded in the early 1990s faced many challenges associated with a brand new business model and no real support infrastructure. Companies such as Brooktree, and Chips and Technologies had to create the script. Yes, there were a few silicon foundries. However, the design rules, the process and the models used for the design were not well documented and not thoroughly verified. The fabless companies had to hire many people to interface with the foundries, the packaging, assembly, and test houses. They also had to hire many designers for the custom design of standard cell libraries and their ICs. The designs had a high level of "hand-crafting." It was not unusual to have an in-house staff of 50–100 or more.

CY 2005	IDM	Fabless
ROE	10.5%	14.8%
ROA	6.4%	12.0%

TABLE 1.6 Comparison of ROE and ROA for IDM's and Fabless Companies (*Courtesy of GSA.*)

> In the mid 1980s it was very difficult for entrepreneurs to get access to silicon foundry capability. Larger companies were able to get limited access to some captive silicon IC manufacturers, especially in Japan. Yet process related information was considered highly proprietary and it was very difficult for companies to get access to necessary design information. After the commercial foundries were launched, they were used initially for low cost second sourcing of older generation products from IC manufacturers. It was very difficult for the foundries to capture new design starts. The foundries then invested in the required infrastructure to provide rules, models and mask making capabilities. In addition, heavy investment in capital, process development, efficient and low cost manufacturing have allowed foundries to offer competitive process technologies, cost effective manufacturing and a support infrastructure to enable the fabless business model.
>
> John Luke
> Former president, TSMC-USA

In the second half of the 1990s, the industry was really getting hot with the dot.com business booming. Numerous IP providers were making available standard cell libraries and memories. The EDA industry was stepping up to the challenges associated with increased transistor densities per chip, increased performance, and the need for increased designer productivity. Foundry technologies were being introduced more aggressively, and they were becoming more competitive with in house processes at the IDMs. There was more talk in the industry about the need for re-usable IP and platform based design methodologies, but these were nascent efforts at the time. The operations infrastructure to manage the flow of products across the oceans and the Far East countries required much human interfacing and tracking. High volume demand for ICs used in cellular phone and consumer applications offered expansion opportunities for supply chain partners. Many advances were made in the packaging, assembly, and test areas. Low cost, non-hermetic packaging with tighter wire bond pitches became common.

In the first 5 years after the turn of the century, there was a significant escalation of demand for cell phones and graphics products. The foundries continued to invest heavily in facilities and process development. They also increased investment in design infrastructure such as libraries, memories, and IP to facilitate design efforts of fabless companies. The EDA industry and the foundries stepped up to provide Process Design Kits (PDKs) and reference design flows to facilitate efforts at the fabless design houses. As the transistor packing density increased further, it became clear that a more focused effort on Designing For Test (DFT) and Automatic Test Pattern Generation (ATPG) were essential for cost effective production of ICs. As designers included more analog functions on chip, there was increased pressure on the foundries to provide high quality Spice models suitable for analog design. Solder bumping and flip chip assembly processes became available for high pin count packaging. Multi-chip, multi-tier, or 3D-stacked packaging of multiple silicon chips within the same package became available as a production technology. This approach became available to fabless companies for high-volume production the first time.

Since 2005, the 65 nm and 45 nm technology nodes have been introduced. Leading fabless companies have already committed many new designs to these technology nodes. The industry is entrenched in the nanometer era. However, the literature is abundant with discussions of process and design challenges, as well as warnings about the need for co-design techniques for integrating process and design issues. The cost of design at these nodes is escalating exponentially, driven primarily by the cost of design verification and the mask cost. It is believed that this increase in design cost will lead to a stratification of the industry. Large fabless companies, and those servicing very high unit demand will be the only ones that could justify committing to the use of such advanced technologies. While there is much concern and chaos in the industry about this, in my view there are many opportunities for technical and business innovations. An example of such innovation is the use of "More than Moore" approaches such as those offered by innovative SiP and other packaging methods. Possible modifications to the funding and business models are discussed in Chap. 10. In the meantime, the industry has many technical challenges associated with device leakage, power dissipation and process variability associated with the small dimensions. These are discussed in Chaps 5 and 6. Techniques for Design For Manufacturability (DFM), Yield (DFY), Test (DFT), and Quality (DFQ) are being discussed in the literature and the press very frequently. It is believed that guidelines and EDA tools that help manage some of these and other issues will be made available as the industry moves forward.

One of the big challenges facing the supply-chain industry is related to the increased cost of designing at the leading edge process nodes. It is predicted that there will be fewer early adopters of technologies at the 65 nm node and beyond. It is also predicted that there will be a longer lead time between the early adopter and the mainstream users, and that there will be fewer "mainstream" users of technology nodes at or below 65 nm. These assertions are depicted pictorially in Fig. 1.16. Some of these ideas were first presented by Walter Ng of Chartered Semiconductor. Whether this phenomenon will manifest itself in lower wafer volumes and slower volume ramps on new technlogy nodes needs to be tracked.

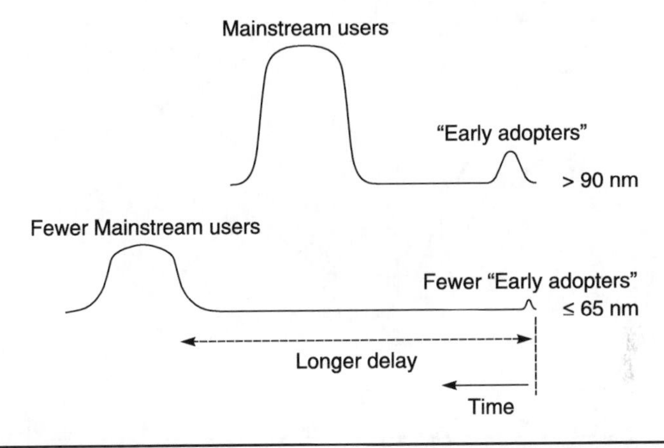

FIGURE 1.16 Pictorial representation of early adopters and mainstream users of deep-nanometer process technology nodes.

So, why will the industry continue to invest in the leading edge nodes? To off-set the high cost of design, there must be high volume demand for the IC. As long as there is high volume demand for the ICs, scaling principles (Chap. 5) will dictate the use of newer technology nodes. Experienced and large fabless companies can make these trade-offs. There is a discussion of such trade-offs in Chap. 7. If you are an emerging fabless company, one guiding principle is to use the most mature technology that allows you to meet the design functionality, cost, and performance goals. This approach is a major change from the historical trend in the industry, one that always required the use of the latest technology node.

Another challenge for fabless companies and the industry is the need for increased, tight coupling between the fabless company and the foundry especially on nanometer tecnology nodes. This is driven by process issues such as process variability and leakage that need co-design solutions (Chap. 5). As discussed in Chap. 10, there is also a trend towards a virtual re-integration of the distributed supply chain. The leading fabless companies are already experiencing this need in the context of stacked packaging of die from multiple suppliers. Tight coupling between the fabless company and its customers is emphasized in Chap. 2.

1.2.3 Evolving Funding Models

Figure 1.17 shows the amount of funding and the number of fabless fundings from 2000 thru 2006. Since 2001, there seems to have been a reduction in annual funding from about $2.5B to under $2B. Data from GSA for 2007 indicates 130 fundings for a total of $1.8B.

As shown in Fig. 1.18, there has been a significant decline in the number of fabless startups since 2000. Unlike the early 1990s, when

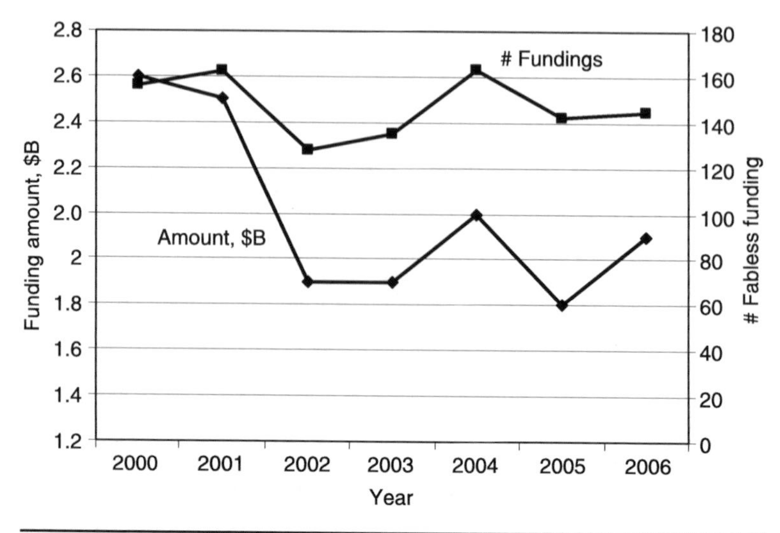

FIGURE 1.17 Fabless funding amount and the number of fabless fundings. (*Courtesy of GSA.*)

funding was difficult to come by because it was an unproven model, there was a funding boom in the late 1990s. In the dot.com era, if you had an idea, had a team of experienced principals and a business plan, it was relatively easy to get funded by venture capitalists. As the dot com bubble burst and as a result of the September 11th attacks in the U.S. in 2001, securing funding got quite difficult. It is postulated

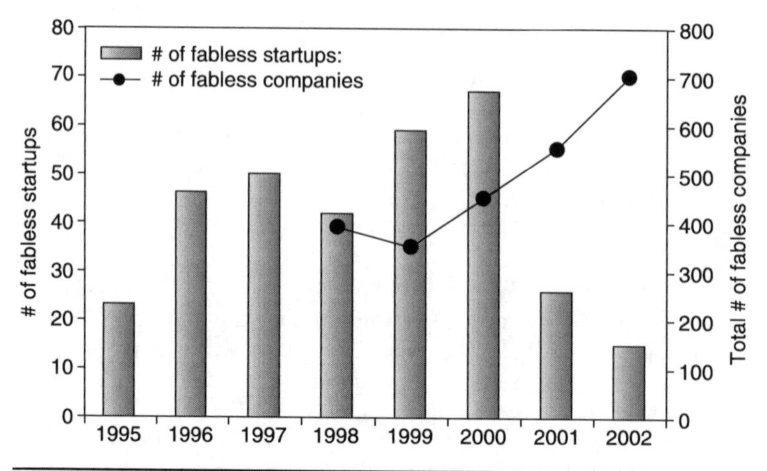

FIGURE 1.18 Number of fabless startups has declined significantly since 2000. (*Courtesy of GSA.*)

here that the new startup funding model will evolve over the next few years. As discussed in Chap. 10, new funding models will emerge as the industry finds ways to deal with the deep-nanometer node technical and business challenges.

1.2.4 The Fabless Support Infrastructure

Ecosystems of support have evolved in various parts of the distributed supply chain. In this section there is a summary of some of the salient elements of these eco-systems.

In the foundry industry:

- Scaled new process technologies are continuing to be introduced every two years in spite of the challenges outlined in Chap. 5. Some customers choose to shrink their designs to intermediate "half-nodes."
- Process variants for expanding applications of semiconductors.
- Design support infrastructure to enable IC designers.
- Coordinated solutions of process/design/packaging co-design, appropriate for nanometer process technologies.
- Improved models that represent accurate behavior of the devices, the parasitics and the interconnects in a circuit simulator.
- DFM methodologies to improve product yield and to capture increasing variability effects at the deep nanometer nodes.

In the EDA industry:

- Improved simulation tools
  - enhanced design kits;
  - improved verification methodologies and tools.
- Improved design productivity methodologies and improved tools for power and leakage management, timing analysis, and DFM assessments.

In the package/assembly/test/socket/load board/qualification industry:

- Continued scaling of wire bond pitch.
- Continued package size (footprint and volume) reductions.
- Continued introduction of new, low cost package families.
- Continued advancement of chip stacking and 3D packaging.
- High volume manufacturing of bumping technology with tighter pitch.
- Introduction of environmental friendly, "green" packages.
- Continued enablement of IC performance improvement.

In the operations area:

- Improved and seamless tracking of and visibility into product flow in the supply chain. Most systems are web-based and provide customers easy access.
- Rapid availability of reports on the material in process.
- Minimal customer involvement in shipping, customs and other handling issues.

1.3 Key Points

- The maturing semiconductor industry has been driven by Moore's Law. While there are indications of a slowing of the growth rate of chip complexity and performance improvement due to technical and business challenges, there is still a continued reduction in cost per function. Creative techniques and leveraging "more than Moore" approaches are expected to drive the industry further.
- The fabless approach is here to stay. Growth rate of the fabless companies, their potential gross margins, and returns on equity and assets are superior to that of the most vertically integrated companies.
- The cost of design and fabrication at the leading process technology nodes is escalating at a rapid pace. Companies need to ship very high product volumes in order to get a reasonable return on these investments. Some key elements required for success if IC companies are: proper product definition, product differentiation and design verification, proper selection of the technology and methodologies used, and aggressive, yet achievable implementation plans.

CHAPTER 2

The Big Picture

In this chapter let us discuss some strategic issues related to the marketing and positioning of the IC product with customers, and a typical execution schedule. First there is a discussion of the electronics market place, market opportunities, and challenges. This is followed by a discussion of the importance of early customer engagement and an illustration of the value chain. Typical development schedules for the fabless IC company as well as for the end customer will then be discussed and illustrated with an example.

2.1 Electronics Markets

The semiconductor industry has been a fertile ground for nurturing electronics innovations. Over the last 40 years, various different markets have driven the growth of the industry and have provided the impetus for innovations. As one contemplates forming a new fabless semiconductor company, it will become clear that a thorough understanding of the company's market is essential in making correct decisions, both technical and business related. Successful, mature fabless companies are continually aware of market demands and trends. The first part of this Chapter will focus on current market drivers and trends.

There are numerous sources available for gathering market information. The GSA (GSA, http://www.fsa.org), recently renamed GSA (http://gsaglobal.org), is a good place to start. Most of the well known sources listed here offer subscription services.

- Databeans (http://www.databeans.com).
- Gartner Dataquest (http://www.gartner.com).
- IC Insights (http://www.icinsights.com).
- iSupply (http://www.isupply.com).
- Semico (http://www.semico.com).
- Semiconductor Industry Association (SIA) (http://www.sia.com).

In 2006, the worldwide semiconductor business totaled $247.7B. According to the SIA, the semiconductor industry manufactured 90 million transistors in 2003 for every man, woman and child on earth. In 2010 it is expected that the industry will manufacture 1 billion transistors per person! [2.1]

Table 2.1 shows a breakdown of the 2006 semiconductor revenue into six market segments. The top three markets with the largest revenue as well as growth rate are computers, consumer products, and wireless products.

In the previous chapter it was shown that 20% of the worldwide semiconductor revenue in 2006 (Fig. 1.14, Table 1.5) is attributed to the fabless market segment, which is growing much faster (26%) than the IDMs (6%). Figure 2.1 shows a breakdown of the end markets serviced by public fabless companies. This data shows that the largest markets currently serviced by public fabless companies are networking/telecommunications, consumer and data processing [1.23].

While the industrial and automotive sectors comprise a small fraction of the overall fabless IC market, they have decent growth predictions (Table 2.1). Therefore these market segments could provide opportunities for innovations and new applications of fabless semiconductors. The military/aerospace market segment also represents a small fraction of fabless revenue. Although this market is characterized by low volumes, there are opportunities for "niche" solutions and innovation.

According to George Scalise, SIA President, "2006 was the 'year of the consumer' in the electronics industry" [2.2]. Semiconductor technology has been used to improve functionality and lower the cost of cell-phones, MP3 players, and HDTVs, which has resulted in significant increases in sales.

Semiconductor revenue by end market	CY 2006 ($B)	CAGR (2006–11)
Computer	89.1	10%
Consumer	53.5	11%
Communications (wireless)	39.6	14%
Communications (wired)	23.1	8%
Industrial	21.9	6%
Automotive	20.0	8%

Source: GSA, Databeans 1/2007

TABLE 2.1 2006 Semiconductor Revenue Partitioned by End Markets and the Predicted Compound Annual Growth Rates CAGR (*Courtesy of GSA, Databeans 1/2007.*)

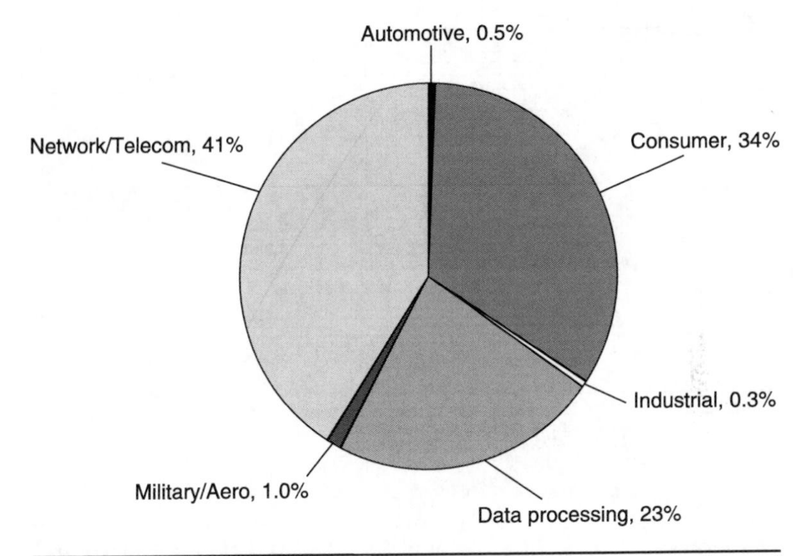

FIGURE 2.1 Breakdown of 2006 public fabless semiconductor revenue into end markets. (*Courtesy of GSA.*)

Another point of note is that, according to iSuppli, the semiconductor content of electronic systems—as measured by cost—has been increasing steadily and now stands at 21.6%. The market size can be gauged by the following summary of volume shipments in 2006 [2.2]:

- cell phones—over 1 billion worldwide;
- personal computers—over 235 million worldwide;
- MP3 players—over 34 million in the U.S.

Yet another important point to keep in mind is that the time to high production volume could be very short, especially if you hit a "jackpot!" Fig. 2.2 shows how quickly unit volumes have reached 1M units for various product families [2.3]. It took over 21 years for the black and white TV to hit the 1M mark. Recently, the DVD player took just over a year. The Apple iPod took almost two years [2.4]. The iPhone is expected to sell 1M units in one quarter [2.5].

In December 2006, Gartner identified the top three markets with the highest growth rate over the next 5 years for ASSPs [2.6]:

- digital cellular handsets;
- LCD and plasma TV;
- digital media players.

A summary of the information regarding the top ASSP markets is shown in Fig. 2.3.

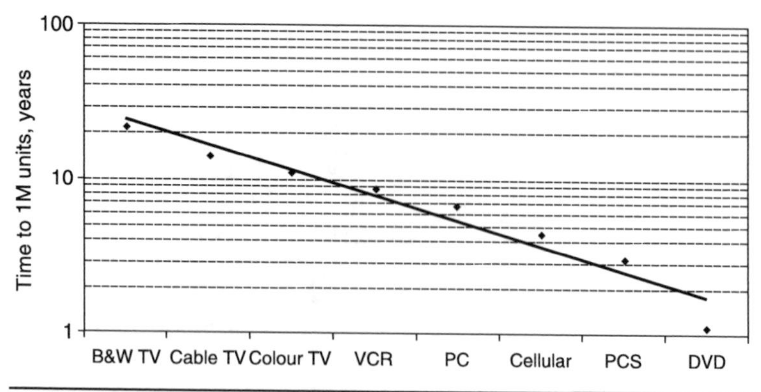

FIGURE 2.2 Time to 1M unit shipments for various electronics products. (*Courtesy of CSM, Semico Research.*)

Successful integration of new features into products is found to be a key to capturing market share. Selecting the right features becomes a special challenge in a highly unpredictable market such as consumer electronics. The problem gets exacerbated further by the fast moving convergence of multi-media features into hand-held products such as media players, enhanced cell phones, video games, and cameras. In the cell phone marketplace, the most popular features for integration are cameras, Bluetooth, digital audio, and GPS capability [2.7]. It is predicted that video capability in portable media players will increase their demand from about 40M in 2006 to 183M in 2010 [2.7].

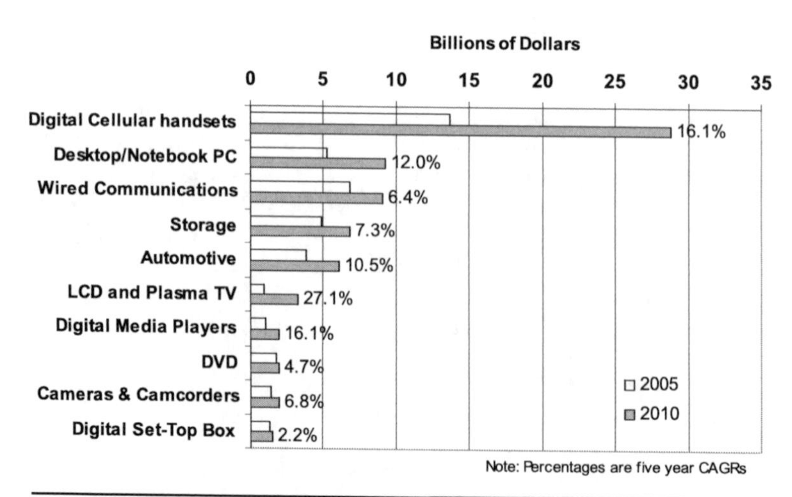

FIGURE 2.3 Top ASSP markets, showing CAGR percentages. (*Courtesy of Gartner Dataquest estimates, November 2006.*)

In some products, "go fast" performance features have less importance than features such as long battery life and long operating life. Form factor, looks and convenience are important considerations. Sometimes these "human factors" become dominant considerations, as demonstrated by the popularity of Apple's iPod.

2.2 The Global Opportunity

In recent times, we have all observed globalization both on the manufacturing and the consumption side. Prahalad and Hart have discussed the "Fortune at the bottom of the pyramid" in an article originally published in 1998 [2.8]. They postulate a tremendous opportunity as companies reach out to the 4+ billion people in "the bottom of the pyramid" in developing countries around the world. Currently this population segment spends very little (under $20 per person) on electronics products annually. They are potential candidates for the procurement of electronics devices such as cell phones, TVs, and low-end laptops. Prahalad and Hart argue, as represented in Fig. 2.4, that the top of the pyramid represents the current market consumption of feature-rich and performance driven devices sold in the developed countries. The average consumer in these developed countries spends 10–15 times more on electronics devices per capita than at the bottom of the pyramid. Even within this top group, product volumes generally go up as lower price points are achieved. Rick Tsai, President and CEO of Taiwan Semiconductor Manufacturing Company, the world's largest independent foundry, recently used a similar analogy in a keynote speech [1.19]. There appears to be a significant expansion opportunity in the consumer, wireless, and computer markets both at the top and the bottom of the pyramid.

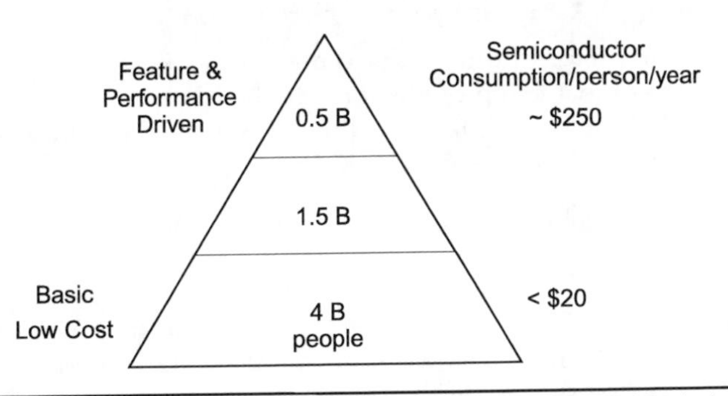

FIGURE 2.4 "Fortune at the bottom of the pyramid." (*Courtesy of Prahalad and Hart [2.8].*)

2.3 Some Challenges in Today's Electronics Marketplace

- **Short market windows and short Time To Market (TTM).** Development delays can be a killer because they can lead to missed market windows and significantly reduced market share.
    - For example, the four primary market window for cell phones, and other consumer electronics products are: Christmas time, the Chinese New year, graduation, and similar hot-selling market windows.
- **Short product life cycles**. Some cell phone and consumer electronics products could have a life cycle as short as 6–12 months.
- **Low cost** is *king*. An annual reduction of ASP by 15–30% is common.
- **Low power dissipation** in ICs has become an extremely important factor since long battery and operating life are important customer considerations.
- **Form factor**, size, shape, looks, and sleekness of the products is important to the consumer. The demands could be different for different age groups and in different geographical locations around the globe.
- **Ever increasing chip complexity** (this can also be an oppurtunity).
- **Escalating development cost** at the leading edge.
    - Careful consideration must be given to potential market size and Served Available Market (SAM) assumptions when selecting the product, the technology and the implementation approach, as is discussed in later chapters.
- **Software**, application software, user interface and convenience are increasingly more important. Hardware and software co-design is becoming very important.
- **Complete solutions.** Customers are looking for the fabless IC company to provide not only the ICs, but also a reference platform, test suites, software, and other items to make it a near "turn-key" solution.

Successful fabless IC companies must understand these challenges and consider solutions in a proactive manner. This is important in order to gain product acceptance and market share. To be successful, the fabless company must understand the customer side of the value chain as well as the manufacturing side of the value chain, as is discussed in the next section.

2.4 Understanding the Value Chain

There is much talk in the literature about the design and manufacturing supply chain. There is also much discussion of the manufacturing supply chain in later chapters of this book. In this section the importance of the customer side of the value chain is emphasized. Some of the challenges in today's dynamic marketplace can be managed via meaningful partnerships with the IC company's customer and the customer's customer [2.9].

A simplified view of the traditional fabless IC value chain is shown in Fig. 2.5. It is assumed that the fabless IC supplier contracts with a "turn-key" supplier, for example the wafer foundry, for management of the manufacturing supply chain. While there could be some direct communication between the IC company and the package/assembly/test providers, the turnkey supplier manages the placement of orders, shipments and deliveries. It is also assumed that in this model the fabless IC supplier deals directly with the EDA tools providers and the chip design work is completed internally. Depending on the location of product inventory, the fabless IC supplier fulfills orders by shipping product to the OEM (original equipment manufacturer), either from its own facilities or arranges for the product to be "drop-shipped" from the turn-key supplier. Product is then shipped from the OEM to the retailer. This value chain shows a fairly "serial" flow of the IC product. Table 2.2 lists some examples of typical fabless value chains. The first three are examples of products shipped by the three largest fabless IC companies. The last two are examples of companies that are fabless but are shipping electronics products under their own logo and brand name. In this model they combine the roles of the fabless IC company and the OEM. Many variants are possible to this simplified model and the specific companies mentioned.

A more streamlined customer side value chain, one that is better suited for integrated development and for reducing time to market, is shown pictorially in Fig. 2.6. In this model, the fabless IC supplier

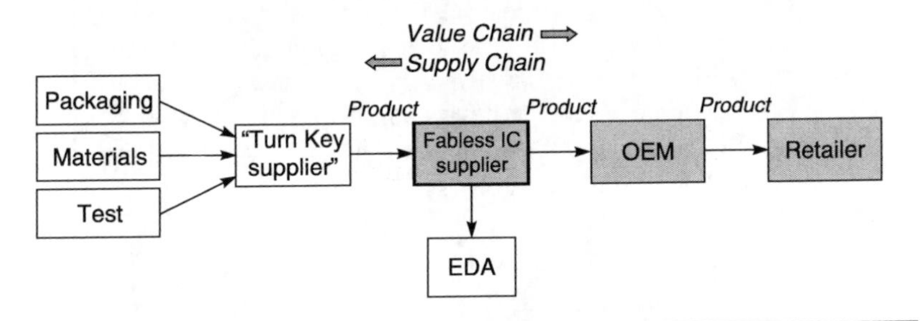

FIGURE 2.5 Simplified representation of the traditional fabless IC value chain.

Product	Fabless Co.	OEM	Retailer
Cell Phones	Qualcomm	Samsung	Verizon
802.x WLAN	Broadcom	Cisco	Circuit City
Graphics cards	nVidia	HP	Best Buy
Xbox	Microsoft	Microsoft product group	Best Buy
Cell phones	Nokia	Nokia product group	AT&T

TABLE 2.2 Customer Side Value Chain Examples

manages the interface and the orders directly with the fab, package/assembly, and test suppliers. In order to reduce development time and the introduction of new technologies, the mature fabless IC company establishes working relationships with equipment, materials, and other suppliers as well as consortia for advanced research and development. The fabless IC company also establishes relationships up the value chain such that it can get in synch with the customer's customer. This is important for understanding the retailer's near term and long term product plans, including plans for new feature introduction, product introduction schedules, and other important product roadmap considerations. Such open communication can also help the OEMs and the retailers understand the implications and benefits of new IC technology capabilities such as lower power dissipation, form factor, and on-board memory.

Partnering with our customers and their customers helps us get a head-start towards long range development of new features and standards.

Partnerships with suppliers down the value chain allow us to help drive new technology developments to meet our IC product requirements. This results in time to market benefits and enables timely introduction of our IC products on new technology nodes. We are careful not to drive custom process technologies. In this way the process technologies are available to all fabless IC companies, thus keeping the wafer availability high and costs low.

Behrooz Abdi
President & CEO, Raza Microelectronics
Former SVP & GM, Qualcomm CDMA Technologies

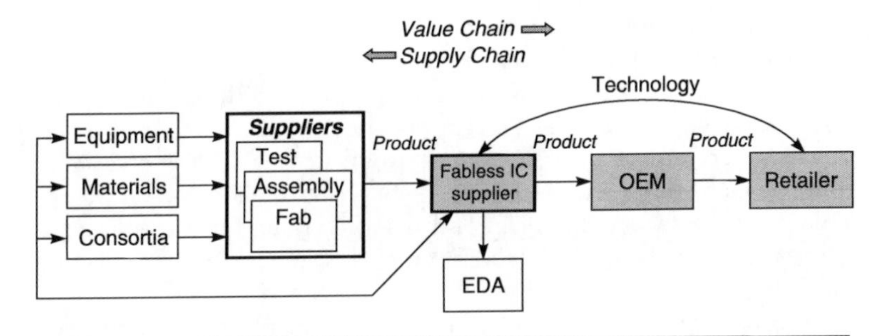

FIGURE 2.6 Streamlined value chain to reduce time to market and product introduction schedules.

While establishment of such partnerships could be a problem for the emerging fabless company, serious efforts to establish communications with the customer side of the value chain are highly recommended. Such open communication does have the possible risk of giving away one's idea. However, such relationships up the value chain could have significant potential upsides. Not having such relationships can be a severe limiter, as can be testified by many frustrated start-ups.

> As a young fabless IC company we focused our energy on developing a specialty DSP for use in digital cameras supporting the MPEG4 standard. While the IC performed well, the customers wanted us to supply complete prototypes of assembled units including the application software. The prototype and the software development kit we provided was not sufficient. This caused us serious workload and skill set issues late in the development cycle.
>
> Brian Fitzgerald
> Former CEO
> ChipWrights

2.5 Customer Needs

In the 1990s the fabless IC company would develop an IC and if it was a chip that could add value to the OEM's system, it would be integrated into the OEM's product. The hardware and software integration of the IC was the responsibility of the OEM. For example, the Chips and Technologies, graphics and video accelerator ICs for PCs, were integrated into the motherboard by the developer. In order to manage short product life cycles and short lead time to market, OEMs are now requiring the fabless IC company to provide kits including reference designs, hardware and application software.

Another important area is the burden on the fabless company that comes from the build-ahead to satisfy short lead product times. This can be a serious financial and resource issue for the emerging company. There is further elaboration of this in Sec. 2.6.1 and also in later chapters.

OEMs are also requiring major component/IC suppliers to supply high quality parts with predictable failure and reliability specifications.

2.6 Overview of Fabless Company Development Activities

Figure 2.7 captures a high level view of the major development activities at a fabless IC company. Chip level "system" architecture/design/simulation and verification activities get started soon after the establishment of the company, and continue throughout the IC development cycle with increasing levels of refinement. Details are discussed in later chapters. The ideas that were the basis of the company's formation must be demonstrated in hardware. This is usually accomplished through implementation using one or more FPGAs (Field Programmable Gate Array). Such a demonstration leads to the creation of a reference design, which can be used for demos to customers. In addition to the hardware implementation, there usually is the need for an early version of the application software. These system level activities are represented in the top part of Fig. 2.7. As shown in the bottom half of Fig. 2.7, the ASIC design activities usually start in parallel. The ASIC design is delivered to the fab foundry via a "tapeout" process for the creation of masks and the complete IC process implementation. The ICs are then assembled in packages and tested. The first samples arriving back at the fabless company must be thoroughly bench-tested to verify that they function and operate as designed. A handful of Engineering Sample (ES) parts can then be made available to one or more lead customers. This starts a validation process with the customer(s). After the initial testing, the IC supplier can make available parts that have been more thoroughly characterized for manufacturability and some early look at reliability. The parts are usually designated QS (Qualifiable Samples). Some suppliers may call them Customer Samples or Shippable Samples. If all goes well, the customer is usually ready to place orders for early production parts, "ramp-up" and full production parts. The schedule within the fabless company and its relationship to the schedule at the OEM is discussed in the next section.

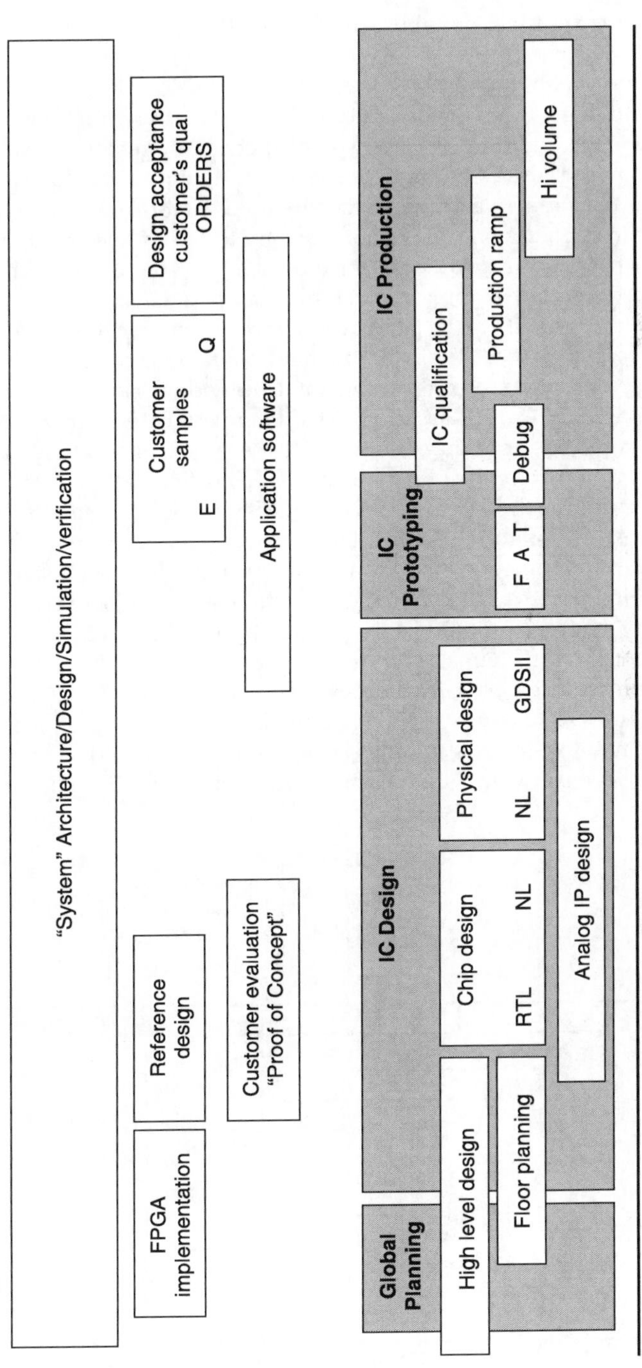

FIGURE 2.7 Overview of development activities at a fabless IC company.

2.6.1 Typical Schedules and Examples

In Fig. 2.8 is shown a possible schedule for the development of an ASIC. Actual schedules can vary depending on the maturity of the design, the capabilities of the design teams, the complexity of the design and many other factors. The ASIC design activity is shown to last about five quarters, at which point there is a "tapeout" milestone—sending of the GDSII tape to the silicon foundry. The foundry fabricates the glass/quartz masks that are used to fabricate the silicon IC. Upon completion of the wafers, they are tested for electrical test ("e-test") parameters that determine if the silicon process was within established limits. The foundry customer can then choose to electrically test functionality of the parts in wafer form—this is called "wafer sort" or "wafer probe." The wafer is then sent to an assembly house that dices and assembles the functional individual chips into the desired package. The ICs are then tested using test equipment called ATE (Automated Test Equipment). After the proper inspections the "prototype" parts are then ready to send to a customer. In this chart, a 3–4 month lead time to the delivery of ES parts is assumed. If the parts meet functionality, the fabless IC house must start to thoroughly characterize the functionality and performance of the design, start a qualification process to verify reliability, and start activities to assess and verify manufacturability of the parts. Having achieved the appropriate benchmarks, the parts are usually deemed fit for shipment to the customer(s) as QS. Software development activities progress in parallel with benchmarks for the shipment of α, β, and production levels. In many of today's complex applications there could be multiple releases of software within each of these categories. Plans for the

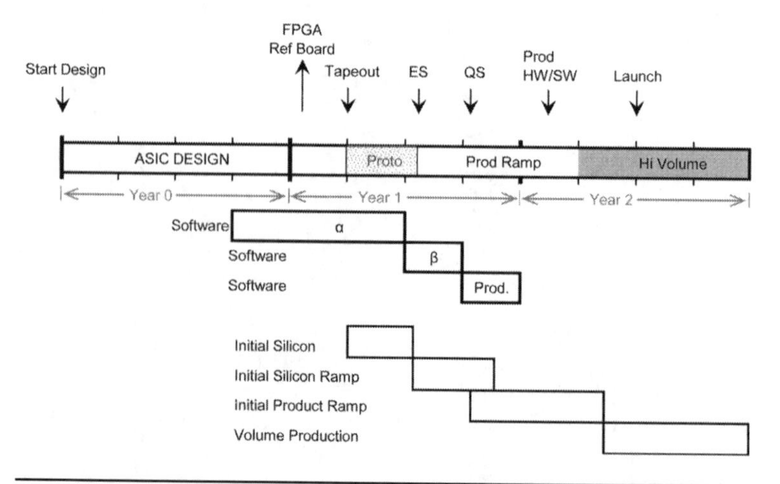

FIGURE 2.8 Typical ASIC development schedule at a fabless IC company.

build up of silicon and packaged parts need to be developed closely with the OEM customers. This is important because the lead times for silicon are usually long and the customer normally wants parts delivered right away—more later.

Overall, this typical development cycle shows a lead time of nine quarters from the start of ASIC development to the launch of a product by the retailer. This schedule also reiterates that for a new start-up the tapeout benchmark is usually only the half-way point for time to revenue. Some designs could be taped out in two to four quarters from the start of design activity, versus the five quarters shown in Fig. 2.8. Such accelerated execution is possible if the design is a re-implementation of a previous architecture or a previous design, and is not too complex.

Now let us discuss the relationship between this ASIC development schedule and the product development schedule at the OEM, and the interdependencies.

As said before, the OEM's product development cycle is generally shorter than the IC development cycle. In Fig. 2.9 the product

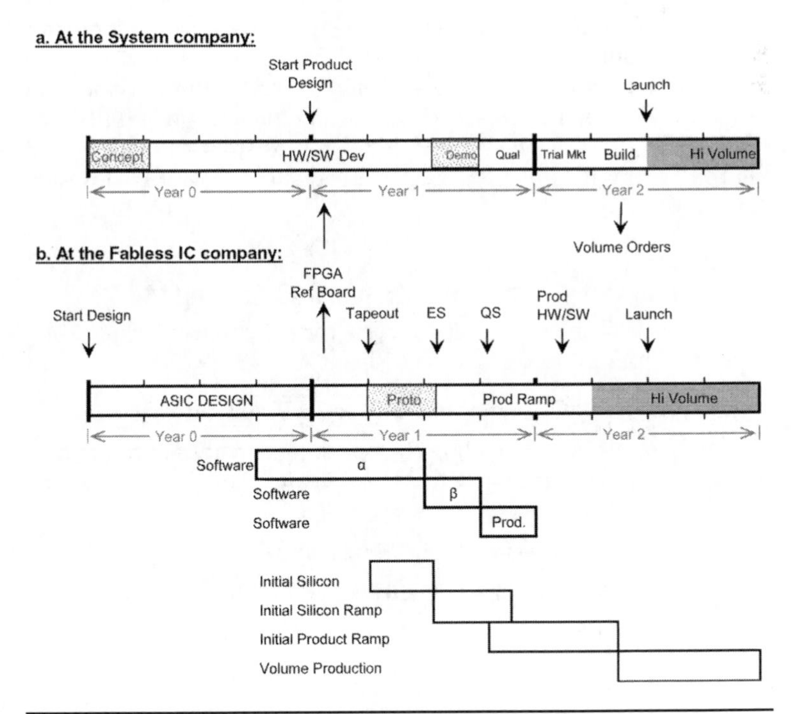

FIGURE 2.9 Product development schedule at the OEM (a) compared to the ASIC development cycle at the fabless company (b).

development is shown to be six quarters long. An initial, strategic "roadmap" discussion between the OEM and the ASIC developer is appropriate as the fabless company launches its design effort. Periodic updates can be invaluable in keeping the two in synch. Of course this is making the assumption that there is a working partnership between the two parties. This is easier to do for an established company and can be a challenge for the brand new fabless IC company—they will need to have some important connections into the OEM or possess an overwhelming idea that the OEM "must have" as a feature in a future product.

It is important for the fabless IC supplier to provide reference boards and software, and possibly other materials as the OEM kicks off their system product development cycle internally. The OEM will validate use of the reference boards on bread-boards and evaluate the product in detail when ES parts are made available. If all goes well, this is followed by a qualification phase and a possible trial marketing run. Successful completion then usually triggers the issuance of an order for initial build quantities. The challenge for the fabless IC supplier is that the demand is usually for parts deliveries in a short period, typically 4–8 weeks. This usually means the fabless company has to build parts at risk and have them in the pipeline. This could cause a serious financial burden for the fledgling start-up company. Following a build cycle of 4–8 weeks the (system level) product could get launched about six quarters after the start of the development program. Note that these schedules assume the design is functional on the first pass. Depending on the complexity of the design and the technology (Chap. 4.4) it may be prudent to factor in a design "re-spin" in the schedule.

2.6.2 An Example—Microsoft's Xbox360 Gaming IC

One example of the development cycle for the chip set for a leading edge consumer product follows [2.10]:

- 3-chip set (CPU, graphics processor, IO processor)
- The chip set has leading edge packing density, approximately 500M transistors, which is greater than predicted by Moore's law.
- CPU, approximately 165M transistors
- 90 nm SOI (Silicon On Insulator) CMOS Technology
- Start to tapeout 15 months
- Beta development kit 20 months
- Product launch 26 months

Shown in Fig. 2.10 is a block diagram of the CPU (Central Processing Unit) [2.11]. Figure 2.11 is a die photo of the CPU die [2.11].

> At the onset of the Xbox 360 chipset design start, we knew we were planning an aggressive schedule and ramp. Flawless, relentless, and close coordination with our service partners made the difference between success and failure.
>
> Bill Adamec
> Senior Director
> Semiconductor Technology Group
> Microsoft XBox

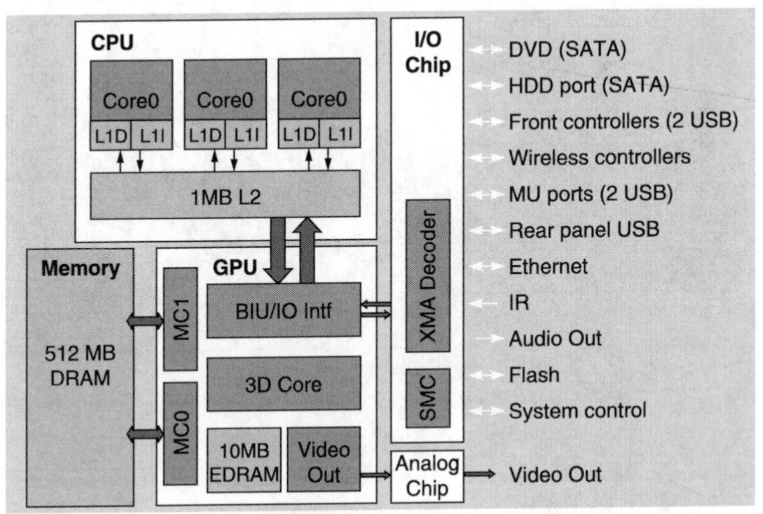

BIU: Bus Interface unit
MC: Memory controller
HDD: Hard disk drive
MU: Memory unit
IR: Infrared receiver
SMC: System management controller
XMA: Xbox media audio

FIGURE 2.10 Block diagram of the Microsoft Xbox CPU launched in late 2005. (*Courtesy of Microsoft [2.11].*)

FIGURE 2.11 Die photo of the Microsoft Xbox360 CPU. (*Courtesy of Microsoft [2.11].*)

2.7 Key Points

- Customers now demand more completer solutions from the IC supplier. This includes software, reference designs, test suites, and the like.
- Establishing relationships on the customer side, as well as the supply side of the value chain are essential implementation of ICs.
- The ASIC design and development timeline is only a fraction of the overall time to production revenue.

CHAPTER 3

Lifecycle of a Fabless IC Company

3.1 Getting Started

It all starts with an idea! A good idea for an IC product with a healthy market demand, an experienced entrepreneurial core team, and a little bit of money are some key elements required to launch a new company. The core team usually has a technologist, a sales/marketing individual, and a business person.

Launching a new fabless IC company was relatively easy in the late 1990s, but has become a little more difficult since 2001. There are many possible roadblocks and impediments that keep the success rate relatively low, as mentioned in Chap. 1. Some of the key strategic elements for success are discussed in this chapter. Other issues related to technology selection, design infrastructure and operations issues are discussed in later chapters. Although the focus here is on a fabless startup, the principles discussed here can also be applied to the launch and execution of a new IC development program at a mature fabless IC company.

3.1.1 Lifecycle Overview

Figure 3.1 is a simplified timeline of the typical fabless IC company life cycle. The development schedules outlined in the previous chapter fit within this broader timeline. An idea, a core team of entrepreneurs and some funding from "friends and family" gets the new company going. The technical member of the team usually starts validating the idea while the marketing and business team starts to sound out possible advisors, customers, and funding sources. Soon they have to go after some seed, or "angel" funding. This is usually around $0.5–1M and could last 9–12 months, depending on how frugal the team is with cash. By the time the CEO (Chief Executive Officer) starts working on the first round of funding, usually called

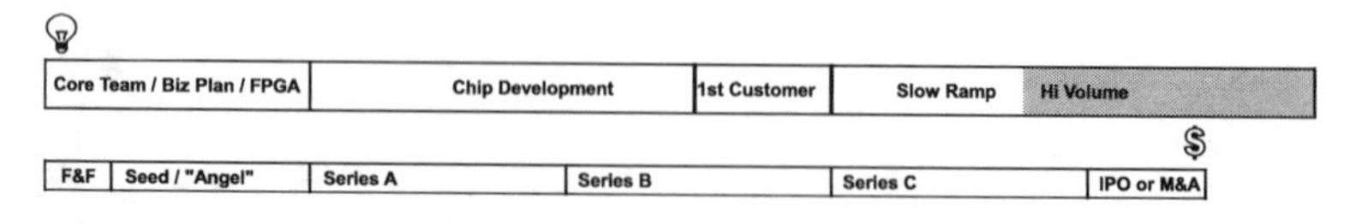

FIGURE 3.1 A simplified overview of a startup fabless company's lifecycle.

"Series A," "the technical idea should be reasonably well demonstrated and a business plan developed. Chip development gets started in earnest as Series A funding becomes available. For simplicity this timeline shows discrete transition points; in reality there are overlaps and discontinuities. Overall activities in the development cycle were discussed in Chap. 2, see Fig. 2.7. There is also more detailed discussion of development activities later in the book. In the traditional model, upon completion of engineering samples the search begins in earnest for locking in the first customer. In the new model, following the recommendations in Chap. 2, the fabless company should already be engaged with one or more key customers. The customer orders initial quantities of pre-production and then production parts. This high level view assumes a success story. In the meantime, the company will need to raise funds every 9–15 months, depending on the amount of funding in each round and the cash "burn-rate." As is discussed later, possible exit strategies for the emerging company are to go public, get acquired, or scale up to be a large company.

How many rounds of funding and how much cash gets raised varies widely. Table 3.1 shows [1.23] the number of fabless company funding events in 2006. This information is tabulated by the various rounds (one thru nine) or Series A thru 1. Nearly 70% of the funding events were for series A, B, and C. The average amount of funding for the first three rounds was $10M, $14M, and $16M. Note also that there was at least one company that raised nine rounds of venture funds.

It is also interesting to note that the number of funding events and the amount of funds varies over the years. These trends are shown in Fig. 3.2. The overall trend is a slight decline in the amount of funds made available, although the total number of fundings has stayed fairly flat. This indicates a downward trend of the average funding amount. Therefore, the startup CEO must expect more challenges and competition in securing funding. The conclusion is that the new startup needs to have better value propositions for the investor—a strong business plan, a strong execution plan, and improved opportunities for investors to receive returns on their investment.

3.1.2 Product Positioning

Proper definition of the product is a key to market positioning and market acceptance, and eventual success. Understanding the trends

Round	Series	Number of fundings		Total ($M)		Ave funding ($M)
1	A	18	12%	183.5	10%	10
2	B	40	28%	554.5	29%	14
3	C	40	28%	623.9	33%	16
4	D	12	8%	307.2	16%	26
5	E	5	3%	72.8	4%	15
6	F	1	1%	15.0	1%	15
7	G					
8	H					
9	I	1	1%	15.0	1%	15
Undisclosed		28	19%	118.3	6%	4
		145		1890.2		13

Source: GSA

TABLE 3.1 Summary of Funding Events in 2006 Categorized by the Funding Round (*Courtesy of GSA.*)

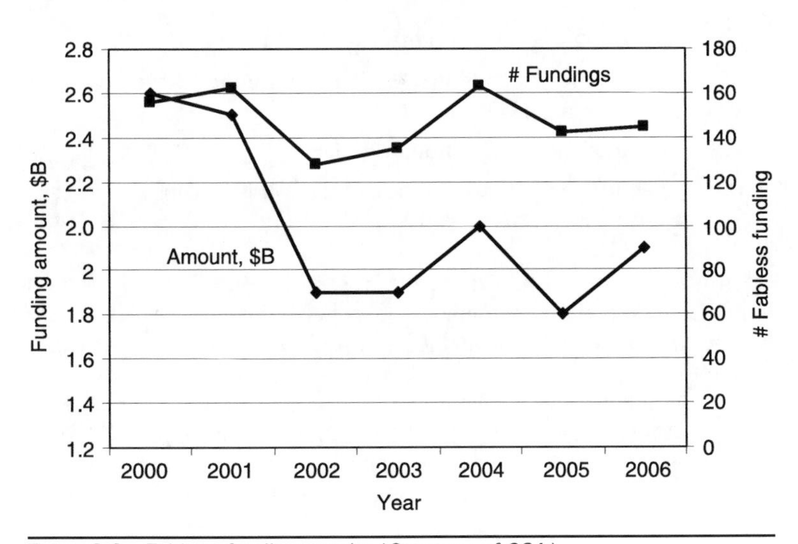

FIGURE 3.2 Fabless funding trends. (*Courtesy of GSA.*)

in the marketplace for the proposed ICs, the customers 'products and the end customers' needs is extremely important in the fast moving electronics market. An aggressive but realistic schedule, an implementation plan and commitments, together with a history of meeting benchmarks, are keys to establishing credibility with customers. These can be achieved with an all-star team and a streamlined execution plan. There have been many great technical ideas that fail because of poor execution or slow market penetration.

The class of IC product being developed at the fabless company can affect its overall lifecycle. Here, I have classified the IC products into two major categories based on standards, technology, market, and the customer base. Each of the two categories is sub-divided into three or four IC types:

EXISTING standard, market, and customer base

- Super integration, e.g., chip sets for PCs (Chips and Technologies, early 1990s).
- Problem solutions, e.g., incorporating an algorithm to fix OLED (Organic Light Emitting Diode) display degradation.
- Evolutionary enhancements, e.g., cost reductions and feature enhancements.

NEW standard/technology, market, and customer base

- New or emerging standard, e.g., WiMax, 802.11x, H.264.
- New features/capabilities, e.g., c.link for multimedia (Entropic), and cameras in cellphones.
- New interfaces, e.g., Bluetooth, USB.
- "Revolutionary" capabilities, e.g., MEMS sensors, special sensors on cell phones.

In the existing standard category, getting to market quickly and gaining market share are the key end-goals. Incorporation of the new IC into the customer's product is usually a "drop-in" into an existing system or socket.

In the new standard/technology category, creating the market is the operative goal. The new standard/technology category is characterized by a longer IC development cycle time, due to evolutions of the specifications and features. Integration of the new IC solutions into customer end products mean a longer development cycle. Moreover, there may be a need for the creation of a new alliance or a new usage standard. These efforts take time and usually cause a slower ramp up of production, and therefore a longer period for TT$ (Time To $) for the emerging IC company.

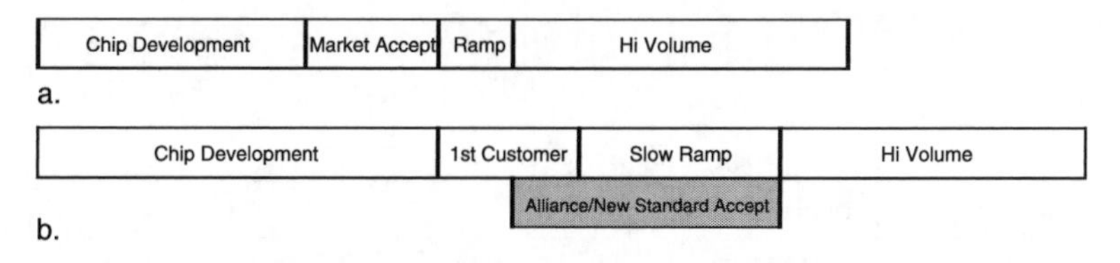

FIGURE 3.3 Conceptual timelines for (a) existing and (b) new standards, markets, and customer base.

Conceptual timelines are shown in Fig. 3.3 for IC products serving an existing market (a) and one that needs a new market created (b).

Early engagement and communication with the customer facilitates a quicker evaluation and validation of the new IC, and results in improved TT$ for the startup.

> The fabless company must identify the first potential customer early in the lifecycle. It is important to have the customer's inputs in shaping the product requirements. Your product will have a much better chance for a design win in the customer's product. Besides, investors and the suppliers will likely get a favorable story when they perform due diligence with the customer about viability of your product and potential volume.
>
> Jim Fiebiger, Ph.D.
> Member, Board of Directors
> Actel, Mentor Graphics, QLogic,
> Pixelworks, Power Integrations

3.1.3 Strategic Decisions—Fabless or IP company?

Should the start-up with the new idea sell/license the IP or be a fabless IC company? This is a strategic question that must be pondered by the executive team. There are definite pros and cons, as discussed here.

Unlike the early 1990s, today's distributed supply chain and the availability of "best-in-class" suppliers has made it relatively easy to get access to design and manufacturing capability. So it is relatively easy to launch a fabless IC company. As elaborated throughout this book though, there are many challenges in becoming a successful fabless company.

A successful fabless company has the potential to achieve a much higher revenue stream than being an IP company. This is illustrated in Table 3.2 through a simplified calculation of the revenue and profit

	Fabless IC company	IP company
Unit price	$10	$10
Royalty rate (p.u.)	n/a	5%
Unit volume/year (M)	10	10
Annual revenue ($M)	100	5
Profit margin (%)	30	80
Net annual profit ($M)	30	4

TABLE 3.2 A Comparison of Revenue and Profit of a Fabless Company and an IP Company

in the two cases. Simplified assumptions include a $10 selling price for the part, a 10M unit annual ship rate, a generous 5% royalty rate, and a much higher profit margin for the IP company. Note that a conservative profit margin of 30% has been assumed for the fabless company—numbers in the 30–50% for successful companies are not unusual. Using a higher profit margin for the fabless IC company will skew the analysis further in its favor.

So the choice between becoming a $100M fabless company versus a $5M IP company makes entrepreneurs generally opt for the fabless IC company model. However, as discussed throughout this book, there are many more risks and headaches in implementing the fabless company.

The IP licensing model could work in a couple of different ways to implement the new idea, as outlined here.

- "Hard" IP block, which is delivered to a licensee as a design GDSII (Graphic Data System II) tape. The IP block needs to be customized for the specific process selected by the licensee. Customers could be system companies, an IDM, or a large fabless company. The IDM could incorporate the IP as a standalone chip and sell it either under their logo (more likely) or as a different brand. The system or fabless company will likely incorporate the block into a larger IC.
- "Soft" IP core, which is delivered as an RTL (Register Transfer Level) description of the IP. The customer then incorporates the new IP into their own design, instantiating it into the right process using a standard cell library of their choice. The customer base is the same as for the hard IP.

The following discussion explains the path for a fabless IC company.

3.1.4 Feasibility Assessment

A structured process for assessing the feasibility of the new idea and the associated decision-making is illustrated in Fig. 3.4. The process

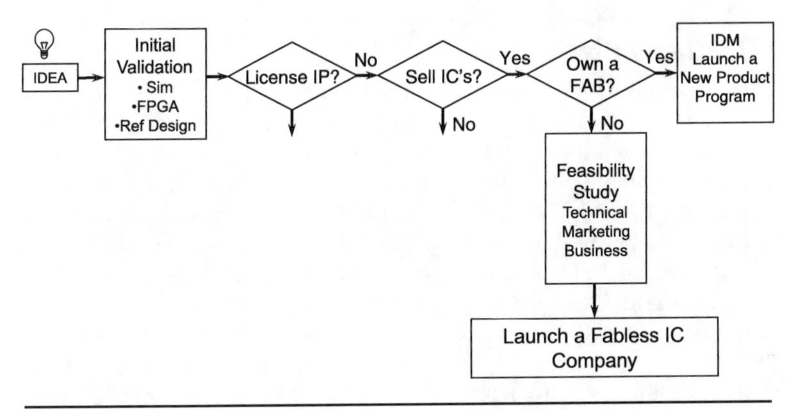

FIGURE 3.4 Process for feasibility assessment.

includes choices such as selling ICs, licensing IP, IDM, or fabless. After an initial validation of the idea through simulations, an FPGA and a reference design implementation, the executives must assess the IP supplier versus fabless company questions discussed in the previous section. If this process occurs within an IDM there are likely defined processes for next steps in developing new products internal to the company. The budding fabless company, though, must conduct a thorough technical, marketing and business study before launching a fabless IC company. In doing so, the principals must compare the strength of the idea in the marketplace, assess core competencies of their team relative to the proposed implementation path and align the company vision as they develop the business plan.

Some of the key questions to be addressed in this feasibility assessment are:

- Technical feasibility assessment
  - Can this be done?
  - Can this be done within cost goals?
  - How much software development will be required?
- Marketing assessment
  - Is there a Market? TAM (Total Available Market)? SAM?
  - Who is the competition? Size it!
  - What is the sales strategy?
  - What are potential volume projections?
- Business assessment
  - Can funding be obtained?
  - How feasible is the business plan?

3.2 Business Plan

That a business plan is a key requirement when approaching investors for requests of funding is well known. The business plan development process requires the executive team to flush out the technical, marketing, and strategic plans, the business issues, and the alignment of the company direction. We are all familiar with startups with brilliant technical ideas that were unsuccessful because of poor execution. Good execution starts with a good team and a good business plan. Overly aggressive business plans showing high rates of return for the investors, especially from a team without an established track record, will have a problem getting funded. I recommend realistic, yet aggressive schedules and financials. A "holistic" view of the program that provides a realistic schedule for product revenue is recommended.

The business plan provides an overview of the company's objectives and the management's plan to achieve the objectives. Information included must outline the technical aspects of the product, how it is differentiated, the target market, and customer base, the market size and served available market, the execution team, the revenue plan, and the margin projections.

The outline of a sample business plan is included in Appendix A.

Many companies pay very little attention to the planning beyond the first GDSII tapeout. And yet, that benchmark represents only the half-way point to realizing any return for investors, as illustrated in Chap. 2. For products that require a new standard and market, the first tapeout could represent only the first 20–30% of the TT$.

Of course, the investors are looking to reduce TT$, improve returns on their investments and increase the company's multiples ratios. However, the fabless company CEO must present a plan that is realistic for the achievement of the stated objectives. It must show realistic development schedules and an execution plan as a fabless IC supplier that aligns the market, the IC capabilities, customer requirements, cost goals, and provides confidence to investors about the team's ability to accomplish the goals. Key questions to be addressed are related to the fabless company's ability to develop and market the new IC:

- within development budget;
- in the right timeframe;
- within the unit cost budget, both initially and with expected price erosion;
- Is the right team available?

3.3 Funding Process

There was an initial discussion of the lifecycle of a new fabless company and its relationship to the funding process in Sec. 3.1. The following funding steps were outlined there.

- Initial funding to get the project off the ground. A few hundred $K from "family and friends" usually gets the principals going.
- Around $0.5—1M, mostly "angel" funding gets one a few key "hands-on" staff. This funding could last 6–9 months.
- By the time the CEO is out looking for the first round, or "series A" funding there should be some tangible results—either simulations or a breadboard reference design as a proof of concept. Typical funding amount is $10M. These funds are usually good for 9–15 months depending on the "burn rate."
- Summary of the 2006 data from fabless fundings tracked by the GSA indicates that the average Series B and Series C fundings were $14M and $16M respectively.

In the late 1990s, during the "dot com" days, almost any reasonable idea with a decent business plan got funded. The success rate of companies was not very good—there are many anecdotal stories about unsuccessful fabless companies. Subsequently, business plans are now scrutinized much closer with serious due diligence.

These days venture capitalists are also looking for rapid return on their investments and are usually looking to own a larger share of the fabless company.

Upon acceptance of the proposed business plan presented by the CEO, a "term sheet" for private placement of funds is generated by the investor(s) agreeing to fund the emerging fabless company. The term sheet usually includes a description of the amount, the shares and type of stock, voting restrictions, and stock transfer restrictions. Also included are considerations such as exit strategy and liquidation, dividends, and dilution. Appendix A shows the major elements of a typical term sheet.

3.3.1 Government Funding Options

As pointed out in Sec. 1.2.3, securing commercial venture capital (VC) funding has become difficult in recent years. One alternative source of funding worth considering is from a government and/or defense related agency. Such a source could be a good match for principals from university and research organizations and for startups with killer ideas that may have a relatively long incubation period (3+ years to commercialization). The basic principles outlined in this chapter are applicable as one embarks on seeking government funding. However, there are additional limitations that must be considered. Markets for government related ICs generally have low product volumes and have long product support cycles. Also, it takes special working relationships to engage with such agencies. Clearly it will be important to get advice from individuals experienced in working with government agencies.

A combination government and commercial strategy may be appropriate for some ICs. Executing such a strategy also has additional challenges since it may be restrictive on factors such as the technologies used and export regulations.

In the United States, it is entirely feasible to fund a fabless semiconductor startup effort with government funding in the form of research contracts from agencies such as the Department of Defense and the Departments of Homeland Security, Transportation, Energy, Agriculture, Justice, and several others. The initial challenge of obtaining such government funding for your startup is no different than that of securing funding from venture capitalists. As in the private sector the first step for the startup is to do your market research. Who are your most likely "customers" within the government and what problem does your technology solve for them? To discern this requires a great deal of legwork, discussions, and relationship building. If you take this route the most valuable asset that your enterprise will have, aside from your IP, will be a "champion" for your technology or solution within one or more government agencies—building such relationships needs to be your goal and the best way to do this is through some initial successes that make everyone look good, your company and your government "champion."

One way to start building a relationship with federal agencies is through the SBIR (Small Business Innovative Research) or STTR (Small Business Technology Transfer) programs. These are "set-aside" programs for small businesses with less than 500 employees. A good starting point to learn about these programs is the Small Business Administration web-site, see http://www.sba.gov/SBIR/indexsbir-sttr.html. Solicitations are issued several times each year and there is an opportunity to speak directly with the government program directors at the early stage of the solicitation process. The vast majority of program monitors will be remarkably frank and open about the problems they are trying to solve and their technology needs so this is the best opportunity to start to build inroads to the agencies. Don't simply submit proposals in response to printed solicitations! Make sure that you are on target and not just wasting your time by first talking to the program monitor. The SBIR/STTR program is multi-phase with Phase 1 awards ranging from $75–100k for an initial 6–12-month project. Successful Phase 1 projects may lead to Phase 2 projects that range from $500k to $750k spread over 2 years typically. Program "options" and Phase 3 awards can increase the total funding into millions of dollars. Your goal should be to develop a partnership with the program monitor and to become a valued technical resource to help them address their agency's needs and perhaps even to provide advice and input for future solicitations. Such an approach can reduce the sometimes long solicitation cycles.

Another funding avenue is to respond directly to BAAs (Broad Agency Announcements) periodically issued by the various agencies. The general programs are not limited to small businesses and draw submissions from both universities and large businesses, and therefore are more competitive. However, funding levels can be much greater and if you have a unique technology that solves the agency's problem a small company can be very successful in this arena.

The start-up should regard the SBIR/STTR program not simply as a potential funding source but rather as a way to introduce your company to the government and to begin building a long-term relationship through which you can align your company goals and technology roadmap with government needs. This cuts both ways. Although the commercial aspirations of your company may not align precisely with the needs of the government agency you should work to identify (and maximize) the overlap of your technology roadmap with the government agency's needs. Likewise, program monitors may not fully understand the capabilities and potential of the technologies that are out there (yours especially!) and they may realign their goals and expectations in light of the capabilities that your company has to offer.

Developing the government as a customer can be a good way to start a fabless semiconductor company. However, it is important to keep your initial goals in mind and to have a good understanding of the limitations of this approach. The goal within the government contractor arena is to have your technology make the transition to an "acquisition program of record," which refers to a specific platform or device being built for the government. One significant advantage may be the ability to leverage IP blocks and other infrastructure across the government and commercial projects. Keep in mind that the government wants your business to be a commercial success, especially in order to assure that your company will be around to sell the needed products to them in the future.

The great trap in the government funding approach is to lose sight of your start-up's commercial goals. For the fabless semiconductor startup it is especially important to avoid the pitfall of perennial pursuing agency solicitations, since your ultimate success will depend on your reaching mass markets. By partnering with the government and maintaining an active dialog throughout the process, the government can be an invaluable ally to help in the early stages of your startup.

(Continued)

> Ultimately your company will require additional funding for commercialization and marketing of your technology. If you utilize the available government funding opportunities to navigate the early stages of your technology development cycle, you will find your startup in a much stronger position, with a much higher company valuation when you go back to VCs for funding three to four years into the development of your enterprise. During this timeframe if the fabless IC company is able to cultivate a customer willing to commit to future purchases or to guarantee loans it may even be possible to avoid the need for substantial equity funding altogether.
>
> Professor Mark Bocko,
> Chair, ECE
> University of Rochester & President and CEO
> ADVantage Imaging Systems, Inc.

Now let us describe the overall IC development and implementation cycle. Remember—besides having a differentiated IC product, execution is the key to success of the company. Making the right choices and tradeoffs is very important for meeting the goals and schedule set forth in the business plan. For the founders of the fabless semiconductor company, product delays may lead to the need for additional rounds of funding which lead to further dilution of ownership in the company [3.1].

3.4 Development Cycle—the Four Phases

The IC development and implementation lifecycle has been partitioned into four phases, as shown pictorially in Fig. 3.5. This representation is similar to Fig. 2.7.

The four phases are:

- global planning;
- IC design;
- prototyping;
- IC production.

3.4.1 Global Planning

This is a short phase prior to the kick-off of the ASIC design. Many companies have a tendency to jump into ASIC design without sufficient due diligence. If properly implemented, this phase can be very valuable not only to define the baseline execution plan, but also to document and estimate some sensitivities to variables. It turns out that almost every design ends up with more gates and features on the

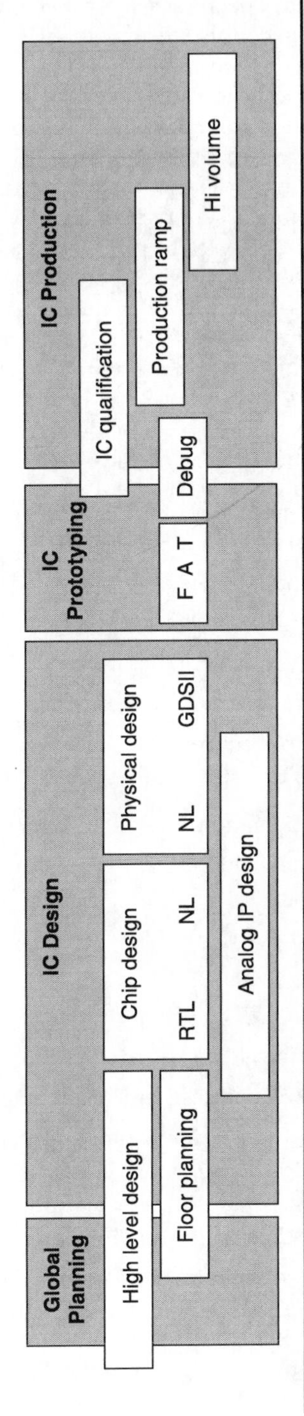

FIGURE 3.5 Four lifecycle phases of IC development and implementation.

IC than originally planned. The die size almost always ends up larger than planned, package pin counts are almost always greater and test times end up being higher than planned. Some experienced CEOs have an add-on fudge factor to estimates they see from the team! In Chap. 4 data is presented from a study that substantiates the fact that a majority of IC designs miss the planned schedule. By investing in a more diligent planning cycle that includes sensitivity analyses, executives could have a much better set of inputs as they refine the business plan. Sensitivity analyses to product cost and schedule as the chip complexity grows, the die size grows, the pin count grows, and the test time increases could prove invaluable. Such due diligence will also help make better trade-offs as incremental requests are made for inclusion of new feature enhancements that marketing and sales "must have."

The following is a list of major activities recommended in the global planning phase. The list is by no means all inclusive.

- Implementation method—ASIC or COT (customer owned tooling) or ? (Chap. 4).
- Design partition parameters and trade-offs:
  - gate count;
  - memory—how much and what kind;
  - (IP) blocks required;
  - package type, pin count;
  - die (or chip) size.
- Technology selection and trade-offs (Chap. 5).
- Supply chain partner identification and early commitments of support (Chap. 6):
  - ASIC supplier;
  - foundry;
  - library, memory, IP;
  - package assembly, test.
- Cost estimation, both unit cost, and development cost (Chap. 7).
- Overall program plan development, including resource estimates (Chap. 9).

3.4.2 IC Design

During this phase, IC design activities are started and gain momentum. Presumably decisions have been made regarding the implementation method and an execution plan. Most startups will at least do the front-end design, from specification through RTL for a digital chip internally.

Chapter 6 will provide details of what is needed for this activity internally. While the initial focus is on design implementation, the latter half of this phase should include increasing operations planning activities. The major focus of these operations activities is the preparation for the prototyping phase. Culmination of this phase occurs when the design database is transferred to the wafer fab for mask making and silicon prototyping.

The following is a list of major activities in this phase.

- design tools and flow selection, acquisition and set-up;
- IP and supplier selection;
- personnel hiring to implement the design;
- design for test and production test methodology;
- supplier relations and contracts;
  - IP, libraries, memory, foundry, design services, packaging, ...
- technology features and cost trade-offs;
  - MiM capacitors, V_T options, ESD implant, deep n-well to name a few;
- shuttle prototyping and/or engineering lot prototyping arrangements;
- characterization plan;
- qualification planning;
- package design;
- test/failure analysis/debug planning;
- operations and manufacturing planning and manpower planning;
- quality planning and document control;
- refinement of cost estimates;
- forecast and order placement for the prototyping.

3.4.3 Prototyping

The major activities in this phase revolve around the generation of masks, fabrication of silicon wafers and the delivery of sample, prototype parts, and the beginning of design validation. If the design functions as planned, ES (engineering samples) parts can be made available for customers to validate. However, if there are problems found during the testing, there needs to be debug activity to determine causes of the problems. Appropriate remedies and action plans need to be put in place on rapid schedules. There should also be an increased focus on preparations for ramping up of production.

A list of the major activities during this phase is as follows:

- coordination of database acceptance at the mask shop and the foundry;
- specification of any special processing instructions to the foundry;
- review of electrical test data from the foundry;
- coordination of packaging and assembly of the prototype parts;
- start of product reliability qualification;
- bench test validation of the design;
- ATE test program debug and parts testing;
- validation of yield and cost assumptions;
- failure analysis and debug, if required;
- firming up of supplier commitments and contracts;
- forecast and capacity planning.

3.4.4 Production Ramp and High Volume Manufacturing

This phase usually does not start in earnest until customers have accepted the design at their facilities in their products, and the IC qualification is well under way under the coordination of the fabless company. There could be some pre-production material that gets built at risk early in this phase. Such material could be used to supply QS parts to customers. Customers will likely build small quantities of their product for reliability testing and product validation.

The major activities in this phase are as follows:

- completion of ASIC reliability testing;
- yield characterization;
- refinement of test program for production test;
- yield enhancement program;
- test/failure analysis and debug, as required;
- validation of yield and cost estimates;
- forecast and capacity planning;
- second sourcing plans for high volume products.

3.5 Roadmap of Products

In formulating the company's strategic and business plans, it is important to have a roadmap of products. Start-ups generally focus on a single IC product. While this is right from a focused execution

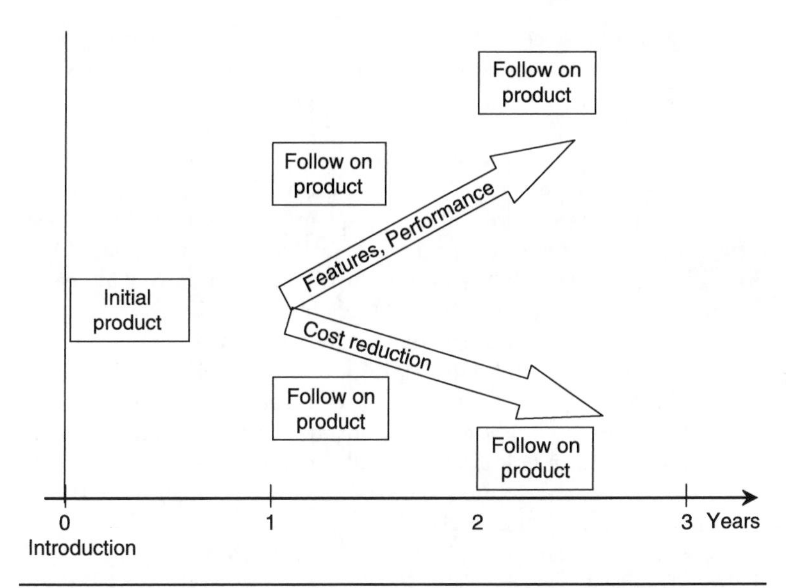

FIGURE 3.6 Typical representation of the IC roadmap of a fabless company.

perspective, the roadmap is important to have for the following reasons. In today's dynamic marketplace there is continual demand from customers for feature and performance enhancements and for price reductions. There is also an internal need to reduce cost in order to improve profit margins. The roadmap will be a vehicle to demonstrate the fabless IC company's understanding of the customer needs and will represent a systematic way to meeting customer expectations. The roadmap approach also helps avoid the "kitchen sink" syndrome on the first design. I refer to the kitchen sink syndrome as the mentality to include any feature that comes along, even if it means recycling the design in the design phase. Figure 3.6 is a schematic representation of a fabless IC company's roadmap showing two tracks—one for feature enhancements and one for cost reductions.

> Establishing your credibility early on is another very important factor. For this I suggest laying out a plan with the customer that includes a staged product introduction. Getting them first articles on schedule is extremely important, even if the initial samples don't have all the "bells and whistles" of the final product.
>
> Jim Fiebiger, Ph.D.
> Member, Board of Directors
> Actel, Mentor Graphics, QLogic, Pixelworks,
> Power Integrations

Some important considerations to keep in mind are as follows.

First and foremost, customers are looking for more complete solutions. The tendency then is for "feature creep"—to keep making modifications to the chip specifications to include more and more features on the present ASIC. There is also the case of "creeping excellence"—engineering enhancements that make sense on a case-by-case basis, but may not be consistent with the big company picture. Both these type of "enhancements" cause delays and "churn" during the development cycle. As a result, there are schedule re-commits, cost increases and increased frustration among the engineering teams. The availability of a well thought out roadmap is a possible way of limiting the number of iterations of the initial design. In the extreme case, it may be feasible to plan the initial design to never go into high volume production—it may be more important to get the initial IC out there as a proof of concept, and without all the "bells and whistles," but on time. Without a roadmap such a strategy will have difficulty in getting acceptance at the customer.

Second, customers are looking for long term relationships with their suppliers. It may be more important to establish credibility by delivering to commitment and establishing a plan that gets the customer the features they need per the roadmap.

A roadmap for cost reduced products is also very important. Customers always want cost reductions. Historical end product cost reductions of 20–30% per year generally translate to lower IC ASP's, as depicted in Fig. 3.7, for the initial product, P0, and follow on products P1 and P2 introduced in years 1 and 2. Whether the company's cost reduction strategy is shared with the customer is left up to the judgement of the executive team. I recommend sharing a

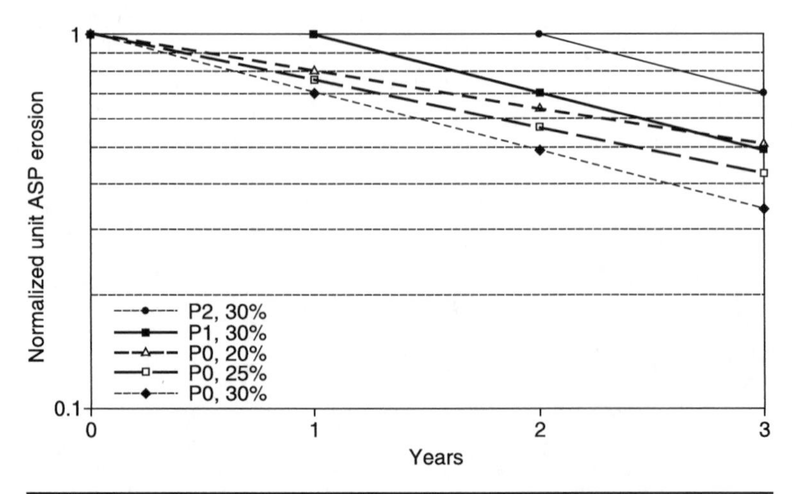

FIGURE 3.7 Typical IC ASP reduction curve for products P0, P1, and P2, introduced in years 0, 1 and 2 respectively with a 30% annual erosion. Also shown is 20% and 25% annual erosion on P0.

"high-level" cost reduction roadmap with the customer. And then driving internally to execute on more aggressive plans in order to preserve and increase profit margins.

Here are some examples of features and performance enhancements and cost reductions that could be included in the roadmap of products:

- Feature enhancements:
  - software;
  - new interfaces—USB 1.x, 2.x, HDMI, . . .;
  - new standards—802.11b,.11g,.11a, 11n, and others.
- Performance enhancements:
  - 1Mbps to 10 to 100 to 1 G to 10G Ethernet.
- Cost reduction:
  - wafer cost reduction;
  - package, assembly and test cost reductions;
  - technology scaling/shrink to the next process node.

3.6 Exit Strategies

A typical start-up that has raised three or more rounds of funding has three major options for an exit strategy that brings winning returns to investors:

- stay private, scale up the company at least for some years;
- go public via an IPO (Initial Public Offering);
- M&A (Merger and Acquisition)—merge with or get acquired by a larger company.

There are many examples of companies in each of these cases. A favorable "valuation" of the company's worth by analysts is a key to determining the feasibility of an IPO or an M&A. In years past, revenue run rate was the major factor in determining valuation using "multiples" ratios. Nowadays profitability is the primary consideration, although revenue is also important.

Table 3.3 is a summary of the number of IPOs and M&A activities over the last few years [1.23]. The number of IPOs is split between total semiconductor IPOs and fabless IPOs. Values of the M&A activities are also listed.

3.7 Long Range Strategies

While the start-up fabless company worries about managing cash flow and getting that first product design win and an exit strategy,

	IPOs		M&A	
	Semiconductor	**Fabless**	**Number**	**Value ($B)**
2003			53	2.7
2004			56	2.9
2005	9	9	59	4.3
2006	8	2	60	13.4

Source: GSA

TABLE 3.3 Summary of Number of IPOs and M&A Activities over the Last Few Years (*Courtesy of GSA.*)

successful companies must embark on an ongoing long range strategic plan and direction. The initial exit strategy is only the beginning of a treadmill that focuses on factors like growth and scaling up revenue, moving up the value chain, margins and profitability, the product portfolio, staff, and locations. A listing of the top 10 fabless IC companies sorted by 2006 revenue is shown in Appendix A. The GSA maintains a database of fabless IC financials [1.23]. A more extensive listing of fabless IC company revenues is also available [1.24].

QLogic started out as a fabless semiconductor supplier in the early 1990s. As our revenues grew to around $250M we chose to move our offerings up the value chain. By supplying board and box level products, and eventually software, our revenue has grown at a CAGR of 19% over the last 5 years. For fiscal year 2007 (ended March, 2007) we had a revenue of $586.7M. Our margins are now over 65%, and are much higher than as a fabless IC company. And yet, fabless semiconductors are our core strength.

H.K. Desai
President and CEO, QLogic Corporation

3.8 Key Points

- Product definition and positioning can be crucial in determining the overall time to revenue. For example, a new IC that requires the establishment and acceptance of a new industry standard can take an extra few years.
- Key elements of a strategic planning cycle include a feasibility assessment, a business plan, funding options, and long-range vision.
- A roadmap of products incorporating a phased introduction of feature/performance enhancements and cost reductions is highly recommended.

CHAPTER 4

Selecting the Implementation Approach

4.1 FPGA, Gate Array ASICs, Semi-custom ASICs

Entrepreneurs have many choices for the implementation approach to be used in converting their ideas into an IC. Selecting the optimum approach can be somewhat difficult because the choices depend on many factors such as the application, chip content, performance, schedule, cost, volume projections, and the teams' experience. Figure 4.1 illustrates the choices. The assumption is that the IC being developed is for a user's, or their customer's, specific application, and is therefore an ASIC. ASICs can be separated into two major categories. The first category consists of devices that can be purchased off-the-shelf and can be programmed by the user to fulfill the required functionality. The second category consists of ASICs that are supplied by an IC company that either has an internal fab (IDM) or is a fabless company. For the next level of categorization, based on the methodology used in the design of ASICs, the industry has used the terms "custom" and "semi-custom." Note that ASICs developed at vertically integrated system companies for their own internal use will have a similar set of sub-categories.

As the name implies, custom integrated circuits have traditionally been "hand-crafted" and require more designer intervention. In the very early days, *all* ICs were handcrafted. As the number of components per chip escalated, the need for automated design tools and methodologies was recognized and this led to the birth of the EDA industry. Nowadays, most chip designs use EDA tools extensively and are not really hand crafted. Therefore the boundary between custom

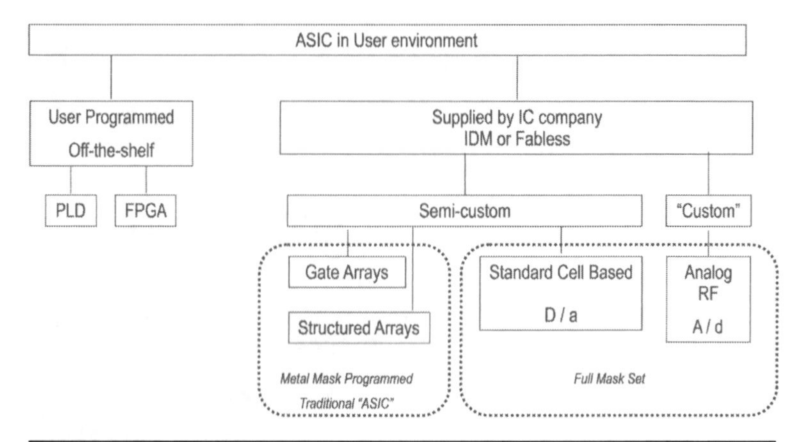

FIGURE 4.1 Schematic representation of the various types of application specific ICs.

and semi-custom has become quite fuzzy. Included in this custom category are:

- Large microprocessors with mostly digital content, with some analog circuit blocks.
- Chips with RF functions and those with mostly analog functions.

Semi-custom ASICs are split between "standard cell" based and array based chips. Standard cell based ASICs are the major focus of this book. They allow designers to incorporate many analog, memory and other IP blocks in addition to a large number of digital logic gates ("D/a"). Some ASICs could have mostly analog functions along with some standard cell-based digital circuits ("A/d"). Both the custom and the standard cell ASICs require a complete set of masks to personalize the chips. It is estimated that there are approximately 3000–4000 new standard cell based ASIC design starts every year. While the number of designs has decreased significantly in recent years, it is estimated that there are many more gates being packed in the new designs overall.

Gate arrays and structured ASICs allow personalization of specific customer requirements in the top metal interconnect layers. The base layers of these chips have a "sea of gates" pre-defined. Therefore, such devices can be fabricated in about half the cycle time and are less expensive to develop. These devices have limited packing density and performance. Gate array devices became popular in the 1980s and 1990s as vehicles for system designers to implement their solutions in silicon. As the demand for embedded memory and re-usable IP blocks grew, it brought out another limitation of gate arrays—lack of an ability to embed memory and other IP blocks. This led to the offerings that included fixed memory blocks. More recently, suppliers have created "structured array" offerings that have pre-defined blocks which are targeted at specific market segments. An example is shown in Fig. 4.2.

Structured arrays could be worth further consideration and may be the right choice for the fabless start-up [4.1], if the following conditions are met:

- The application of the ASIC is a mainstream application for which the supplier happens to have the right structured array base already developed.
- The volume projections are relatively low.
- Low cost is important, but is not a major consideration.
- In using this option, the unit cost will be higher than in the standard cell approach, but the development NRE (Non-Recurring Engineering) cost will be lower and the development time will be shorter.

Another major reason for the decline in the use of gate arrays is the emergence of FPGAs. The density capabilities and user-friendliness of FPGAs have made them a favorite with system designers looking to validate new ideas into silicon. FPGAs now play a crucial role in the validation of design ideas in hardware. There are over 90,000 FPGA designs implemented each year. In addition to being chosen for design validation, the use of FPGAs may also be a good choice for the fabless IC company that only ships low volume products where cost and performance using this solution is acceptable. The major benefits are a very low NRE charge and quick time to market. A detailed discussion of the FPGA development and implementation methodologies can be found in the literature and is not included in this book. A good place to start a literature search is the leading suppliers' web sites. Leading FPGA suppliers are Xilinx (http://www.xilinx.com) and Altera (http://www.altera.com). This book however, can provide useful insights to the fabless company in managing the operations aspects of the company even when using an FPGA as their product.

As mentioned before, the industry has traditionally used the term "ASIC" synonymously with array based devices. In its heyday, there were over 10,000 such ASIC design starts per year. This attracted many suppliers. Over the last 5–10 years, however, there has been a significant drop in the number of such ASIC devices. The drop has been associated with the "death" of the ASIC industry. Complicating this perception is the over 90,000 FPGA designs that are implemented annually. In reality, keeping the broader ASIC definition in mind, it is my contention that as long as there are creative and entrepreneurial designers, so will there be ASICs. They fulfill a very important need of system designers to be creative and to get their ideas into high volume IC products. It should also be noted that although the number

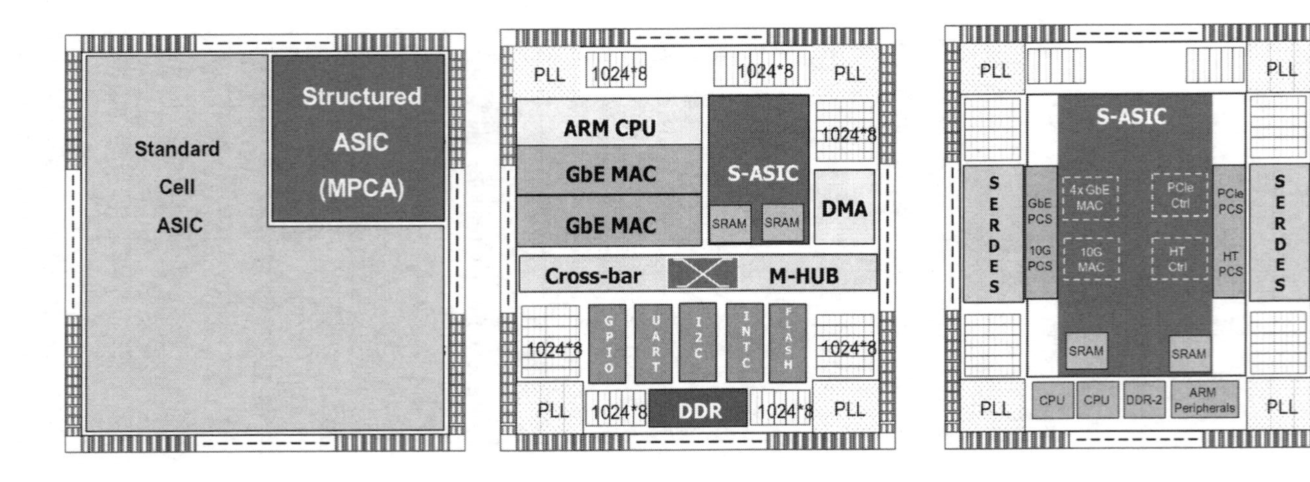

FIGURE 4.2 Examples of structured array layouts. (*Courtesy of Faraday, http://www.faraday-tech.com.*)

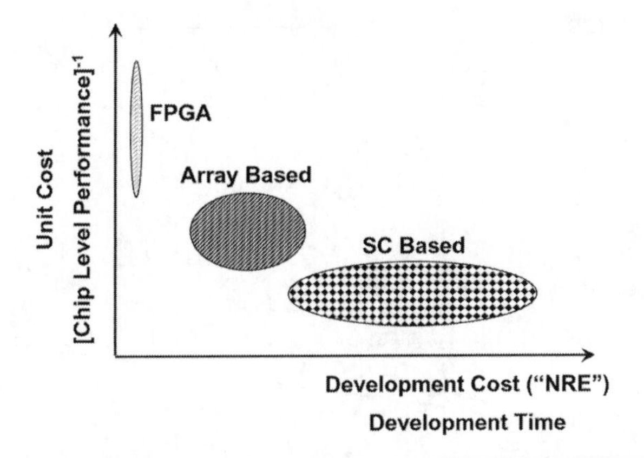

FIGURE 4.3 Pictorial representation of unit cost and development cost of various ASIC types.

of ASIC design starts is down in recent years, many more transistors are designed and shipped today because the average complexity per chip has increased.

4.1.1 Volume, NRE, Unit Cost Tradeoffs

Figure 4.3 is a conceptual summary of the positioning of FPGAs, array based ASICs and standard cell based ASICs in the space that helps illustrate the optimization of unit cost, development cost (NRE) and time, and performance. FPGAs are good for design validation, have a very low NRE cost but have higher unit cost. Array based devices have higher NRE but the unit cost is lower than FPGAs. Standard cell based ASICs offer the lowest unit cost, and the best performance, but the NRE is the highest and development time is the longest.

The chart in Fig. 4.4 can be used to estimate a crossover point when deciding to use high NRE, standard cell ASICs. For example, if the standard cell based design NRE is $10M and the unit cost saving relative to an FPGA or a Structured ASIC is $50, the annual unit volume must be greater than 100K in order to recover the investment in two years, or 200K to recover the investment in one year.

In summary:

- FPGAs offer a powerful solution for design validation and debug. They could be considered for shipping product in low volume applications. If the product is highly successful and there is demand for high volumes, it is possible to convert the design into an array-based or a standard cell ASIC.
- Structured custom arrays should be considered if a mainstream market is targeted and volumes are low.

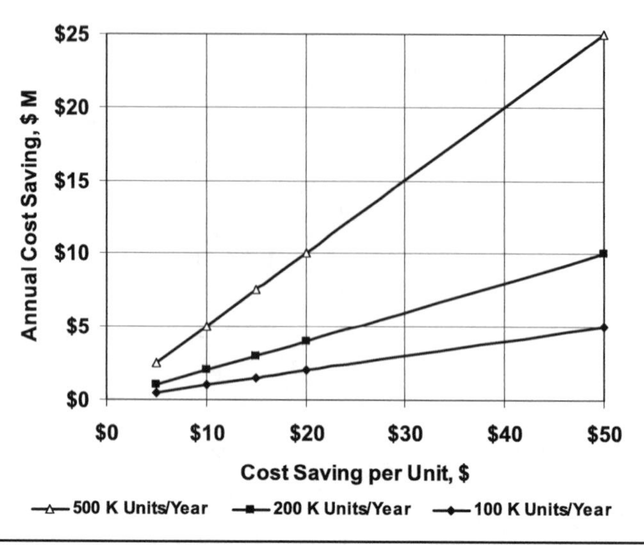

FIGURE 4.4 Unit cost savings from using a standard cell ASIC for various annual unit volumes.

- For high volume applications, a standard cell based design approach is likely to be the right solution. Selecting the right technologies, the implementation approaches and managing the implementation are topics to be discussed in the rest of this book. Sometimes it is worthwhile to invest in an analysis of similar, existing products in the market. Such an analysis will likely provide insight into the methodology used for those products.

4.1.2 Standard Cell Based ASICs

ASICs designed using standard cells offer:

- Highest packing density (transistors per square mm) and performance;
- Lowest chip size and unit cost;
- Most flexibility in integrating memory, analog, RF and other functions;
- Most abundantly available IP blocks, facilitating re-use and reduced development time.

However, they have the highest development cost and the longest prototyping lead time.

Standard cell based ASICs are assumed as the baseline implementation approach throughout this book. Gate arrays and structured arrays are special cases of the considerations offered. Examples of

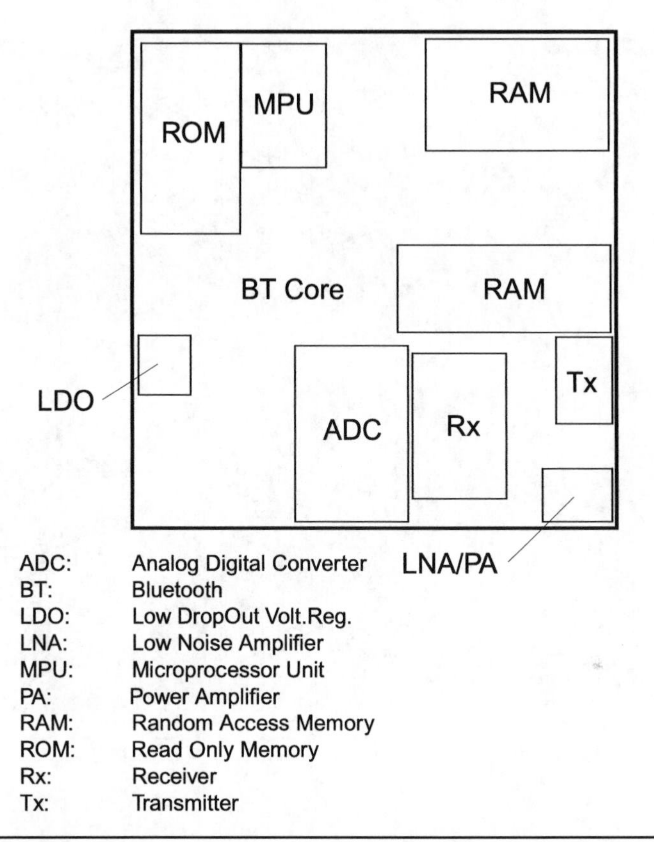

FIGURE 4.5 Block diagram of a standard cell based Bluetooth SoC. (*Courtesy* of *T.I. [4.2].*)

typical standard cell based ASICs are shown in Figs 4.5, 4.6 and 4.7. Figs 2.11 and 5.5 show additional examples.

Notice the logic gate areas, RAM, ROM, and processor blocks in Figs. 4.5 and 4.6. Notice also the use of many analog and RF blocks in these ICs, all implemented in CMOS technologies.

4.2 If You Had a Fab (IDM)

Having determined that the right solution is to design and implement an IC using the standard-cell approach, the next major decision to be made is *how* to implement it? Let us first discuss how semiconductors are developed and implemented at a vertically integrated system company or an IDM, as depicted in Fig. 4.8. All these activities occur in-house. An overview of acitivities follows, with details discussed in later chapters.

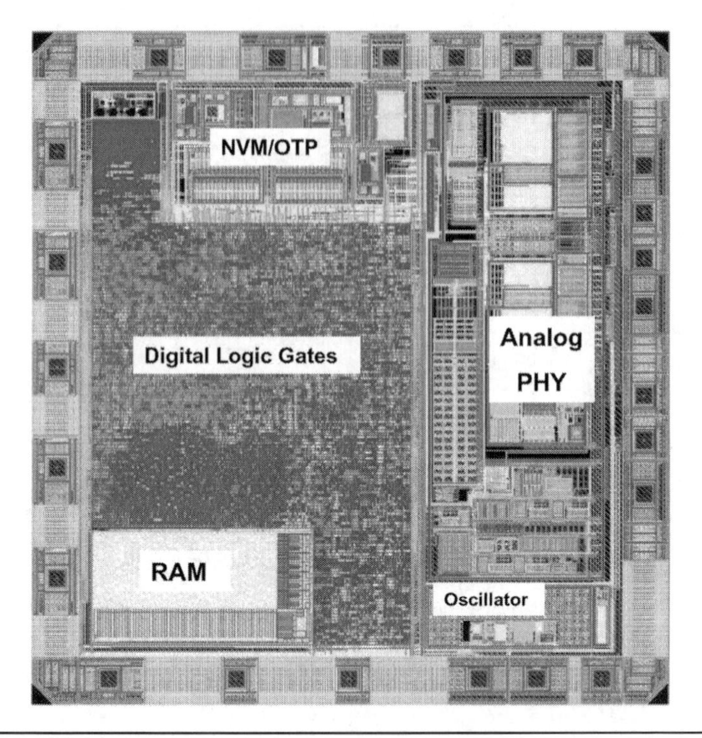

FIGURE 4.6 Twisted pair transceiver IC. (*Courtesy of Echelon.*)

System- and chip-level design activities include hardware and software design and verification, design definition, simulation, and implementation of the chip-level architecture. The outcome is usually a chip-level specification that is used as a basis to begin chip design. One of the first steps is to partition the chip into "bite-sized" blocks, mostly by their function. Some of these blocks may have significant analog content and may require custom design.

The digital logic portion of the chip and the behavior is usually defined via the development of synthesizable RTL codes. By using synthesis tools a gate level Netlist (NL) is generated and the chip goes through a "place and route" operation. The design may need special circuit design and intellectual property blocks such as the standard cell library, memory blocks, pre-designed microprocessors, input/output (I/O) interfaces, Phase Locked Loops (PLL) and a variety of others. Culmination of the physical design and verification phases results in the generation of a design "tapeout," usually in a "GDSII" format. The mask making facility is then able to convert this graphic design tapeout into a format that can be read by a mask writing "MEBES" (Moving Electron Beam Exposure System) machine. In this way, the design's polygons and other shapes get reproduced on the mask.

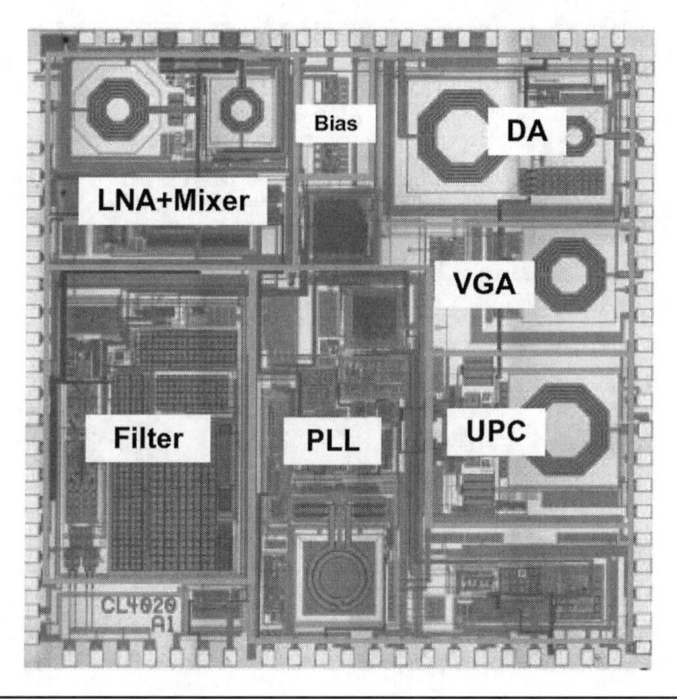

FIGURE 4.7 A 3G TD-SCDMA transceiver with numerous analog blocks [4.3].

Once the masks are made, the wafer fabrication is completed at the wafer fab facility.

Completed wafers are electrically tested at the wafer level ("e-test") to verify that the process parameters fit within specified limits. Sometimes there is also a wafer level test called wafer sort or wafer probe to identify electrically functioning dice. In parallel, the technology development teams works on developing new process technology nodes, new packaging, and assembly technologies. If a special package is required for the specific design, it must also be completed in parallel to the design and wafer fabrication activities. Sorted, or probed wafers are then sent to the assembly operation, which dices up the wafer and mounts functionally good dice in the package. This is followed by a final test that verifies functionality of the IC, and that it meets the product specifications.

The product support activity usually involves test program development, test support, failure analysis, debug, reliability, and other product functionality and quality related issues. The IDM company has the engineering, operations, and business management infrastructure to support all of these activities in house. Business management software plays a key role in facilitating order entry, tracking

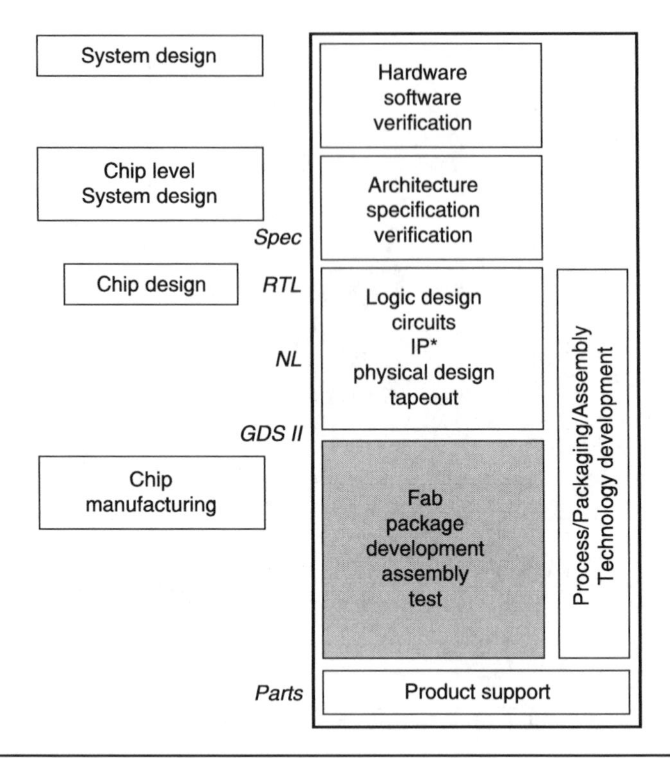

FIGURE 4.8 IC product flow at an IDM or vertically integrated system company.

production material flow, and in fulfilling delivery commitments to customers.

As discussed in Chap. 1, most of the chip manufacturing and many of the chip design functions can now be outsourced to a distributed supply chain. The supply chain partners focus on their individual core competencies. The fabless company has an opportunity to leverage "best in class" partners in the fabrication of their ICs.

A more convenient pictorial view of the IC product flow at an IDM company is shown in Fig. 4.9.

4.3 Fabless Sourcing Models (ASSP, COT, ASIC)

Simplified versions of various alternatives available to the fabless IC company for implementing new ideas are discussed next. While there are numerous variants possible, the major alternatives are illustrated in Fig. 4.10.

Compare these fabless implementation options to the IDM model shown in Fig. 4.9.

The shaded boxes in the figure represent activities at the fabless company. Open boxes represent outsourced activities. Contrast these

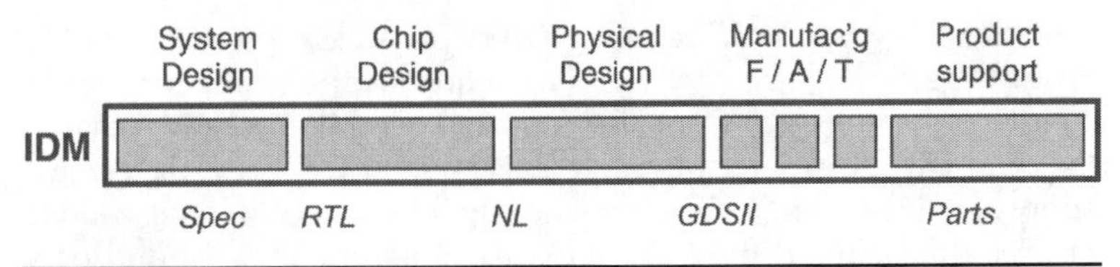

FIGURE 4.9 IC product flow at an IDM or a system company, in a more convenient format.

with the illustration in Fig. 4.9, where all activities are in shaded boxes, indicating that they are performed internally. The IC implementation alternatives are:

- as an ASSP;
- as an ASIC using two possible COT (mask) versions:
  - "Do-it-yourself" (DIY) at a large fabless IC company, that have significant internal COT experience and infrastructure;
  - "Do-it-yourself" (DIY) at a small fabless IC company, that have some internal COT experience;
- as an ASIC using an ASIC supplier, either with or without a wafer fab. This alternative is attractive for the system company or the small fabless IC company with no COT infrastructure.

It should be noted that the models described here are in a simplified form and for illustration purposes. Actual implementations could be a mixture of these models. For simplicity the hand-offs referenced here are for mostly digital (D/a) designs. The dark areas represent

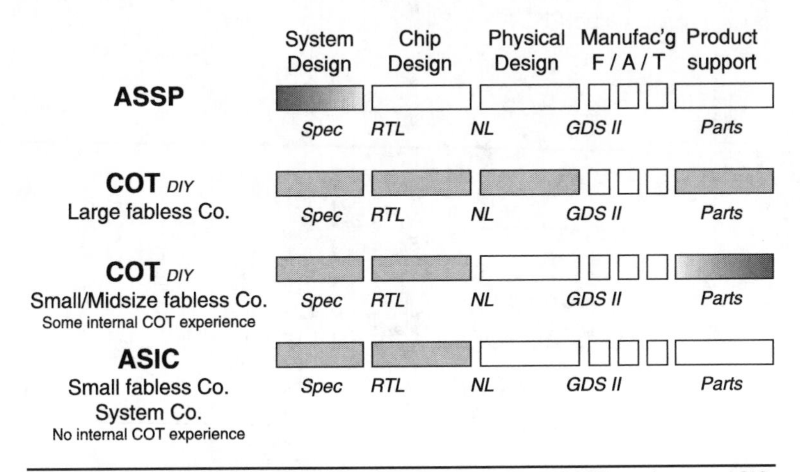

FIGURE 4.10 Product flow for IC implementation as an ASSP, and as an ASIC using either a COT approach or through an ASIC supplier.

activities at the fabless IC company. The clear areas represent activities at third party suppliers. My recommended approaches for fabless companies as they grow are also captured here. It is assumed that as the company grows it acquires internal experience with the distributed supply chain, in the COT model. The larger fabless companies tend to bring in-house the activities key to differentiating their IC products. Examples are the use of internal resources for library development, test, and characterization.

4.3.1 ASSP Sourcing Model

In this model, the parts are usually manufactured by the IDMs and have high volume demand. The architectural design is usually implemented by the fabless IC or system company. The physical design could be implemented either by the system company or the merchant manufacturer. The parts could be marketed by either the systems company or one or more semiconductor companies. An example would be the cell-phone chip sets designed by Nokia and manufactured and sold by Texas Instruments.

The model typically espouses a simple set of interfaces, as depicted in Fig. 4.11, where the fabless company transfers a specification to the IDM semiconductor chip-house, and then sells the resulting ICs as a standard product, either with its own logo or that of the IDM. The marketing and sales channels of the IDM may be available for use by the fabless IC company. The business arrangement will usually include a license to make and sell the IC. The key attributes of this model are:

- Hand off at an IC specification level. Effort at the fabless company is focused on high level design and verification.
- The chip needs to be designed into the system like a standard, off-the-shelf IC.

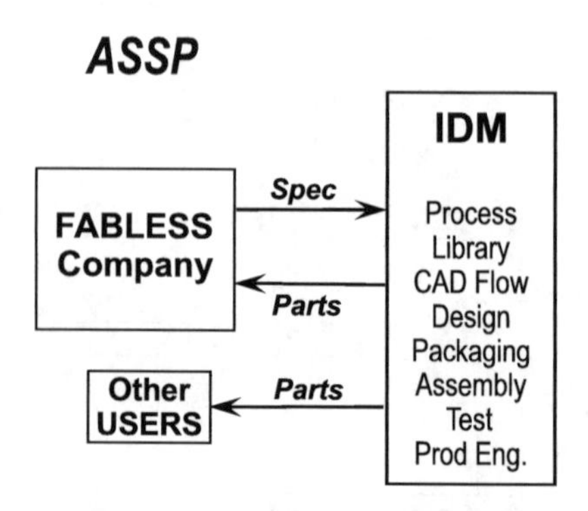

FIGURE 4.11 Relationships in an ASSP sourcing model.

- The chip supplier is typically an IDM such as Texas Instruments, ST Microelectronics, National Semiconductor.
- The supplier assumes responsibility for the IC design.
- The supplier supports the IC as a standard semiconductor product.
- The supplier usually gets rights to the IP and is licensed to sell the part in the open market.

This avenue is attractive to a fabless company that competes at the box level. Differentiation usually comes from system architecture and not necessarily the content of the chip level IP. This model also requires low levels of investment in chip level infrastructure at the fabless IC company.

4.3.2 COT, ("Do-It-Yourself") Large Fabless Company Sourcing Model

An IC implemented using this model is manufactured by best-in-class silicon, assembly and test suppliers based on an architectural and physical design implemented by a fabless IC company. The fabless IC company usually has the COT experience and infrastructure internally to leverage the distributed supply chain. The company builds internal expertise in areas key to the differentiation and characterization of its IC product and for better control of its development cycle time. This model is appropriate for high volume applications and for a company implementing many designs every year. The parts are marketed by the fabless IC company with its own logo.

The model is typically supported through a series of bilateral interfaces to many suppliers, one of which is the silicon foundry that manufactures wafers. The fabless IC company transfers the design to the foundry supplier as a GDSII design database, and buys either wafers, probed Known Good Die (KGD), or packaged, tested, finished units in a "turn key" mode. Such a traditional COT model has been used extensively by many fabless IC companies. The key attributes of this model are:

- Hand off at GDSII level.
- Fabless company does all the chip level design, including the physical design.
- The chip is a custom chip, owned solely by the fabless company.
- The silicon supplier is a foundry; either a "pure play" type (e.g. TSMC, UMC, Chartered Semi, Silterra, …), or a division of an IDM (e.g. TI, Samsung, IBM, …).
- For very high volume parts and/or for limited customers the pure-play foundries will likely provide a "turn-key" service

to manage the assembly and test supply chain. Delivery from the pure play foundry could be wafers, probed dice or finished units.

- For low volume ICs and customers in lesser standing with the foundries, the fabless company needs to engage with package, assembly and test houses and manage the supply chain themselves.
- The foundry provides process design rules and models and the fabless company obtains its own design infrastructure, including libraries, IP, sign off procedures, design flows, etc. Some foundries do make available some portions of the design infrastructure, now being referred to as the design ecosystem.
- The silicon supplier assumes no responsibility for the design, and carries a liability associated only with the manufacturing process.
- The fabless company, who is the foundry's customer, owns the product, the physical database, the mask works and provides product support. Ownership of the mask tooling at the silicon foundry has been likened to ownership of the mold tooling at metal foundries, and hence the term (COT).
- The design and the manufacturing supply chains are disaggregated and many opportunities exist to create difficulties (or "gaps") in the methodologies, logistics and engineering aspects of the IC fabrication. The gaps can be due to business, technical, or operational issues.

This model is attractive to fabless entities that compete at the box or chip level. The company differentiates its product significantly through chip level IP. The model requires investment in chip design and IC product support infrastructure, as is illustrated in Fig. 4.12. The fabless company manages interfaces directly to the foundry, the SATS (Semiconductor Assembly and Test Suppliers), the design tool, library, memory, and IP suppliers.

Unit costs are generally lower than in the ASIC and ASSP models. Therefore this model is an attractive alternative for cost-sensitive parts. If the large fabless company has negotiated the lowest, most favorable pricing from the foundry and the SATS, the unit cost could approach the lowest possible commercially available pricing. Of course, the fabless company must add the internal infrastructure cost, amortized over unit volumes to calculate the total unit cost. Details are discussed in Chap. 7. In the ASIC and ASSP models, the fabless company ends up paying for the internal support infrastructure costs at the suppliers in addition to the supplier's profit margin. This "margin stacking" has been a key to driving the larger fabless companies to invest in portions of the required infrastructure and yet espouse this COT sourcing model.

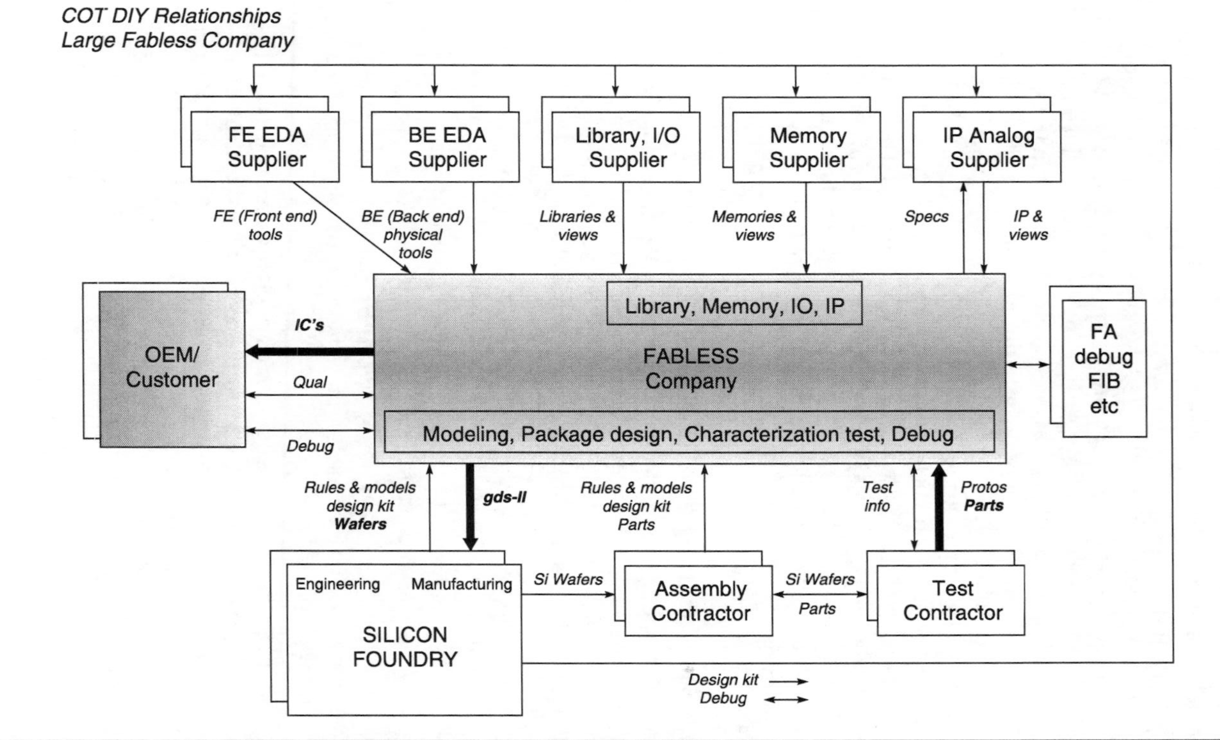

FIGURE 4.12 Relationships in a COT DIY sourcing model at a large fabless IC company.

4.3.3 COT (DIY), Small/Midsize Fabless Company Sourcing Model

Clearly there are variants to the basic COT model. At smaller or midsize fabless companies that do not have all the infrastructure and experience required, one could outsource pieces of the supply chain. For instance, it is possible to outsource the physical design to one of many design services providers. Examples are QThink (http://www.qthink.com) and Global UniChip (http://www.globalunichip.com). Using a design services provider avoids the need to invest in physical design EDA tools and the manpower required to support the design tools and the design flow. At the same time, the company leverages experienced teams at the design services provider. Figure 4.13 represents the relationships in such a model. The model is very similar to that shown in Fig. 4.12 for larger fabless companies. The major difference is the outsourcing of physical design. In addition, the fabless company may also leverage external resources for product support. This will include activities such as debug, test characterization and failure analysis.

4.3.4 Sourcing Using an ASIC supplier

This model is typically supported through a series of bilateral interfaces to a single, dedicated ASIC supplier. The fabless IC or system company transfers the design to the ASIC supplier, and buys the resulting IC as a custom product. The key attributes of this model are:

- Hand off at RTL or gate level netlist for digital chips. Fabless company does the logic design, simulation and verification.
- Analog and other IP blocks could be designed by the ASIC supplier based on specifications from the fabless company.
- The chip is a custom chip, owned by the fabless company. System specific functions and interfaces are usually embedded in the chip.
- The supplier could be a dedicated ASIC supplier such as AMI Semiconductor (http://www.amis.com) or Texas Instruments (http://www.ti.com), with an internal fab. Or, it could be a fabless ASIC supplier such as eSilicon (http://www.esilicon.com) or Open-Silicon (http://www.open-silicon.com). The relationships for such a supplier are illustrated in Fig. 4.14. Most of the time the fabless company's IC product is sole-sourced at the ASIC supplier.
- The ASIC supplier provides the design infrastructure, including libraries, some IP, sign off procedures, design flows, etc. These could be leveraged from third parties, or from internal resources at the fabless ASIC suppliers.

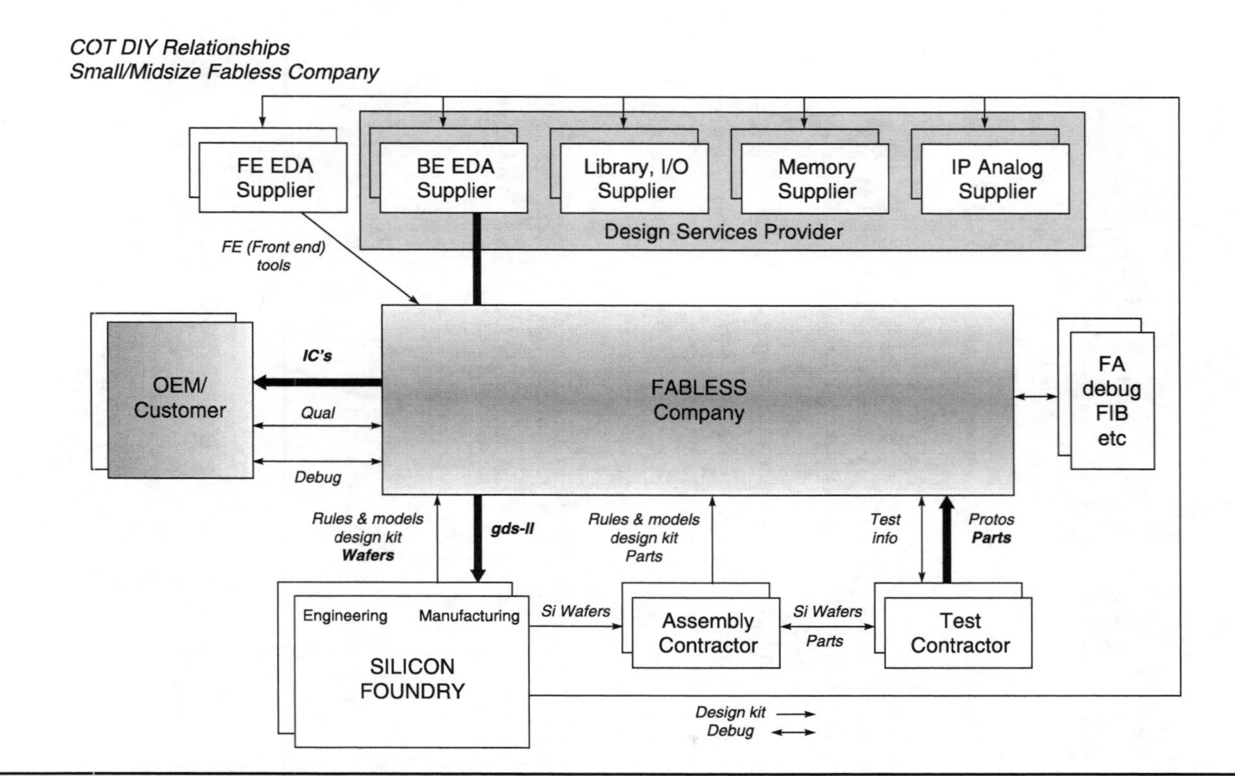

Figure 4.13 Relationships in a COT DIY sourcing model at small/midsize fabless IC companies.

> The high cost and long lead time to develop special fiber channel IP blocks drove our selection of the ASIC sourcing model. Our ASIC partners invest in the IP blocks that can be used by multiple users.
>
> H.K. Desai
> President and CEO, QLogic Corporation

- The ASIC supplier assumes responsibility for the physical design, not the behavior level.
- The ASIC supplier supports the IC as a custom product.
- The ASIC supplier does not own the chip IP and does not sell the product in the open market. While this is true in general, there could be exceptions.
- The ASIC supplier owns the physical database of the design.
- The ASIC supplier is responsible for the delivery of parts built to a forecast from the fabless company.
- The ASIC supplier owns the yield responsibility and the WIP (work in process) inventory.

This approach is attractive to a fabless IC or system entity that competes at the box level, and does differentiate its product through some form of chip level IP. This model does not require investment in chip physical design infrastructure or in IC product support infrastructure, as is illustrated in Fig. 4.14.

One commonly used variant to this model is that of a system company using a large ASIC supplier as described here, with the exception that the design is transferred as a GDSII tape.

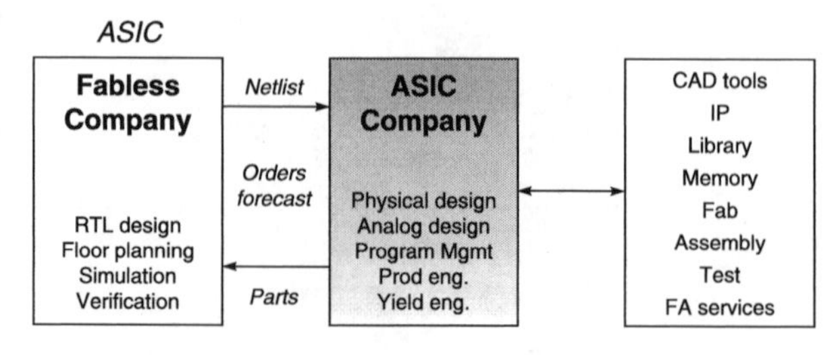

Figure 4.14 ASIC sourcing model relationships.

4.3.5 Sourcing Model Trade-offs

The primary considerations for tradeoff decisions between the various models are as follows:

- **Control and protection of IP**. The COT model allows the fabless company to maintain control of their IP, while the ASSP model offers very little control. The ASIC model is in-between, and is often perceived as semi-permeable IP protection. Thus, sensitivity to IP protection can drive the selection of the appropriate model.
- **Investment in chip level infrastructure at the fabless company**. The ASSP model requires none, while the COT model requires significant investment in both chip design and IC product support infrastructure. ASIC is in-between. Use of the fabless ASIC model reduces the need for in house infrastructure at the system company. Note that the liabilities tend to track the infrastructure. Thus, the profile of capabilities at the system level entity can force a choice of the appropriate model. An analysis of the engineering development manpower and capital costs are presented in Chap. 7. Such a quantification can be used to calculate a per-unit cost adder to the product unit cost in order to get a holistic view of costs.
- **Product unit costs**. The COT model tends to offer lowest unit cost, and ASIC the highest. ASSP tends to be associated with high volumes and hence competitive costs. The manufacturing partner will usually charge a premium over the COT model pricing to cover cost of the supply chain integration services. For companies servicing very high volume demand, unit cost may be the dominant factor, and may drive selection of the model.
- **WIP inventory and yield ownership**. In the ASSP model the manufacturing supplier has complete ownership of the WIP inventory and yields. ASIC suppliers take ownership based on the fabless company's forecast. In the COT model the fabless company takes ownership of the WIP inventory and the yield. The foundries usually have very good defect limited yield. They also provide support for process related yield enhancement, especially for high volume customers. However, responsibility for design related yield issues belongs to the fabless company both in the prototype and manufacturing phases.

Another factor that impacts selection of the sourcing model is the level of control the system company would like to maintain in the design and/or manufacturing phases.

Thus, selection of the appropriate sourcing model needs to be made based on the conditions and constraints specific to the fabless company, such as IP ownership control, volume demand, unit cost targets, margin targets and investment objectives.

For the sake of completeness the discussion in this book assumes the COT model. The other models will drop out of that super-set. By definition, the fabless company outsources the wafer fabrication, assembly, and high volume test. The major strategic decisions are then related to how much of the design work is completed in-house. Another strategic decision will be regarding how much internal support infrastructure needs to be built at the fabless company. These issues are clearly a function of factors such as the volume, the revenue stream, the number of designs and the team's experience. These factors are discussed in later chapters in the context of the operations infrastructure required and the cost.

Some of the infrastructure requirements can be strongly affected by the definition of the product goals, as is discussed next.

4.4 Design Strategies

4.4.1 Designing for Cost Schedule, and Success

A holistic view is important when defining the product specification and the implementation plan. The product must meet the following criteria:

- the market and lead customer's expectations of functionality and performance;
- the unit cost target;
- the schedule for delivery of samples and production.

It has been reported that 85% of ASIC and ASSP design schedules are missed [4.4]. Eight percent of projects were reported to complete as scheduled and 7% finished early. It is estimated that a primary reason for the missed schedules is an underestimation of the design complexity, both hardware and software. The author also observes an overestimation of the development team's productivity.

In addition, here are some other reasons why implementation schedules are missed, especially at emerging companies. At the company implementing their first design, there is much difficulty balancing the limited funds, changing requirements, and limited resources to make thorough evaluation of requested changes. There tends to be much "churn," starts and stops and repeats in the design phase. The top reasons for missed schedules are:

- "Over-Specification" and underestimating interactive effects:
  - One example is the design of a networking chip with a 500 MHz clock [4.5]. The design team created a chip-architecture

on paper and then spent 9 months writing RTL code. Once in the synthesis flow they realized they could not meet the timing constraints. This resulted in the need to use a high performance library with increased NRE cost and a delayed schedule. Then, in the physical design phase, they discovered that the use of the high performance cells blew the power dissipation budget. This resulted in the need for a thermally enhanced package, costing three times the original estimate. The unhappy result was that the product was priced out of the market and was way late.

- There are a number of ways to avoid such situations. In the recommended global planning phase, a holistic investigation of and sensitivity analyses of performance, power, leakage, yield, and cost can help prevent surprises. Negotiating a slightly less aggressive performance spec with the customer, together with a roadmap for a follow on chip, may have been a better alternative.

- Changing features, specifications and requirements.
- Additional, anticipated features. Especially those that are not quite as firm as the rest of the chip can become bottlenecks later in the design cycle.

Very seldom do initial estimates of cost and schedule decrease during the execution phase. Estimates of chip complexity, die size, the unit cost, and the schedule almost always get worse. There are many reasons why this is true—creeping changes of market requirements, engineering enhancements, new supplier information, inflationary factors, more detailed, and refined planning, etc. Factor this into execution plans. Sensitivity analyses and contingency planning can help prepare the team and allow better tradeoffs. Sensitivity analyses of cost are discussed in Chap. 7.

Another fact that is not well documented in the literature, but is a reality, is that most designs go through iterations. This is especially true if there is significant analog content on the die. Even digital designs implemented the first time may require minor changes. The industry has managed to reduce the cycle time for implementation of such changes through changes only in the metal layers—this reduces the cost and cycle time of implementing the changes. Most design service providers will pack in "spare" standard cells in unused space on the dice. In this way metal level changes can include the addition of a few additional gates. More discussion of this in Chap. 6.

One possible approach is a two-stage plan, as discussed in Sec. 3.5. The objective of the first stage chip is to be a "pipe cleaner," which allows the customer to see some samples of a "Rev 0" version of the production chip which may never go into production. This first stage chip could be fabricated using one of the Multi-Project Wafers (MPW) or "shuttles" offered by the foundries (more in Chap. 6). It is important to realize that in at least some situations, and especially with a properly negotiated arrangement with the lead customer, getting the first, "Rev. 0" product out on schedule is extremely important. This establishes credibility with the customer and gets them something real to start evaluating—both hardware and software-wise. Such a plan will keep from loading the prototype of the design with many "bells and whistles," and reduces execution risk. Realistically, one should plan for one metal layer design iteration for a first time implementation of a brand new design anyway. So, there is usually an opportunity to make refinements. A balance has to be made, though, to make sure the design is still competitive. Good communication with the customer is required to establish an open relationship to allow such trade-offs. Such an approach could also be handy in really evaluating the limits of the design specification. By evaluating the real distributions on real silicon, it may be possible to sign up to a slightly more aggressive specification.

Remember also that, to a large extent, today's SoC chips resemble yesteryear's complete PCBs (Printed Circuits Boards) populated with many unique components and ICs. Of course, the functionality, number of transistors per chip and performance is far superior now. There are many considerations that determine optimization of the content and the implementation plan for the ASIC. Some of these considerations are discussed throughout this book, and are listed here to emphasize the need for a holistic view.

- Meet the functionality, performance, unit cost and schedule.
- Be within the company's development budget.
- Selection of a single chip or multi-chip solution.
- Design complexity, tools, methodology to be used.
- Process technology to be used.
- Packaging technology to be used—package type, pin count etc.
- Test methodology and strategy to be used.
- Verification strategy and methodology to be used.
- Implementation methodology—COT or ?
- Outsourcing strategy for design as well as manufacturing supply chain.
- Software development strategy.
- System validation, reference boards, etc.
- Product positioning, competitiveness and differentiation.

4.4.2 Defining the Right Specification—Some Hints

- Must balance market requirements, cost, schedule, risks, and technical ingenuity.
- Selecting the right combination of chip complexity, process technology, packaging technology and test approach are all important considerations. While this can get tricky, opt for slightly conservative goals rather than pushing the limits. Being too conservative can make the product noncompetitive and/or uninteresting to the client. Too aggressive a spec. can mean delays, cost increases and could blow you out of the market!
- Re-use as much as possible from previous designs.
- Leverage as much external IP and design experience as possible without losing control of your proprietary design know-how. Make sure the IP selected has been implemented in high volume on the same technology.
- Avoid custom development as much as possible. Customization of IP blocks, the fab process, packages and test will, almost always, add to cost, schedule, and risk.
- If feasible, incorporate hardware options that can be selected using software changes. This could be a way to mitigate risk of chip modifications.

4.4.3 The Right Design Implementation Targets

There are many considerations in signing up for the IC's implementation targets. As discussed in the previous section the specification must be realistic and achievable both technically and from a business perspective. For example, signing up for an IC specification that can only be met via the use of a 45 nm process technology node is not realistic because the high development cost will likely preclude the emerging fabless company from pursuing the opportunity. Figure 4.15 shows a qualitative trend of increasing complexity, physical effects, and development cost in using leading edge process technology nodes. Selecting the right process, design, and packaging technologies is discussed in Chap. 5.

The major factors to consider in selecting the implementation targets are:

- Specification
- Functionality, performance, reliability
- Design partitioning—complexity, whether a leading edge SoC
- Process technology—which node, which features
- Package type, pin count, single chip or SiP
- Unit cost

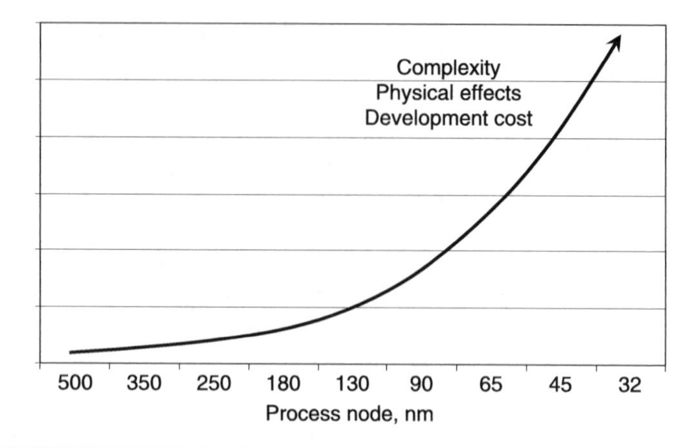

Figure 4.15 Qualitative trend of increasing complexity, physical effects and development cost when using leading edge process technologies.

- Development cost
- Schedule.

One prudent approach is to make an assessment of where the IC definition fits in the "envelope." It is believed that project risks can be reduced significantly by selecting implementation targets that allow accomplishing the project goals without pushing limits. The concept is illustrated using Fig. 4.16. While the industry and the media has

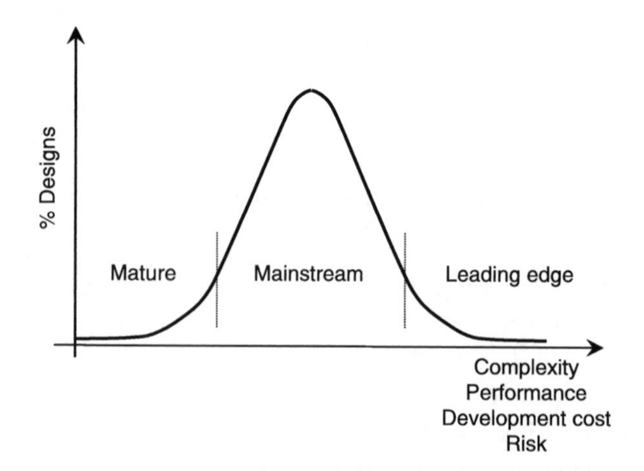

Process Node Maturity								
Mature			Mainstream			Leading Edge		
500	350	250	180	130	90	65	45	nm

Figure 4.16 Stratification of implementation targets into leading edge, mainstream and mature categories. Also shown is how process nodes matched up in 2007.

	Mature	Mainstream	Leading edge
Process	Very stable	Stable	Changing
Design tools	Very stable	Stable	Changing
Library	Very stable	Stable	Changing
Changes	~None	Few	Many
Development cost	Lowest		Highest
CPF in production[1]	Highest		Lowest
Schedule predictability	High	High	Low

[1]*CPF–cost per function.*

TABLE 4.1 Summary of the Stability and Predictability of Various Maturity Levels

focused much on the pizzazz of leading edge process nodes and leading edge design implementations, there are risks associated with riding the leading edge bandwagon. There are also distinct advantages. A judicious choice must be made by the fabless company in selecting the implementation targets. Leading edge designs with the highest complexity will have the highest implementation risks, the lowest predictability and the highest likelihood of missing schedules and exceeding budgeted cost [4.6]. Table 4.1 summarizes a comparison of factors affecting risk and predictability in various areas (process, design, cost and schedule) for leading edge designs, mainstream designs, and mature designs.

The curve in Fig. 4.16 is generic and can be applied in various ways. As shown, it applies conceptually to the overall design. One variant could be applied to the classification of the various process nodes available today. For instance, the process nodes 45 nm and 65 nm are considered leading edge in 2007. Mainstream nodes are 90 nm, 130 nm, and 180 nm. Mature nodes are 250 nm and larger minimum feature size. This is also shown in Fig 4.16. Another variant could be packing density in any one node. For instance, in the 45 nm node, transistor density over 2M transistors/sq mm could be considered leading edge. While this analysis is somewhat qualitative, it is helpful in benchmarking and positioning one's design.

A design complexity summary is shown in Table 4.2. It shows the maximum number of gates/sq mm in thousands of gates, and transistors/sq mm in thousands of transistors for each process node between 500 nm and 45 nm. The design complexity numbers are those that are achieved routinely with a reasonable effort. Higher densities are possible with increased effort, more risk and reduced predictability.

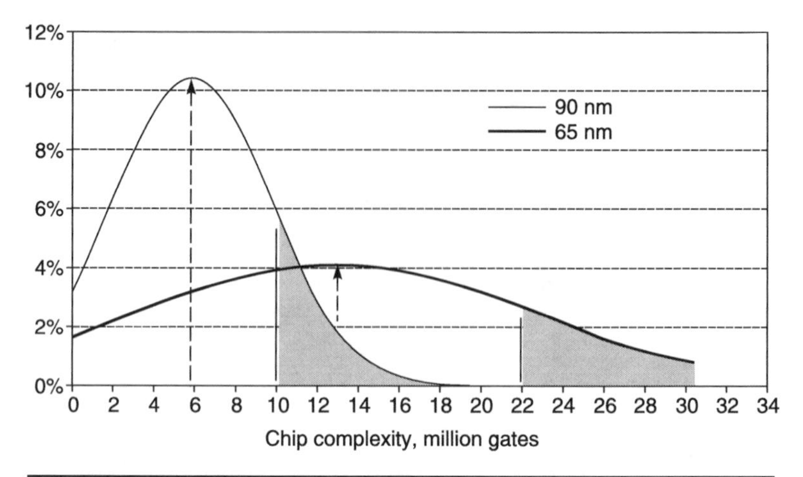

FIGURE 4.17 Distribution of recent design complexity in 90 nm and 65 nm nodes. Leading edge design would be ones in the shaded regions.

The maturity levels of process nodes in 2007 are also shown in Table 4.2.Late in 2008 and in 2009, it is expected that a new leading edge node, 32 nm will need to be included on this chart. At that time, 65 nm will move into the mainstream category.

On counting gates

Count total number of transistors on the chip. Divide by four. This will give you an equivalent number of 2-input NAND gates on the chip. Note that a 6-transistor SRAM cell is therefore equivalent to 1.5 NAND gates. Remember to add an estimate of a gates used in the SRAM block periphery-sense amplifiers, control logic, etc.

Alternatively, divide the total usable area by the area of a 2-input NAND gate to determine an equivalent gate count in that area. EDA tools use such an algorithm to report the number of gates. Sometimes this number is referred to as an "areal" gate count.

The usable (or "Core") area is calculated excluding the area occupied by peripheral circuits from the die area.

Process node maturity								
	Mature			Mainstream			Leading edge	
nm	500	350	250	180	130	90	65	45
K gates/sq mm	7	15	30	60	120	240	425	700
K transistors/ sq mm	28	60	120	240	480	960	1700	2800

TABLE 4.2 Transistor and Gate Packing Density for Various Process Technology Nodes. Also shown is the Node Maturity Level in 2007

As mentioned earlier, within a process node there is a distribution of complexity similar to that shown in Fig. 4.16. For example, for the 45nm process node, design complexity above about 2.5M transistors/sq mm will be considered leading edge.

Now let us look at some real data from a recent study at a leading design services provider. Figure 4.17 shows a distribution of recent design complexities. The average design complexity was 6M and 13M gates at 90 nm and 65 nm respectively. Leading edge designs would be those above 10M and 22M gates respectively.

Leading edge ICs with over 5–10 M transistors have been called SoC. While there is no standard definition and it is difficult to identify a truly complete system on a single chip, the term is used in the context of chips that have multiple functional blocks and could perform as a sub-system. SoC ICs usually incorporate the maximum density of transistors/chip, using the latest available process technology, design tools and methodologies, packaging technology and test methodologies at the time it is implemented.

4.4.4 In-house or Outsource?

Having set a reasonable IC specification and implementation targets, another important decision to be made is the implementation strategy. Should the emerging company make investments in design tools, methodologies, and a design staff? Should they invest in a package design capability, in silicon device modeling and characterization, and in product engineering and test?

In today's sophisticated manufacturing infrastructure, there are many high class providers eager to provide services, especially to fabless IC companies. Finding and managing the right partners could be a challenge. By using their services the fabless IC company can avoid capital and manpower investments. A simple rule of thumb in today's environment is to buy services wherever possible. The exceptions are in areas where the company either has core competency, or wants to protect their IP, or in an area where the company wants to build internal expertise.

> It is important to focus internal design resources in areas that provide differentiation to one's IC products. Investing in internal expertise that extends this differentiation should take priority over resources available through external service providers and ASIC providers. If managed properly, such external resources can also play an important role in transitioning rapidly to the next project on the product roadmap.
>
> Russ Harris
> Senior Vice President
> Echelon Corporation

The following is a list of some helpful considerations that can be of assistance in this difficult decision making process.

The case for "front end" design in-house (recommended):

- better control of IP;
- closer link to the architecture, software and other system groups. Reduced cycle time for tweaks to the RTL and the circuit behavior;
- limited EDA tool investment required. EDA tool costs around $0.25M for a starter set.

The case for "physical" design in-house:

- recommended only if the company has an experienced design team;
- the team has the experience to select IP, libraries and manage both the technical, and business issues with multiple suppliers;
- many design starts are planned every year;
- need extra close connection with the front end design team because there are special timing, performance, cost or power constraints;
- EDA tools cost could be $0.8–1M for a starter set.

The case for analog/RF design in-house:

- recommended if execution of the company's strategy requires key innovations in analog or RF circuit design and layout;
- unlike in the past, there are now numerous talented teams providing sophisticated services. Evaluate their capabilities before launching the internal effort.

The case for outsourcing physical design (recommended starting point):

- avoids EDA tool cost, internal design infrastructure and manpower investments;
- ability to leverage talented teams with up to date experience in implementing designs;
- predictable cost if properly managed;
- the internal team will need to be responsible for the IP and library partner selection and the business relationship. The company will need to have some in-house expertise, maybe one experienced individual.

The case for outsourcing silicon device and modeling expertise (recommended):

- the requirements of such services are usually sporadic. Some support is required during the design phase and some in the prototyping and characterization. In addition, some effort may be required on an as-needed basis when the design is in mass production.

The case for product engineering and test in-house (recommended):

- when taken in a broad definition of this role, it is a crucial role throughout the company's lifecycle. More discussion in the context of an "operations dilemma" in Chap. 6. The right individual with a breadth of knowledge and experience can play a pivotal role in the growth of the fabless company. The problem is finding individuals with such broad expertise.

4.4.5 IP Strategies

In order to get new designs to market in the shortest time, with the lowest development cost, and the highest probability of success, one must leverage available IP. In earlier days, system designers put together systems with standard ICs bought off the shelf and manufactured by IDMs. As more and more of the system functions have been incorporated into the ICs, the design of today's "mega chips" requires the incorporation of off-the-shelf IP blocks. There are many providers of IP blocks. The silicon foundries provide IP blocks, and there are independent providers of libraries, memories, and other special functions. Checking the IP portfolio available at the foundries is a good starting point. There are also a few of organizations set up to facilitate searches of available IP.

- Design and Reuse (http://www.us.design-reuse.com).
- ChipEstimate (http://www.chipestimate.com).
- VSI Alliance (http://www.vsi.org).

Some simple guidelines:

- maximize "re-use" from previous designs and experience;
- maximize use of available IP:
  - soft IP, where the RTL description is used;
  - hard IP, where a design database in GDSII format is available.
- Make sure IP has been thoroughly validated in order to reduce risk to the design. The GSA has recently launched a tool that enables risk assessment of IP (http://gsaglobal.org).

4.4.6 Design Differentiation

The fabless company's ASIC offering has to be differentiated from the competition. Careful consideration must be given to matching the differentiating features of the design with the core competencies at the emerging company. There are many possible differentiating features. Some examples of new features and capabilities are:

- A new standard.
- A new multimedia feature.
- A new version that improves form factor or power dissipation, lower cost etc. This could be facilitated by:
  - a new architecture;
  - a new software feature;
  - a new circuits technique;
  - a new process technology;
  - a new packaging approach.

With the general availability of foundry technologies, off-the-shelf IP, libraries, memory and outsourced design services, it is becoming difficult for fabless companies to differentiate the ASIC product based on these elements. For the emerging company the differentiating features must come from architecture, software, circuit techniques, packaging approaches or a unique way of combining chip-level capabilities, and not from the use of leading edge technologies. In the yesteryears it was possible to differentiate one's product through the use of special process features or the latest process technology node. The use of special process features is now strongly discouraged, unless the plan is to serve a high volume market. The use of leading edge technologies can be expensive and adds risks to the development project. As discussed in later chapters, the emerging company will be well advised to leave the leading edge implementations to the "big boys!"

It is important not to confuse uniqueness with differentiation. Realize that in spite of the use of available IP (standard cell library, I/Os, memories, analog mixed signal (AMS) blocks) and process technology, each ASIC has:

- a unique RTL description, incorporating any architectural uniqueness;
- a unique ATE program;
- a unique DFT implementation;

- some unique, specially designed blocks such as an ADC, DAC, analog front-end.

Differentiation refers to unique features and/or capabilities of the ASIC that enable the customer to put together a better system level product. Overall, in positioning their product it is important for the fabless IC company to ask a few key questions:

- What specific customer problem is the new IC solving?
- How much are they willing to pay for the solution and the IC?
- Who is the competition?

In my view, the start-up IC company should avoid differentiating their IC product only on cost and the use of the latest process technology node. Playing the unit cost game with the larger semiconductor suppliers is a losing proposition for the startup. As is discussed earlier in this chapter and in Chap. 7, the high development cost and risks associated with leading edge ICs can cause serious financial and credibility issues for the startup company. During the product definition and feasibility assessment phases, it is extremely important for the executives to understand the differentiating features of their IC. Acceptable features could be based on a unique architecture, performance, software, I/O interfaces, packaging that leads to an improved form factor, reduced power, and the like. They should also align the differentiation with their team's core competencies. The company will be well advised to focus their precious internal resources on the unique features that are not available externally.

> Emerging fabless IC companies can improve their success probability by defining their product to be differentiated even when using a process technology node that is one to two generations behind the leading edge node. To achieve such differentiation there must be focus on system and software techniques. It is advisable for the startup to focus their energy and efforts on the key ingredients that differentiate their IC, e.g., high speed I/O design.
>
> Behrooz Abdi
> President & CEO, Read Microelectronics
> Former SPV & GM, Qualcomm CDMA Technologies

Larger fabless companies do invest in key areas that enable them to differentiate their IC products. Some example areas are library design, memory and package design, product, and test engineering.

> As a developer of storage processors, we found that differentiating our ICs through architectural features is very important. Our unique implementations allow us to achieve a 10× performance advantage over the leading edge CPUs even though we used process technologies that were two generations behind the leading edge node.
>
> Fazil Osman
> CTO
> Astute Networks

4.5 Key Points

- A design approach using standard cells provides high packing density, flexibility, availability of IP blocks, and the lowest unit cost. However, the development cost is high.
- Various implementation approaches are possible (ASIC, ASSP, COT) with many pros and cons, and tradeoffs need to be made by the fabless IC company.
- Designing for cost, schedule, and success requires definition of the right specification and the right implementation targets. Fitting within the "envelop" of available technologies and methodologies reduces risk and improves the probability of success.
- Careful consideration of sourcing strategies, whether in-house or outsourced, is recommended.
- Your product must solve a real problem and must have differentiation. Do not rely on leading edge process technology as the only differentiator.

CHAPTER 5

Selecting the Technologies

5.1 Introduction

Throughout the history of ICs, process technology has been the driver for reducing feature size and thereby maximizing the possible complexity (number of transistors/chip). This was discussed in Chap. 1 and as shown in Figs. 1.1 and 1.2. Designing on the latest process technology has been a key to producing products with competitive features, performance, and cost. In this chapter let us first discuss the business advantages of scaling when a design is migrated to the next technology node, and when larger diameter silicon wafers are used to manufacture the chip. Later in the chapter there is discussion of available process, design, packaging and test technologies and many hints on how to select the right one for your application.

The drivers for scaling and introducing new technology nodes have been:

- Unit cost reduction through:
  - the reduction of die size for the same chip complexity via a migration to the next technology node (Sec. 5.2);
  - the use of increased wafer size (Sec. 5.3).
- Feature and performance enhancement through additional transistors on chip, new circuit and design techniques, and new circuit/chip/system architectures together with a migration to the next technology node..

Introduction of a new process technology node every 2–3 years has been enabled by the development of process equipment, new materials, new process techniques, new transistor device design, and new process/device integration. Some examples of new process node challenges in the fabrication of ICs are:

- lithography process and equipment;

- thin film process and deposition equipment;
- MOSFET (Metal Oxide Semiconductor Field Effect Transistor) gate and insulator dielectric materials;
- MOSFET drain and channel engineering;
- shallow junction fabrication;
- integration of up to nine layers of metal interconnect.

The need for automated design tools and design methodologies was recognized in the 1980s as the number of transistors per chip approached 0.5–1M. In addition, the importance of design re-use became very significant in the 1990s. It was recognized that as design complexity increased, so did the design time and effort. Re-using IP design blocks became a popular concept and is a must for achieving the increasing chip complexities of present day ICs.

As the number of transistors per chip has increased, so has the number of Input/Output (I/O) pins required to move signals on and off the chips. An empirical relationship between the number of gates (or transistors) per chip and the number of chip I/Os was first proposed by Rent in the 1960s, was published later [5.1–5.3] and became known as "Rent's rule." Packaging technology has moved along rapidly to accommodate the hundreds of millions of transistors that are being incorporated on leading edge ICs.

Established IDM companies with leading edge products have traditionally used their captive wafer fabs and their internal design, packaging, and test infrastructure to get a jump on new technologies. However, there are now only a handful of companies that are able to contain such investments [1.22]. There is further discussion of this in Chap. 10. In the meantime, foundry technologies are catching up with the leading edge IDM suppliers. New business models for foundry engagements are emerging where leading fabless companies are now able to offer new ICs competitively (within one to four quarters) of the very leading edge, captive IDMs—more in Sec. 10.4.4.

While Moore's law has held up and new technology nodes are being introduced every two years, many new challenges are emerging. Some examples are:

- diminishing performance advantage node to node;
- increasing power dissipation due to leakage;
- increasing process variability/manufacturability concerns;
- increasing process and design cost.

The IC industry will likely find solutions for these challenges, just like it has over the last 50 years. Later in this chapter there is discussion of current solutions to these latest challenges.

The industry has been enamored with designing new ICs at the leading edge process technology node. This is generally considered a must. Because successful implementation of leading edge ICs requires the ability to synchronize the latest process technology, the latest design infrastructure (libraries, tools, methodologies), and the latest packaging and test techniques, it requires large investments in development cost. It also adds significant risk to implementation schedules.

For the fabless start-up IC company, this creates a dilemma. Selecting the right technology node becomes an exercise in balancing benefits against additional development cost and schedule risk. These factors are especially important in the current environment where late introduction or the inability to ramp up production could be devastating due to the short-product life cycles. The next few chapters offer valuable hints on how to make well informed and judicious choices.

So, how does one select the right technology node, the right foundry, the right wafer size, the right design, packaging and test technology?

5.2 Considerations to Pick the Right Technology

Important factors in picking the right technology can be categorized into five major areas. Some of the controlling factors within each major area are also listed. The relative importance of the different areas can vary by market segment, the product application, and the competition. Note that this listing provides a simplified view. In reality, some of the listed controlling factors affect more than one of the major considerations. For instance, availability of re-usable IP blocks affects chip size (cost), power, performance, schedule, and risk.

- **Functionality:**
  - technology node and features;
  - design tools and methodologies;
  - availability of re-usable IP blocks.
- **Cost:**
  - process maturity and features used;
  - die size;
  - pin count;
  - reliability requirements;
  - design tools and methodologies.
- **Performance—speed and power:**
  - process node and features;
  - circuit design;

    - design optimization;
    - quality and reliability goals.
- **Schedule**
    - process maturity;
    - design flow and tools maturity;
    - product positioning risk—leading edge, mainstream or mature.
- Risk:
    - aggressiveness of IC specification;
    - meeting customer expectations.

First let's discuss a very important metric called Cost Per Function (CPF). This background discussion will illustrate scaling approaches and will show the benefits of using larger diameter wafers. A discussion of similar concepts was published recently [5.4]. Then let's delve into the available process, design, packaging and test technologies. Chapter 6 is an elaboration of how to implement these technologies. In general, the selection process is an iterative approach that balances the benefits and risks of using various available technologies such that the end result is an IC meeting the expected functionality, cost, performance, schedule, and quality goals.

5.3 Cost per Function (CPF)

As the semiconductor industry introduces new process techniques, equipment, wafer fab facilities, larger diameter wafers—everything that is essential to fabricate ICs at the nanometer scale—the cost of fabricating the wafer continually goes up, as was shown in Fig. 1.3. A key parameter used in managing the business feasibility of ICs and the scaling concept is the CPF. The semiconductor industry has managed to reduce the overall CPF by 29–35%/year [1.14, 1.15] over the years. Within any given process technology node the die cost and CPF are reduced due to the manufacturing and defectivity learning curves. This is shown graphically in a conceptual chart, Fig. 5.1. As the volume of wafer and product shipments ramps up in each technology node, there is a reduction in die cost (and therefore CPF) due to a reduction in wafer cost, a decrease in defectivity and a corresponding improvement in the yield of good dice per wafer. An example trend of defectivity at Intel is shown in Fig. 5.2 [5.5]. Notice the faster reduction of defectivity (sometimes called "defect density") for the newer process nodes. Similar charts can be obtained from the leading foundries. The decreases in defectivity are due to process optimization, the manufacturing learning curve and yield improvement efforts at the wafer fabs. A compilation of Defect Density (DD) trends indicates an average reduction of 19% per year over the last

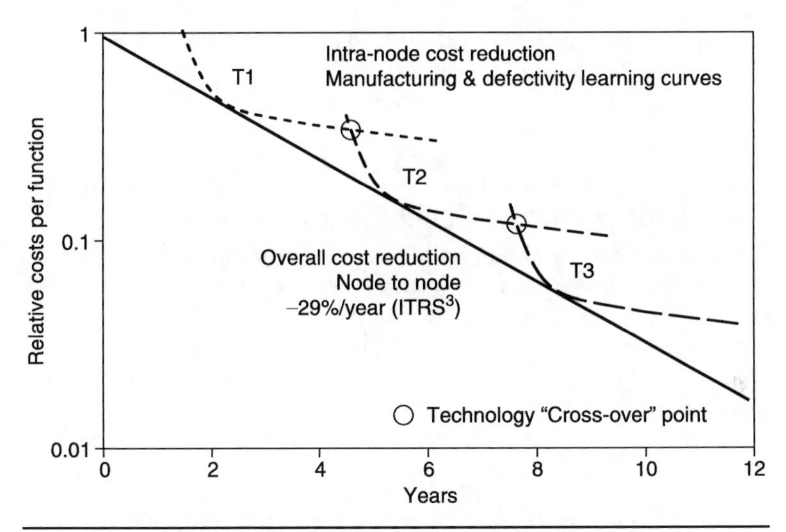

FIGURE 5.1 Cost per function trend and technology "cross-over" points.

35 years [1.15]. In Fig. 5.1, as technology T1 is maturing the technology node T2 gets introduced. Initially the defectivity and the CPF on T2 are higher than in technology T1. Through yield enhancement efforts defectivity drops rapidly as the volume is ramped up in T2. The technology cross-over occurs when the CPF in the newer technology is below the CPF in the older technology. Two cross-overs are shown in Fig. 5.1, one from T1 to T2 and another one from T2 to T3.

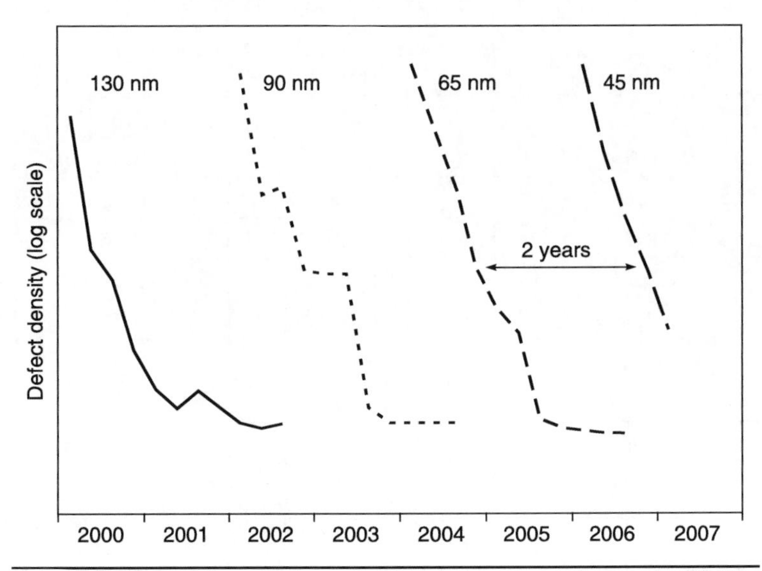

FIGURE 5.2 Defectivity reduction at Intel for various technology nodes [5.5].

Such cost reductions have been a cornerstone in the success of IC economics through the years. Chapter 7 displays the results of detailed CPF calculations. Commonly used CPFs are cost per gate and cost per transistor.

5.4 CPF Reduction from Technology Scaling

An industry target has been to reduce minimum feature size by around 30% at every process technology transition, as discussed in Chap. 1. Table 5.1 shows the various process technology generations or "technology nodes" used since the mid 1980s. Also shown is the scaling factor, k, which is the inverse of Dennard's scaling factor, λ introduced in Chap. 1.

Such technology scaling was achieved typically in the following manner:

1. Introduction of new photo lithography equipment and processes that allowed printing and patterning of dimensions 30% smaller than in the previous generation.
2. Improvements to other parts of the process, e.g., gate oxidation, ion implantation, diffusion, etching, interconnect metallurgy, etc.
3. Engineering and optimization of the transistor device structure and other aspects of the process in order to meet

Technology node, (μm)	Technology node (nm)	Scale factor (k)
1.5	1500	
1	1000	0.67
0.8	800	0.80
0.6	600	0.75
0.5	500	0.83
0.35	350	0.70
0.25	250	0.71
0.18	**180**	0.72
0.13	**130**	0.72
0.09	**90**	0.69
0.065	**65**	0.72
0.045	**45**	0.69
0.032	**32**	0.71

TABLE 5.1 Scaling Ratio for Various Technology Nodes since the Mid 1980s

performance and cost goals, and to enhance manufacturability and reliability.

4. Executing a Linear Shrink of an existing product, thereby reducing the die size per side by a scaling factor, k, such as 0.7. The area of the new die was then approximately one half of the original product. In the early days such an approach was called a "dumb shrink" or an "optical shrink" referring to the way it was implemented. Due to various intricacies of the process, the design rules and device characteristics at shrinking geometries, such scaling became increasingly difficult. The approach became known as an "intelligent laborious shrink" at some companies.
5. A new set of design rules—both physical and electrical—were usually used to design new products that took full advantage of the new technology capability. While the shrink approach was able to get an initial product out in the new technology node, the "Re-Design" approach was necessary to maximize performance and minimize cost of products in the new node.
6. In addition, the new technology usually had some new features aimed at increasing the packing efficiency, design productivity and device performance. Some examples are:
   - increasing the number of metal interconnect layers,
   - self-aligned polysilicon gate structure,
   - oxide isolation,
   - shallow trench isolation (STI),
   - multiple MOSFET transistor types with different gate 'turn on' threshold voltage (V_T) characteristics ("enhancement," "depletion," "high" to name a few),
   - standard cells,
   - EDA tools, and
   - re-usable IP blocks.

Now let us discuss migration of designs from one node to the next using either the "Linear Shrink" or the "Re-Design" approach. To illustrate the "Linear Shrink", consider Fig. 5.3(a) which depicts a square die with dimension y and having N transistors, in technology node T1. A simple shrink of the die into technology node T2 would reduce the die size by the scale factor k, where $0 < k < 1$. It should be noted that this scaling factor corresponds to the factor $1/\lambda$ used by Dennard in his papers [1.16]. Table 5.2 (a) is a summary of the resulting scaling parameters as well as typical values for such a scaling. Although the cost to process the wafer in the new technology node increases by a factor C (typically a 20% premium), the maximum number of available Gross Die on the Wafer (GDPW) increases by k^{-2}. Therefore, the die cost and the CPF reduces to Ck^2 or 60% of the cost in the technology node T1, for $k = 0.7$. This initial analysis assumes the new technology is processed using the same wafer size, and that the yield is the same in both technologies.

(a) LINEAR SHRINK			
Constant Wafer Size Constant Yield			
	T1	**T2**	**Typical**
Technology Scale Factor	1	k	0.7
Die Size	1	k	0.7
Wafer Cost	1	C	1.2
GDPW	1	k^{-2}	2
Die Cost	1	Ck^2	0.6
# Functions	1	1	1
CPF	1	Ck^2	0.6

(b) RE-DESIGN including Increased Die Size and New Technology Cleverness			
Constant Wafer Size Constant Yield Increase Die Size to Increase Packing Density			
	T1	**T2**	**Typical**
Technology Scale Factor	1	k	0.7
Die Size	1	S	1.1
Wafer Cost	1	C	1.2
GDPW	1	S^{-2}	0.9
Die Cost	1	CS^2	1.4
Cleverness Factor	1	F	1.3
# Functions	1	$S^2F^2k^{-2}$	4
CPF	1	$CF^{-2}k^2$	0.36
CPF reduction/year, 3yr cycle			29%

Table 5-2 (a) Summary of Scale Factors for a "Linear Shrink"
(b) Summary of Scale Factors for "Re-Design"

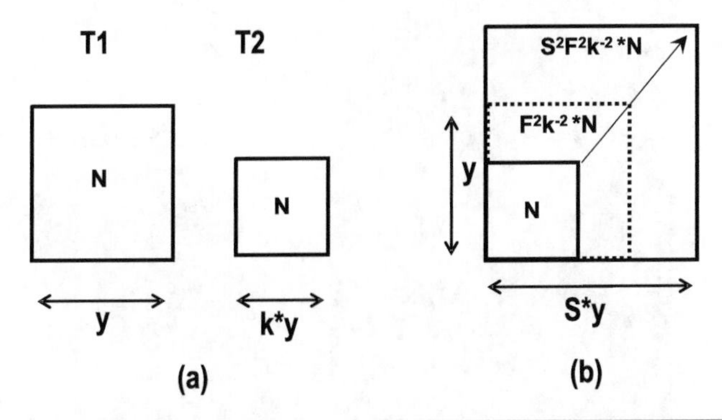

FIGURE 5.3 (a)"Linear shrink" from technology T1 to T2 and (b)"Re-design."

The "Re-Design" approach is illustrated via Fig. 5.3(b) which depicts increased packing density achieved by taking advantage of more aggressive technology features and design rules and a "cleverness factor," F. The number of transistors packed in the same size die increases by a factor F^2k^{-2}. Further increases in packing density result from the use of larger die sizes. Manufacturing enhancements of the process, the equipment and the clean room environment result in lower defect densities. This allows the fabrication of larger die with acceptable yields in the new technology node in spite of the tighter geometries. The increase in the maximum allowed die size is represented by the factor S. For simplicity, a square die is assumed and "die size" represents one linear edge of the die. Table 5.2(b) summarizes the scale factors and typical values. These typical values show a 29% annual reduction in CPF and a four-fold increase in functions over a 3 year period, which is consistent with Moore's Law [1.10, 1.12, 1.13] and the ITRS 200.

Such a scaling methodology has been reported by Intel for their 80 × 86 microprocessors. Figure 5.4 shows the migration of the 8086,

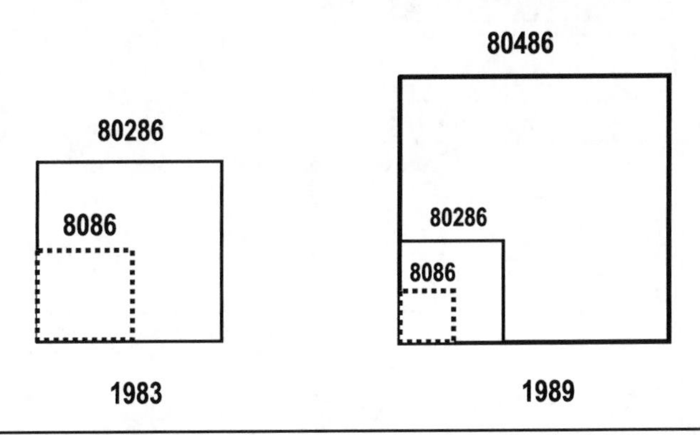

FIGURE 5.4 Technology scaling methodology reported by Intel [5.5].

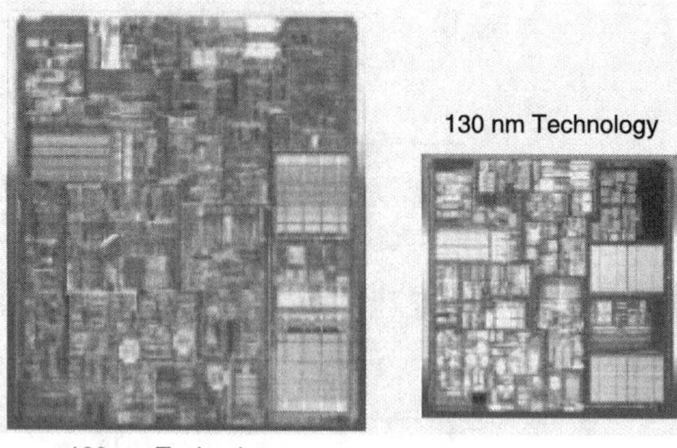

FIGURE 5.5 Intel's Pentium 4 die fabricated in 180 nm and 130 nm technology nodes (*Courtesy of Intel [5.7].*)

80286, and the 80486 processors with increasing transistors per chip [5.6]. For example, in 1989 the 8086 and 80286 microprocessors were shrunk and fitted into an area that was a fraction of the area in their previous implementations. The 80486 was then introduced in the new node with a larger die size and having 4 times the number of transistors of the previous processor in the previous node.

Figure 5.5 shows side by side pictures of the Pentium 4 microprocessor die implemented in 180 nm and the 130 nm nodes [5.7]. Intel has recently presented their "Tick Tock" product model [5.5], which is illustrated for the 65 nm, 45 nm, and 32 nm nodes in Fig. 5.6.

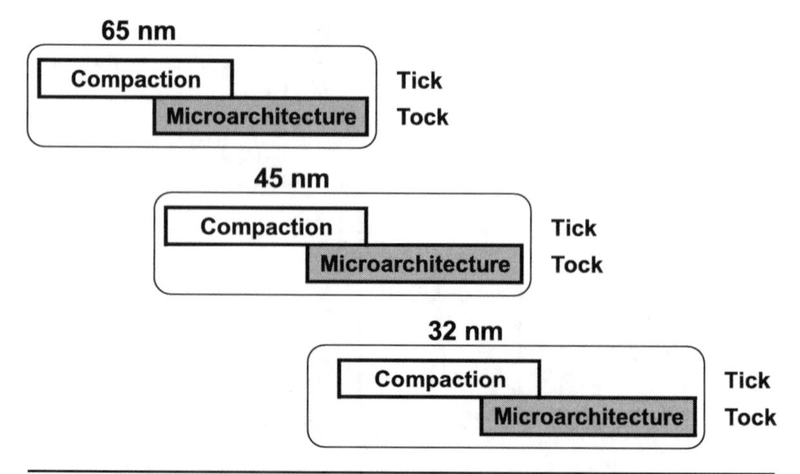

FIGURE 5.6 Intel's "Tick Tock" product model for introduction of new products in new technology nodes.

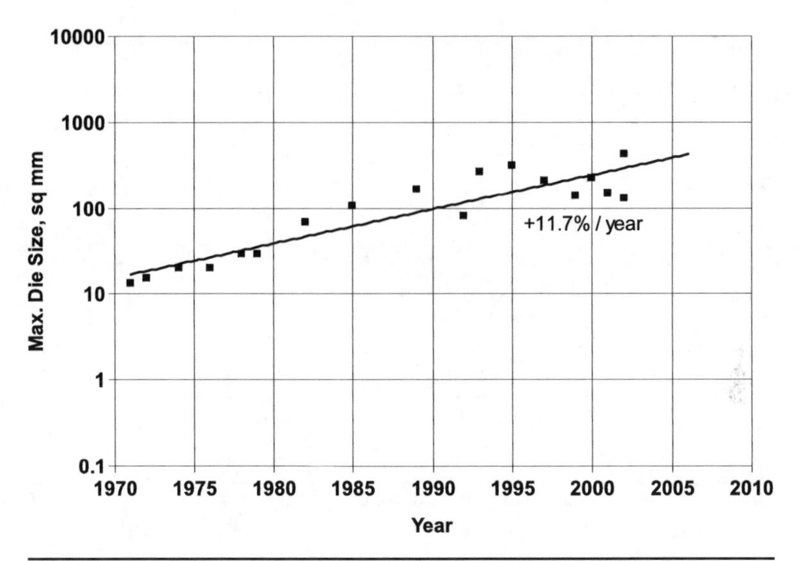

FIGURE 5.7 Die size of leading Intel microprocessors [1.15].

They use "compaction" or "Tick" to introduce the initial product in a new technology. This appears to correspond to the linear shrink approach addressed in this book. The re-design addressed here corresponds to the "micro architecture" or "Tock" approach suggested in the recent Intel publication. The compaction design is a derivative from the previous technology node while the micro-architecture version is a new design with full capabilities of the new technology node and includes architectural and circuit enhancements. The new design version also likely includes additional transistors on a larger die size.

Figure 5.7 shows a plot of the die size of Intel's microprocessors. The trend represents an annual increase of approximately 12% in die size. The maximum allowable limit is set by the size and quality of the optics in the lithography equipment. Currently the maximum field is approximately 22×22 mm. This upper limit causes some flattening of the die size trend shown in Fig. 5.7. The fabless IC company must select the optimum die size for their chips using models such as the ones discussed in Chap. 7.

5.5 Die Cost Reduction by Increasing Wafer Size

The industry has successfully increased maximum wafer size [5.8] from 50 mm (2″) diameter to 300 mm (12″), as shown in Fig. 5.8. Figure 5.9 shows pictures of completed wafers corresponding to the steps in Fig. 5.8. The wafer diameter steps result in an increased diameter

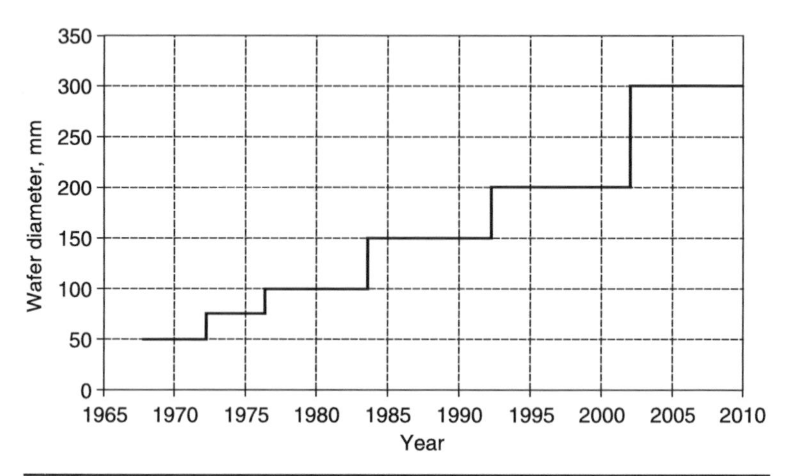

FIGURE 5.8 Silicon wafer diameter increase over time.

ratio by a factor of 1.33 or 1.5. An increased number of gross die per wafer results from the use of larger diameter wafers, as shown in Fig. 5.10. The available silicon area is either 1.78 or 2.25 for the two different diameter ratios. The actual ratio of GDPW is generally higher and is a function of the die size, as shown in Fig. 5.11. This is due to improved die-stepping algorithms that maximize the number of full die. Larger diameter wafers also allow a reduction of the number of partial die around the perimeter of the wafer; this effect is

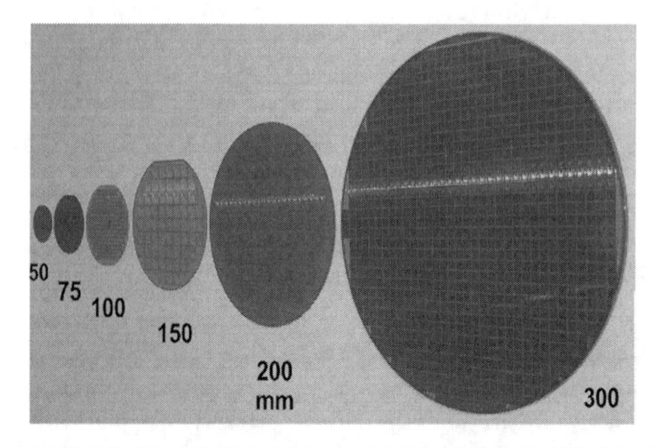

FIGURE 5.9 Silicon wafers with various diameters from 50 mm to 300 mm. Note that the round wafers appear oval due to the picture's perspective view.

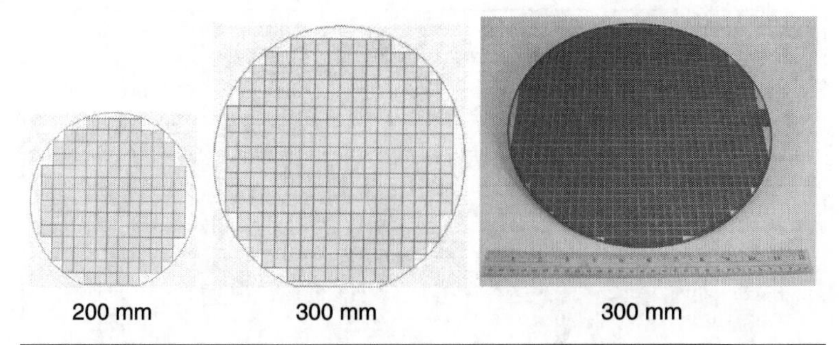

Figure 5.10 Increased gross die from a wafer diameter increase in the same technology. Also shown is the picture of a 300 mm diameter wafer. The wafer appears oval due to the picture's perspective.

more dominant for larger die sizes. Therefore, manufacturing on larger diameter wafers offers an improved economy of scale. As discussed in Chap. 7, modeling the number of GDPW on a wafer can be tricky. Estimates are illustrated in Chap. 7.

As mentioned earlier, the use of larger diameter wafers does increase wafer cost. However, there is a reduction in the die cost and CPF. This is because the relative GDPW increase exceeds the relative wafer cost increase. Early on in the introduction of a new wafer size, a 70% increase in wafer cost is reasonable [1.15]. In mature production, the cost to process a larger diameter wafer could be 30% higher. Notice that these increases are less than the GDPW ratios of 1.78–2.25 discussed previously. Some examples are illustrated in the accompanying sidebar.

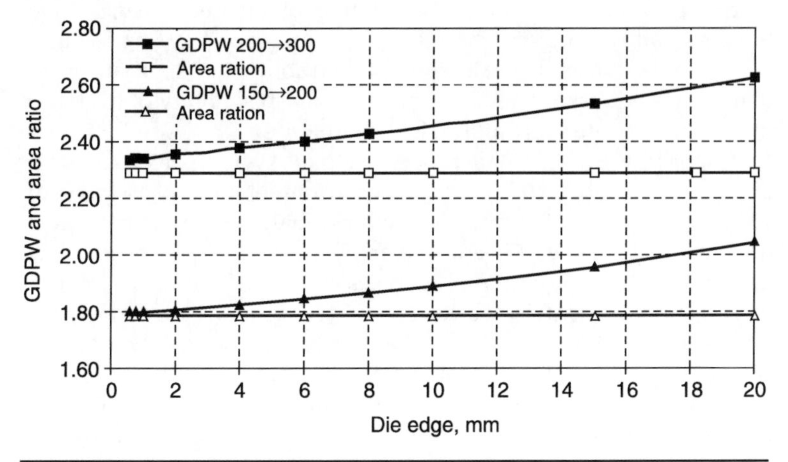

Figure 5.11 GDPW increase as a function of die size for two different wafer size transitions (200 to 300 and 150 to 200 mm diameter).

Relative die cost on larger diameter wafers = W/g,
where W is the relative wafer cost for the larger wafer and g is the relative GDPW

As mentioned earlier, the range of values is 1.3–1.7 for W and 1.78–2.25 for g. Therefore, the range of relative die cost is 0.5–0.9, a 10–50% die cost reduction when using larger diameter wafers.

A couple of examples for a mature and a relatively new technology are shown here:

- For a 10 mm die in 0.8 μm technology processed on 150 mm and 200 mm wafers, W = 1.35, g = 1.95. Therefore, die cost on 200 mm wafers = 69% of die cost on 150 mm wafers.
- For a 10 mm die in 130 nm technology processed on 200 mm and 300 mm wafers, W = 1.75, g = 2.45. Therefore, die cost on 300 mm wafers = 71% of die cost on 200 mm wafers.

5.5.1 Available Process Nodes

The following Table 5.3 shows a listing of currently available process technology nodes at commercial foundries, shown by wafer size. The latest process nodes are developed and introduced on the largest diameter wafer available, both at the equipment manufacturers as well as at the foundries. Of course the foundries make decisions about wafer diameter based on fab capability and on demand. It would be wise to check with the foundry representatives for the latest availability. When a new wafer diameter is introduced, it generally requires a new production wafer fab. The new wafer fab is generally ramped up on the most advanced process node in stable manufacturing at the time. As will be noted in Chap. 10, the high cost of building 300 mm wafer fabs is restricting the number of IDM companies that can afford to make such investments. This will limit the availability of the latest process nodes internally at the IDM companies.

In addition to these main process technology nodes, it has become popular to consider "half-nodes." For example, the process node at 150 nm is called a "half-node"—it offers a half step to the next technology node, between 130 and 180 nm. Some foundry customers like to complete their design in one process node and then use a linear, optical shrink to actually fabricate their devices in the half-node. This allows them to take advantage of a unit cost reduction because of a smaller die and also achieve some performance improvement. Typical linear shrink factors are 10–15%, instead of the 30% for the full node. In this way the process technology can be "squeezed" to operate as an evolutionary process, without major changes. Migrating design using the smaller shrink factor is less of a challenge, especially if the

Wafer Diameter (mm)	Technology node (nm)	Technology node μm
300	45	0.045
300	65	0.065
300	90	0.09
300	130	0.13
300	180	0.18
200	130	0.13
200	180	0.18
200	250	0.25
200	350	0.35
200	500	0.5
200	800	0.8
150	500	0.5
150	800	0.8
150	1000	1

Table 5.3 Summary of the Currently Available Process Nodes at Commercial Foundries, Sorted by Wafer Size

company plans ahead. Success in implementing such shrinks requires pro-active considerations that allow execution at the half-node with no design rule violations and no yield or reliability issues.

5.6 Foundry Financials and Technologies

Now let us review revenue information about the top commercial foundries [1.23].

- TSMC (Taiwan Semiconductor Manufacturing Company) (http://www.tsmc.com).
- UMC (United Microelectronics Corporation) (http://www.umc.com).
- Chartered Semiconductor Manufacturing Company (http://www.charteredsemic.com).
- SMIC (Semiconductor Manufacturing International Corporation), (http://www.smics.com).

The information was derived from the company annual reports and was consolidated by the GSA [1.23]. In 2006 the $19.3B dedicated foundry revenue was split as shown in Table 5.4. TSMC has maintained its position as a leader in the dedicated foundry business over the years.

	2006	
	Revenue ($B)	Market share (%)
TSMC	9.74	50.5
UMC	3.44	17.8
SMIC	1.465	7.6
CSM	1.415	7.3
Others	3.24	16.8
Total	19.3	

Source: GSA, Company Annual Reports

Table 5.4 Dedicated Foundry Revenue and Market Share for 2006 (*Courtesy of GSA, Company Annual Accounts.*)

Figure 5.12 shows the fraction of 2006 revenue sorted by process technology node for each of the foundries. This data shows that 40–75% of each foundry's revenue was from the 130/150/180 nm nodes. Note that in 2006, three years after the introduction of the 90 nm node, the revenue contribution from the 45, 65 and 90 nm nodes together was only 10–30% at the commercial foundries. The revenue contribution from 250 nm and older nodes ranged from 20–40% at three of the foundries. SMIC has been focusing only on the 180 nm and more advanced nodes.

One key piece of information here is that 70–90% of the foundry revenue contribution in 2006 came from the "mainstream" and "mature" process nodes, as defined in Fig. 4.16. Another key piece of

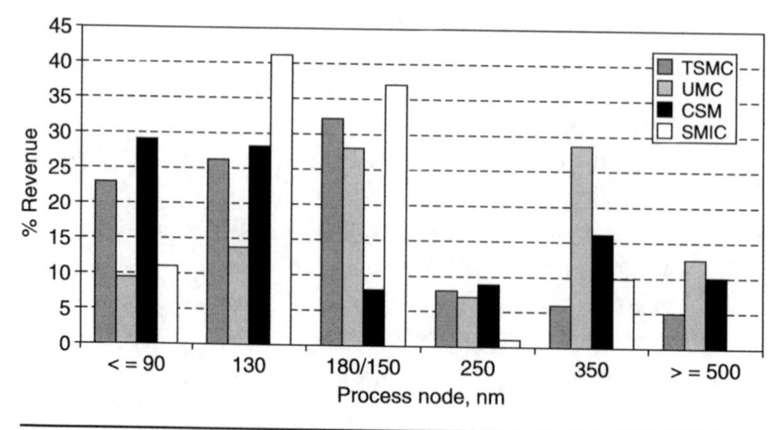

Figure 5.12 2006 Revenue contribution of various process nodes for the top four commercial foundries (*Courtesy of GSA.*)

information is that it takes two to three years for high volume production to ramp up on the latest technology node. The emerging start-up fabless company should keep this in mind when selecting the right process node for their design. Although the industry and media are focused on the latest process node, there is a lot of "horsepower" in the mainstream and mature nodes!

The data in Fig. 5.13 shows similar results. While the leading edge component of revenue has grown from 0% in 2004 to 10% in 2005 to 20% in 2006, approximately 80% of TSMC's revenue came from the mainstream and mature technology nodes.

5.7 Process Technologies

5.7.1 Foundry Methodologies

In the 1970s and the 1980s, internal wafer fab capabilities offered designers an opportunity to make adjustments to the process so they could derive unique advantages in their ICs over their competition. Examples are a "tweak" or "push" of design rules to allow greater packing density, or an extra ion implantation to allow an extra V_T device. These tweaks resulted in differentiated features such as lower cost, lower power and/or higher performance than offered by their competition. However, introduction of such process tweaks caused an escalation of process recipés in the wafer fab. In the extreme cases there were hundreds of recipes with a limited volume per recipe. Most of the time this required manual intervention in the process to insure accurate processing, and yet there were many opportunities for misprocessing. It was difficult to get economies of scale in manufacturing. In the 1990s the commercial wafer foundries offered fixed baseline process technologies with little opportunity to make adjustments.

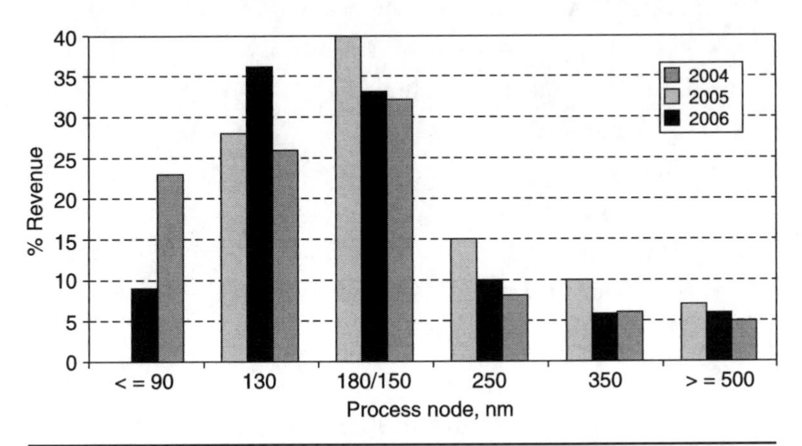

FIGURE 5.13 2004–2006 revenue contribution of various process nodes at the leading commercial foundry, TSMC. (*Courtesy of GSA.*)

This required a modification of the standard operating procedure for designers used to process customization; yet the foundries were able to demonstrate superior quality of processing, high yields and consistent quality. Motivated by the need to attract more designs into their foundries, the foundries offered improved design infrastructure. This has now led to a design ecosystem, as discussed later in this chapter.

> For more than 20 years, TSMC not only has built a world-class foundry with multiple fabs, but a business model as well. This business model of being a pure-play foundry for manufacturing and design services has enabled many semiconductor companies to focus on their differential designs without having to additionally invest in their own wafer fabs. Through process technology expertise and close collaboration, we have helped our customers bring their best and brightest product. Ideas to market faster with the highest degree of quality and reliability possible.
>
> Rick Tsai, Ph.D.
> President, TSMC Ltd.

Over the years, foundries have become very responsive to designer needs. The foundries have aligned their offerings with the popular market segments. They have aligned technology features most important to designers in the market segments. Technology platforms are offered for the following various market segments.

- Communications—wired and wireless,
- Consumer—gaming, multimedia,
- Computers—PCs serveres, ...
- Automotive,
- Power ICS,
- Display Driver ICS,
- CMOS Image Sensors,
- Microcontrollers,
- RFID.

It is interesting to note that the three C's (communications, consumer and computers) usually drive the leading edge technology platforms while the other applications use mainstream and mature process technologies.

Of course, in supporting so many different platforms and applications, there has been an escalation of the number of process recipés and offerings supported by the foundries. Availability of automated MRP (Material Resource Planning) and other systems to manage material flow, recipés and data are must. Foundries have done very well in managing outstanding process quality, yield and reliability. This is all good news for the fabless IC designer.

5.7.2 Start with CMOS

CMOS technology, introduced in high volume manufacturing in the early 1980s, has entrenched itself as the mainstream technology. Its competitors used to be Bipolar, BiCMOS, GaAs, and SOI. There is some discussion of these alternatives in Sec. 5.8.

As CMOS processes have been scaled to use smaller geometry lithography, their performance has increased. This is illustrated in Fig. 5.14, which compares the F_T (unity current gain "cutoff" frequency) of a CMOS transistor [5.9, 5.10] versus that of a SiGe BiCMOS device [5.11]. It is indeed true that the same F_T can be achieved with a larger geometry node in SiGe. However, the higher performance advantage of the SiGe process is offset by a higher cost. The enormous investment and high volume manufacturing in CMOS processes provides a compelling argument for their increased usage because of a better IC cost-performance.

So, starting with CMOS as the technology of choice is a safe bet. There must be a compelling reason for considering an alternative technology.

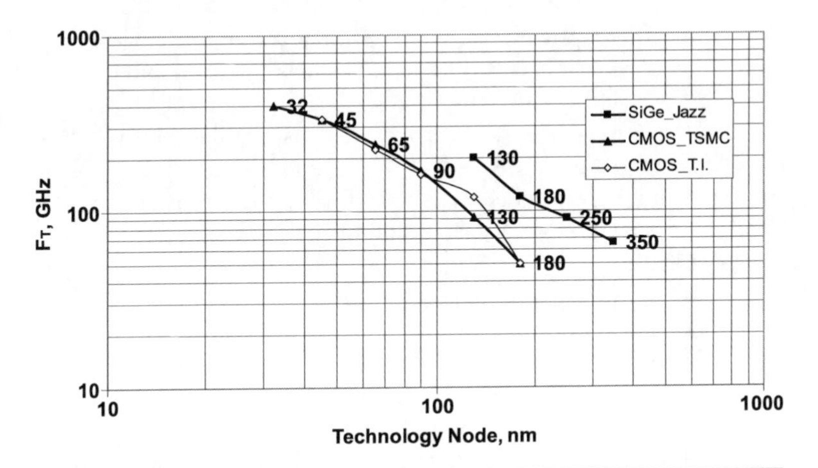

Figure 5.14 Comparison of cutoff frequency of SiGe and CMOS devices for various process technology nodes. Data points for 32 and 45 nm nodes are projections.

As discussed in previous chapters, there are many process nodes available. There are also many process variants and options within each node, as discussed here.

The next step is to decide which process technology variant and which node to use. Digital logic technologies have been used as the driver for new process technology nodes at the foundries. Appendix B has a table that summarizes the roadmap and the many process enhancements that have been implemented in the last decade. The information is a compilation of details available from the four major dedicated foundries [5.12]. It has also been cross-checked with information from Texas Instruments [5.10]. Enhancements have been made in lithography, gate engineering, metal interconnects, and dielectrics to allow scaling of ICs. The roadmap provides a summary of the major changes introduced at each node. The industry has an excellent track record of innovating and executing solutions to overcome technical challenges. A summary of the major features of the most recent five technology nodes is as follows.

- 180 nm:
  - risk production started in 1999;
  - power supply voltage for internal core using minimum devices is 1.5 volts. I/O devices designed to support 1.8, 2.5, and 3.3 volts to allow compatible interfacing with other ICs at the system level;
  - embedded Static Random Access Memory (SRAM) cell size as small as 5 square micron;
  - process uses a retrograde well, shallow trench isolation and a cobalt salicided polysilicon silicon/silicon dioxide (SiO_2) gate materials;
  - lithography using 248 nm optical wavelength. Some critical mask levels used Phase Shift Masking (PSM) and Optical Proximity Correction (OPC);
  - interconnects were fabricated using aluminum based metallurgy;
  - inter-metal dielectric has a dielectric constant, K, of 3.6. For reference, K is around 3.9 for SiO_2.
- 130 nm:
  - risk production started in 2001;
  - power supply voltage for internal core using minimum devices is 1.2 volts. I/O devices designed to support 1.8, 2.5, and 3.3 volts to allow compatible interfacing with other ICs at the system level;
  - embedded SRAM cell size as small as 2.5 square micron;

- process uses a super-steep retrograde well, shallow trench isolation and cobalt salicided polysilicon silicon/silicon dioxide (SiO_2) gate materials;
- lithography using 193 nm optical wavelength introduced;
- interconnects fabricated using copper based metallurgy;
- inter-metal dielectric has a dielectric constant, K, of 2.9.

- 90 nm:
  - risk production started in 2003;
  - power supply voltage for internal core using minimum devices is 1.2 volts. I/O devices designed to support 1.8, 2.5, and 3.3 volts to allow compatible interfacing with other ICs at the system level;
  - embedded SRAM cell size as small as 1.2 square micron;
  - process uses a super-steep retrograde well, shallow trench isolation and cobalt salicided polysilicon silicon/silicon dioxide (SiO_2) gate materials;
  - strain silicon techniques introduced to enhance the carrier mobility of the transistors;
  - lithography using 193 nm optical wavelength introduced;
  - interconnects fabricated using copper based metallurgy;
  - low K inter-metal dielectrics having a dielectric constant, K, of 2.6 introduced.
- 65 nm:
  - risk production started in late 2005/2006;
  - power supply voltage for internal core using minimum devices is 1.2 volts. I/O devices designed to support 1.8, 2.5, and 3.3 volts to allow compatible interfacing with other ICs at the system level;
  - embedded SRAM cell size as small as 0.5 square micron;
  - process uses a super-steep retrograde well, shallow trench isolation and nickel salicided polysilicon silicon/silicon dioxide (SiO_2) gate materials;
  - strain silicon techniques used to enhance the carrier mobility of the transistors;
  - lithography using 193 nm optical wavelength introduced;
  - interconnects fabricated using copper based metallurgy;
  - new low K inter-metal dielectrics having a dielectric constant, K, lower than 2.6 introduced.

- 45 nm:
  - risk production started in late 2007/2008;
  - power supply voltage for internal core using minimum devices is 1.1 volts. I/O devices designed to support 1.8 volts to allow compatible interfacing with other ICs at the system level;
  - embedded SRAM cell size as small as 0.3 square micron;
  - process uses a super-steep retrograde well, shallow trench isolation and nickel salicided polysilicon silicon/silicon dioxide (SiO_2) gate materials;
  - enhanced strain silicon techniques are used to further enhance the carrier mobility of the transistors;
  - immersion lithography using 193 nm optical wavelength introduced;
  - interconnects fabricated using copper based metallurgy;
  - extreme Low K (ELK) inter-metal dielectrics having a dielectric constant, K, lower than 2.5 introduced.

Figure 5.15 shows cross sections of the MOSFET gate and the interconnect layers for devices from the 130, 90, 65, and 45 nm nodes. Notice the narrowing gate length. Intel's technology dates are slightly more aggressive than the foundry risk production dates. As discussed in Sec. 5.9.1, Intel has announced the use of a Hi-K/metal gate transistor in 45 nm technology, while the foundry industry will likely adopt

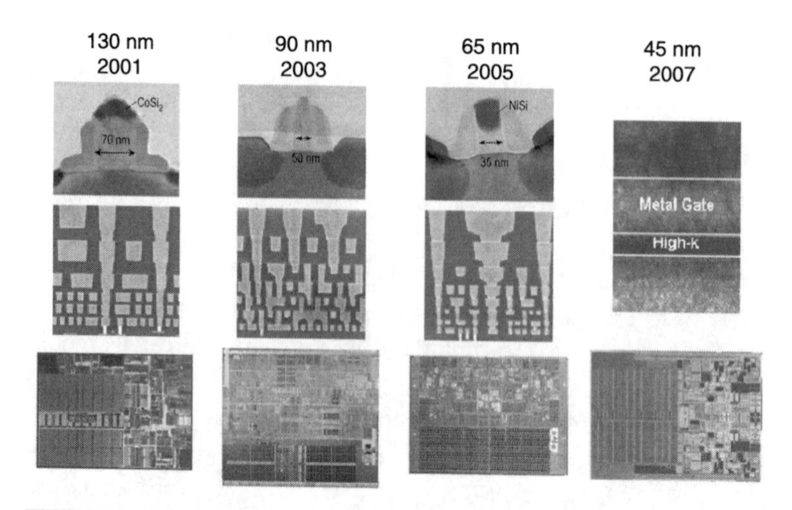

FIGURE 5.15 Cross sections of the MOSFET gate and the interconnect layers for 130, 90, 65 and 45 nm process nodes published by Intel [5.5].

a Hi-K/metal gate solution in 32 nm, the next process node. Notice also the nearly flat metallization layers and the vertically stacked via connections between metal layers. Some of the challenges encountered on the latest process nodes are discussed in the next section.

The foundries now provide rich portfolios of process technologies and the associated design support ecosystem aligned to the major market segments. So, a good place to start is with the process technology offerings for the fabless company's selected market segment. The fabless company must articulate their design requirements and align them with the foundry process offerings discussed in the next few sub-sections. A word of caution to the reader—the information provided here can be very dependent on the specific foundry of choice. The information is dynamic and usually gets updated periodically. My purpose is to provide a reference framework for the reader that should be helpful in making initial selections. Detailed updates must be obtained from the foundry being considered for the specific IC implementation.

5.7.3 Aligning Design Requirements

The fabless company should have the following global parameters and information in hand. Keep in mind that technology selection is an iterative process that usually converges to the right choices for the process, the various process options, the design, the package and test considerations. Of course, cost and schedule are also of prime importance.

- the market segment, for example, wireless for cell phone applications;
- the chip block diagram and specification including "order of magnitude" approximations for the following:
  - complexity (number of gates or transistors);
  - memory required and type (SRAM, DRAM, Flash, OTP);
  - embedded IP requirements (microprocessor, DSP);
  - analog blocks required (Data converters, ADC, DAC, PLL, etc);
  - interface requirements (USB, HDMI, etc);
  - pin count;
  - performance—clock frequency, gate delays;
  - power dissipation, both standby and operating;
  - external voltage limits, both for power supply and signal inputs and outputs;
  - cost;

- reliability requirements;
- volume projections.

5.7.4 Logic Processes

As mentioned earlier, the digital logic processes are usually the first to come online at the foundries. As the foundry technology leader, TSMC "sets the bar" for technology capabilities and offerings. The other foundries usually follow with similar offerings. Some foundries designate these as G (Generic) or GP (General Purpose). Starting at the 130 nm node the foundries introduce additional versions of the technologies to support Low Power (LP), High Performance (HP) and other applications. Adjustments are made to the available threshold voltage (V_T) devices, the power supply voltage for the chip core ($V_{DD\,Core}$), external power supply (V_{DD}) for I/O interface, memory type and density, and metal interconnect options to support various applications.

As an example, at the 65 nm process technology node, TSMC offers five versions [5.9]:

- GP — General Purpose
- LP — Low Power
- ULP — Ultra Low Power
- LPG-G — High Speed with low power consumption
- LPG-LP — High Speed with ultra low power

Many of the trade-offs in these offerings are related to the core voltage (either 1.0 or 1.2 volts) and the available V_T for the transistors. There are four possible transistor types:

- Low V_T
- Medium Low V_T
- Standard V_T
- High V_T
- The available I/O options support either 1.8 volts, 2.5 volts, or 3.3 volt signals external to the IC.

The SRAM comes in high density, ultra high density, and high current versions. Dual port SRAMs are also available.

The Back End of Line (BEOL), which represents processing beyond the contact mask can have many metal interconnect variants including high density and high performance versions. This is achieved by adjusting the metal thickness and the dielectrics used in the process. There are also different versions of the top layer metal thickness to accommodate different requirements of analog and RF applications.

There are many other variants possible, and therefore process technology selection requires a careful and diligent assessment.

5.7.5 Packing Density

Now, turning to the packing density, the maximum number of gates per square mm was discussed in Chap. 4 and is shown in Table 5.5. Unfortunately, there is no standard way to quantify the packing density. Sometimes the minimum 2-input NAND gate is laid out. The reciprocal of this gate area is one popular technology metric; the number gets calculated as the maximum possible gate density per square mm of die area. In reality, the actual packing density achieved is approximately 50–70% of the calculated maximum. The packing density numbers shown in Table 5.5 represent actual (and realistic) packing density achieved.

The SRAM cell size is used as a Figure Of Merit (FOM) for various technology nodes, as shown in the roadmap table in Appendix B. Note that in the 45 nm node, a six transistor SRAM cell can have an area under 0.3 square micron. Up-to-date and specific estimates for SRAM blocks (or macros) can be obtained from the foundry or the services provider. Note that there is usually a trade-off of SRAM cell area against yield and sensitivities, especially at minimum operating voltage (V_{min}). Such sensitivies are becoming especially important at the nanometer nodes.

5.7.6 Analog/Mixed Signal/RF Processes

CMOS processes that are used to incorporate analog or RF functions along with digital logic, memory and other functions must have the option to fabricate passive devices. The following devices are commonly used:

- resistors;
- capacitors—either MiM (Metal insulator Metal) or MOM (Metal Oxide Metal);
- inductors—usually built using a thick, top layer of metal;
- diodes;

nm	500	350	250	180	130	90	65	45
K Gates/sq mm	7	15	30	60	120	240	425	700
K Transistors/sq mm	28	60	120	240	480	960	1700	2800

TABLE 5.5 Maximum Packing Density in Various CMOS Process Technology Nodes

- bipolar transistors—both NPN and PNP; these are usually parasitic devices that are available in the CMOS process;
- MOS Varactors.

In addition, some applications require a high voltage to drive off-chip. There may also be a requirement for a deep N-well implant for improved isolation. And some processes may require a special V_T, although most of the newer process nodes have usable, standard V_T's.

Accurate process models for both active and passive devices, and PDKs are important for good design. This requires the foundry to have quality characterization information available, including noise and other parameters.

Analog and RF designers have successfully used the 180 nm process node to implement many designs with mostly analog functions. Digital designers have successfully incorporated analog functions such as Data Converters (ADC, DAC) and Phase Locked Loops (PLLs) on large digital chips. Single chip SoCs that incorporate an analog front end and RF functions on a single chip may become a commercial reality on one of the newer process nodes such as 45 nm and 32 nm.

5.7.7 Embedded Non-Volatile Memory (NVM)

NVM memory elements can be programmed in the field and have the ability to retain the programmed information even when the power is turned off. Traditional process technologies used for the fabrication of embedded "Flash" devices require approximately 30% additional steps compared to the baseline, digital logic process. While there are many applications that could utilize the NVM capability, the increased cost premium makes this a non-viable option for many customers. The increased process complexity also causes lower yields. Product level requirements for data retention and endurance (on/off) cycles are also difficult to test and to satisfy. However, some of the foundries have indeed made available NVM technologies on the 500, 350, 250, and 180 nm nodes. Due to some of the issues mentioned, NVM technology nodes lag behind the leading logic process nodes. Some foundries are developing a 130 nm NVM process.

In the meantime, NVM technologies using the standard, single poly, logic process have now been introduced. This approach does not require any extra mask processing, but is limited in the memory density. These memory blocks consume a relatively large area. Therefore, limited amounts of such NVM are feasible. Some possible applications that could use these small amounts of NVM storage are:

- analog trimming;
- gamma correction;
- HDMI decode;
- chip ID and encryption;

- memory repair;
- RFID.

There are three versions that are available at this time:

- OTP (One-Time Programmable);
- MTP (Multi-Time Programmable);
- Electrical fuse.

Third party suppliers are working with the foundries to make the memory blocks available for inclusion on designs from the 350 nm node down to the 45 nm node. It turns out that there are different approaches to achieving the OTP. The two common approaches are oxide rupture and floating gate. The fabless company planning to use NVM devices is advised to check out the latest availability and reliability information both from the foundry and the third party supplier.

5.7.8 Embedded DRAM

DRAMs (Dynamic Random Access Memories) have been a major driver of the semiconductor industry as a whole. The main memory inside laptops and desktop computers is usually 1, 2, or 4 Gigabytes and is all made up of DRAM chips. Stand-alone DRAM chips have traditionally offered much higher density and lower cost per bit than SRAMs. However, their performance is not as fast as SRAMs. DRAMs also require special processing, have special process constraints, and require refresh circuitry. Therefore, SRAMs have been the storage elements of choice for random access memory embedded within ICs. Many companies have tried to incorporate DRAMs into logic ICs. As in the case of the embedded NVM, the incremental cost of merging a traditional DRAM process with baseline CMOS digital processes has been a barrier. Recently, the development of a cost-effective MiM capacitor has resurrected the feasibility of fabricating ICs with embedded DRAM blocks. The MiM capacitor replaces the traditional trench capacitor to store charge in the DRAM cell. The advantages for embedding the DRAM on-chip versus using an off-chip DRAM are:

- Lower cost due to a smaller block size than SRAM. A 50-70% area reduction has been shown to be feasible [5.9].
- Improved bandwidth because a wider bus interface DRAM is possible.
- Lower power dissipation because the chip IO's do not have to drive the external buses.

As shown in Table 5.6, embedded DRAM (eDRAM) offers a significant improvement in performance, as measured by latency when compared to external DRAM. Note that the large latency shown

Memory	Area/Mb (Relative)	SoC cost adder/ Mb (Relative)	Latency (clk cycles)
eSRAM	1	1	1
eDRAM	0.4	1.25	2 to 3
External DRAM	0.15	0.5	20 to 25

TABLE 5.6 Trade-offs between eSRAM, eDRAM and External DRAM [5.13]

in this example for external DRAMs may not be so high for all systems. However, the embedded solution is definitely more expensive due to the additional processing steps and lower yields compared to an embedded SRAM (eSRAM) fabricated in a standard logic process. However, because of the significant area advantage that the eDRAM has over eSRAM, the use of eDRAM could be appropriate for some applications. Careful trade off analyses are recommended before embarking on the use of such memory blocks.

Manufacturing issues have been managed and embedded DRAM products are now in production at one or more commercial foundries.

5.7.9 Power ICs

There are at least three types of ICs in this category—power management ICs, high voltage display driver ICs and power converters (DC-DC, AC-DC, others). These ICs may use mainstream CMOS processes that have been modified to support high voltages. This is usually achieved through the use of thicker gate oxides and some device re-engineering and recharacterization. Another possibility is to use a "BCD"—Bipolar CMOS DMOS process. This is a more complex process but offers much higher maximum voltage capability.

Power management ICs incorporate the following functions for use in applications such as cell phones [5.9]:

- battery management;
- display driver;
- audio amplifier;
- power regulators;
- power switches.

Traditionally these ICs have used 350 nm or larger process nodes. Current usage is at the 250 and 180 nm nodes.

Display drivers for liquid crystal display (LCD) or OLED displays require ICs that incorporate the following functions:

- frame buffer (SRAM);
- timing generation;

- gamma adjustment;
- CPU interface;
- source driver;
- gate driver.

These ICs typically require three different voltages and high packing density for the buffer memory. When fabricated using a 180 nm node, the voltages could be 1.8, 5 and 20 volts. Such high voltage technologies are under development at the 130 nm process node.

5.8 CMOS Challenges

In Chap. 1 there was discussion of how Moore's law and scaling theory has driven the semiconductor industry. The literature has many publications predicting the end of CMOS scaling and that it is hitting a brick wall. There has indeed been a slowing down of traditional scaling since the power supply voltage has not scaled proportionately, violating the assumption of constant power density. As the minimum feature size is approaching atomistic levels and quantum mechanical limits, two primary problems have emerged at the top of the list for most technologists [5.14]. The two problems are:

- Increasing standby power dissipation due to:
  - increasing density of devices on chip;
  - rising sub-threshold leakage current;
  - increased gate leakage due to tunneling.
- Increasing variability of device characteristics.

In keeping with IC development history, it is expected that process and device technologists will find solutions to process issues through the use of new materials, processes, and device structures. It is predicted, however, that some issues may be better solved through a close cooperation between circuits, IC design, systems, packaging, and silicon engineers [5.14].

Let us now further discuss some of the challenges. If you are a startup fabless IC company that is not using a leading edge process technology, the information in this section will serve as a reference.

5.8.1 Leakage Power and Active Power

Many applications in the mobility driven marketplace today are requiring longer battery life. Examples are laptop computers and cell phones. The latter require long battery life both in standby and talk modes. The typical power dissipation in these products is shown in

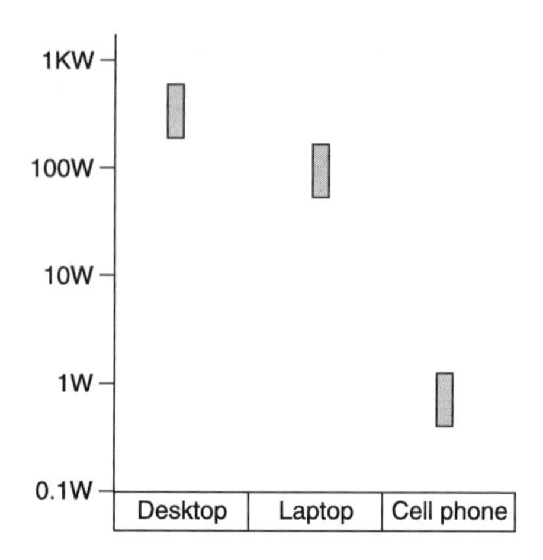

FIGURE 5.16 Comparison of power dissipation in desktop PCs, laptop PCs, and cell phones.

Fig 5.16. Bio-medical applications such as hearing aids require power dissipation that must be much lower.

What makes the task of lowering power dissipation even more challenging for IC designers is the fact that there has been a significant increase in the transistor's sub-threshold leakage in the latest nanometer process technologies. This is in addition to the expected increase in active power dissipation due to higher operating frequencies and the increased number of devices per chip. An excellent treatment of CMOS scaling issues is provided in references [5.15, 5.16]. Figure 5.17 shows the calculation of active power density, due to switching, as a function of gate length. Also plotted is the increasing passive power. The major component of this is the gate tunneling current.

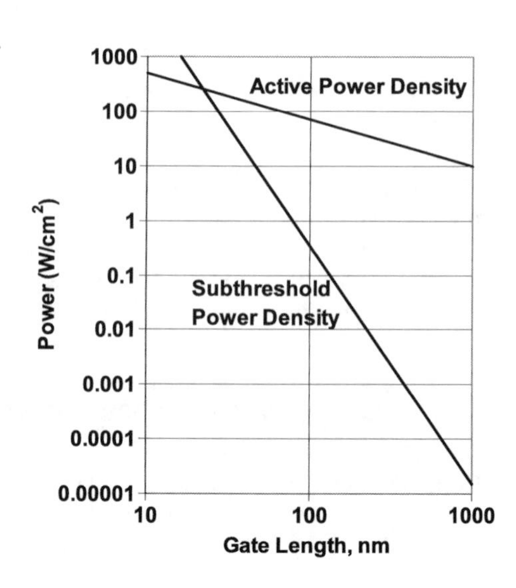

FIGURE 5.17 Rapid increase of subthreshold power in the latest nanotechnologies. (*Courtesy of E.S. Nowack. IBM J. R&D [5.12].*)

The two lines intersect at 20 nm at room temperature. The off-current nearly doubles for every 10°C rise in junction temperature. The sub-threshold power will equal active power at 50 nm gate length. This is why the industry is so focused on finding solutions to manage power dissipation.

The good news is that the industry has a track record of finding solutions to power dissipation issues. The following is a list of major changes made in the industry to manage power [5.17].

- 1970s—a switch to MOS devices from Bipolar.
- 1980s—a switch to CMOS.
- 1990s—scaling down V_{DD}.
- 2000s—power efficient scaling/design innovations.
- 2010s—new materials and device structures.

Many possible solutions are being implemented in the industry. One example of a design related solution that reduced leakage power 40 fold has been reported by Texas Instrument for their OMAP 2420 IC implemented in 90 nm technology with 90M transistors [5.10]. This reduction was achieved by using dual V_T synthesis and multiple power domains. A more complete treatment of design approaches is provided later in this chapter in the context of nanotechnology co-design solutions. A holistic approach for managing power is recommended. Concurrent solutions at the process technology, the design and the architecture levels are being considered in the industry.

5.8.2 Process Variability

Process variability issues have been around since the inception of the industry. What has changed is that variations are a much larger component relative to the nano-scale dimensions in today's processes. While process technologists are working hard to understand and control process variability, the realities of nanometer technologies must be modeled and incorporated into design tools. The problem is that variability can affect performance and yield of products designed at the leading edge. Two references are recommended for the reader interested in delving deeper into this topic [5.14, 5.18]. There are many sources of variations. One way to organize and understand them is as suggested by Dr. Chen [5.14].

- **Global** variations in the process can occur chip-chip, wafer-wafer, or lot-to-lot. These type of variations are well understood by experienced manufacturing engineers and can usually be managed via automated process control methods and tools.
- **Regional** variations refer to variations across the wafer or across the die. If these variations are systematic, they can usually be addressed by making process modifications. If these

variations occur randomly, they can be a challenge to fix. Solutions can encompass on-chip monitors that test and self-correct issues.

- **Local** variations can also be systematic or random.

Examples of systematic variations are:

- Non-uniform surface finish during the CMP (Chemical Mechanical Polishing) step due to local pattern density.
- Ion implant proximity scattering which can result in threshold variations.
- Device current variation dependency on the active area size due to Shallow Trench Isolation (STI) stress.
- Reflectivity variations across chip caused by feature shape. OPCs have been in use since the 180 nm process node to minimize such variations.

Stochastic or random local variation examples are:

- Random dopant fluctuations. This problem arises because in the 45 nm MOSFET gate there may be only 100 dopant atoms. Any physical variation causing a displacement of even a few atoms can cause micro variations of the gate threshold voltage.
- Since sub-wavelength lithography has been pushed to resolve features far smaller than the 193 nm wavelength, the printed features have rough edges. Figure 5.18 shows the increasing gap between lithography wavelength and minimum feature size [1.15, 5.17]. Edge roughness of the printed lines is called LER (Line Edge Roughness). Roughness of the line width is called

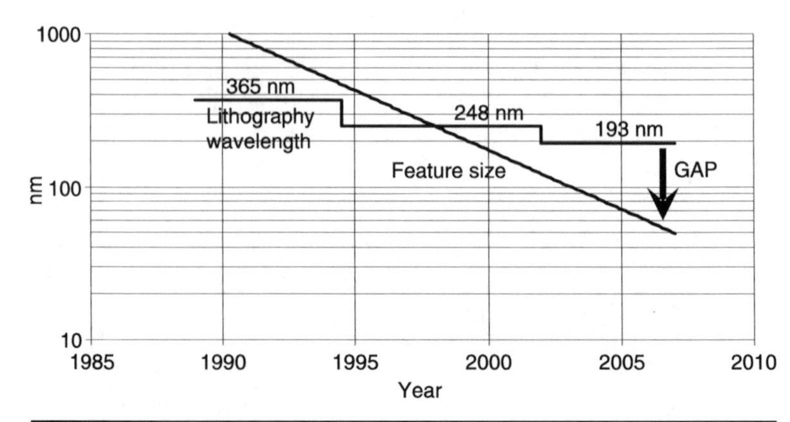

FIGURE 5.18 Increasing gap between lithography wavelength and minimum feature size. (*Source: IC Knowledge [1.15], Intel [5.17].*)

LWR (Line Width Roughness). Characterization and minimization of these effects has been studied extensively by wafer fab engineers and researchers.

- Temperature hot spots and heat flux hot spots have been described by Borkar in reference [5.18].

The effects of variability of parameters such as V_T, I_{dsat} and L_{Gate} on device characteristics have been reported in the literature [5.14]. There have also been reports of on chip test structures to measure, monitor and control variations [5.14, 5.19]. It has also been reported that significant performance hits (~30%) and power increase (~100%) can occur due to threshold voltage variation [5.14].

While there is much focus in wafer fabs to reduce and control variability, it is also necessary to manage the effects in the design process.

5.9 The Design Ecosystem

As commercial foundries and the distributed manufacturing supply chain emerged in the 1990s, there were many gaps between the IC design community and silicon manufacturing. Designers within vertically integrated companies were used to having libraries, memory blocks, models, design tools, process yield, and characterization information readily available. In the early 1990s, the foundries either did not have information available in a usable format to make available to customers, or were not open to sharing such information across customer boundaries. Much progress has been made since then at the foundries, EDA companies, and IP providers to address issues facing customers. What has emerged is an ecosystem aimed at enabling new IC designs. The ecosystem includes solutions to known issues. Some elements of this progress are represented by the following:

- Standard cell libraries, I/O, memories, and some IP is routinely made available by the foundries, or their partners. The business model has evolved to make it convenient for the user.
- IP is available from numerous "third party" IP suppliers. Generally there has to be a business arrangement directly between the fabless IC company and the IP supplier(s).
- EDA tools are available from the four major EDA companies and numerous other specialty suppliers.
- PDKs are provided by the foundries through EDA suppliers.
- A reference design flow is made available to customers by the leading foundries.

A recent article outlined one possible framework of solutions offered by the EDA industry [5.20]. In this "three P" model it is proposed that designers are able to meet design objectives through the capture and application of learning from one technolgy generation to the next. The three P's are:

- Performance—important to the engineer and represents the ability to meet the functionality and performance targets.
- Productivity—important to the middle manager and represents the ability to meet project goals within budget.
- Predictability—important to the senior executive and represents the ability to get the product to market and within budget.

As can be expected, design challenges have grown significantly with every generation of process technology. In general, designers have had to play 'catch up' with new process nodes. The EDA industry has been challenged to make both technical and productivity enhancements. Such enhancements have allowed designers to manage increased chip complexity. The following list illustrates some examples.

- Increased transistor packing density. The increased capability is meaningless unless one can implement designs with it!
- Increased number of interconnect layers. In some ways this made it easier to route connections to the transistors and blocks. However the router needed to be modified to use two, three, four, six, eight, and more layers of interconnects.
- Increased number of IP blocks—memories, analog data converters, processors, etc. In some ways designing today's chips is like designing at the PCB (Printed Circuit Board) level, with many different off-the-shelf integrated circuits.
- Tighter timing and clock skew requirements. Clocks operating at or above GHz clock rates require clock skews that are no more than 10s of picoseconds.
- Low power constraints. Power dissipation in the ICs needs to be reduced in order to reduce thermal constraints as well as to extend battery life and operation time (e.g., talk time in a cell phone).
- Combat process effects such as variability, random dopant fluctuations and well proximity effect.
- Provide solutions such as OPC for pattern distortions when printed on the silicon wafer.
- Combat design challenges such as SRAM stability and smaller "headroom" for analog designs as the power supply voltage has been scaled down.
- Combat reliability challenges such as Hot Carrier Injection (HCI) and Negative Bias Temperature Instability (NBTI) which are aggravated as the device sizes have been reduced.

HCI refers to the effects of changes to the gate oxide/silicon interface caused by injection and trapping of carriers and the generation of unwanted space charge [5.21, 5.22]. These effects result from the high velocity of electrons in an NMOS (N-channel MOS) transistor due to the high electric fields near the drain. The change in the oxide interface charge distribution results in shifts of threshold voltage (V_T) and a reduction of the Drain Saturation current (I_{dsat}).

NBTI refers to a negative shift in PMOS (P-channel MOS) transistor threshold voltage. Prolonged application of a negative voltage applied to the PMOS gate creates interface traps at the gate oxide/silicon boundary. Trapping of holes in these interface traps results in an increased positive charge distribution, causing the negative V_T shift [5.22, 5.23].

The following is a listing of the major design challenges that have been incorporated into EDA solutions [5.20, 5.24]. New solutions are added on a regular basis and there is also continuous improvement of existing solutions.

- **Timing and timing closure** (250/180 nm onwards)—Timing closure encompasses many aspects of the design process and is built around a Static Timing Analysis (STA) environment that is an integral part of the place/route tools, but is also run as a standalone timing signoff process. The primary components include:
  - A robust set of timing constraints that cover all paths from chip inputs to registers, register to register and registers to chip outputs. These constraints must cover all the important aspects of the design, such as clocks with their insertion delay and skew, input arrival and driving cell, output delay and loading, false, and multi-cycle paths, etc.
  - For designs that will be implemented with a physical hierarchy, the chip level timing must be partitioned into block level constraints. Often, this can require multiple iterations of the blocks and chip level timing to refine the constraints.
  - During logic synthesis, additional timing margin must be maintained in order to facilitate timing closure during physical synthesis and place/route.
  - Accurate extraction of RC (Resistance Capacitance) parasitics for interconnect is required to support post-layout STA. The RC extraction is performed for both best-case and worst-case conditions of the interconnect.
  - STA is performed on multiple corners, or settings, for Process, Voltage and Temperatures (PVT), as well as in multiple functional and test modes.

- **Signal integrity** (130 nm onwards):
  - The effects of cross-coupling between adjacent signal interconnects requires crosstalk-aware RC extraction and timing analysis, for both the signal delays and signal glitches (noise) it can cause. Of particular concern is crosstalk on clock nets. Often these are implemented with wider wires, double spacing between the wires and sometimes even shielding around the clock lines. For normal signal nets, the tools can upsize drivers, spread or jog wires, or add repeaters to avoid or fix crosstalk issues.
  - IR drop on power buses becomes more of an issue as supply voltages are lowered and resistance of the metal layers increase. This requires wider and more frequent power straps to be added, along with decoupling capacitors, as well as more rigorous analysis tools. Often one or more of the top layers of interconnect are reserved for power routing, and are implemented as thicker layers of metal. For very large die, flip-chip techniques can help to distribute power more evenly across the die. One must also consider the effect of inductive switching noise on power and clock buses (L di/dt).
  - Electromigration on power buses has been an issue for several generations of technology, but in the most recent technologies, electromigration on signal wires must also be analyzed, especially those with high drive and high frequencies. For signal wires, wider routing and redundant vias help to address electromigration problems.
  - Taken together, all of the above are driving the overall cell utilization to lower levels in the leading edge technologies. This is because additional room must be allowed for the required buffers, repeaters, upsizing of drivers, spreading of wires, etc. For instance, if the cell utilization was in the 70–80% range before place/route in the 130 nm technology node, at 65 nm it may be in the 60–70% range prior to place/route.
- **Active power** (90/65 nm onwards)—To minimize active (or dynamic) power, typical techniques include:
  - Clock gating to turn off the clocks to logic that is not switching and therefore inactive.
  - Careful selection of the clock buffers to be used in the automatically generated clock trees can result in lower power consumption.
  - Voltage islands such that slower logic is placed in one island that is running at a lower voltage and faster logic is

placed in islands running at a higher voltage. These techniques require level shifters for the islands at different voltages. Libraries characterized to the specific voltages are also required.

- **Leakage** (90/65/45 nm onwards)—To minimize leakage, or static power, some techniques commonly used are:
  - Multi-V_T libraries allow lower V_T cells to be used for high speed paths, although this will result in higher leakage currents. Standard or higher V_T cells can be used to minimize leakage power on paths that are meeting timing.
  - Power islands allow a complete set of logic to be powered down when not in use. Retention flops may be included in these islands to maintain important states when the island is powered down, but this requires more complexity in the place/route system to route separate power for these cells. Power islands also require isolation gates to maintain the outputs of the island at a valid voltage state for the circuits they drive. If the power-down is controlled from within the chip, a power gating cell is also required for the power island, which adds not only place/route complexity, but also adds complexity to STA.
- **DFM implementation** (90/65/45 nm onwards).

This is discussed further later in this chapter

- **Statistical timing** (65/45 nm onwards):
  - On-Chip Variation (OCV) techniques are applied as part of the final timing sign-off to counteract process variations that have become more of an issue in recent technologies. Going forward, tools that implement Statistical Static Timing Analysis (SSTA) techniques will most likely be added as part of the standard timing signoff to more accurately address the process variation issues, and hopefully with less overall impact on the design.

5.9.1 Overall Design Flow

The three major phases in the development of a SoC ASIC are listed below [5.24, 5.25]. Also listed are the major operations within each of the phases.

- **Concept phase**
  - marketing inputs;
  - feasibility assessment;
  - functionality, cost, performance assessment.

- **Definition phase**
  - Behavior, architecture and structure definition. An example of the structure is the block diagram shown in Fig. 2.10.
  - Transaction level modeling (TLM) of system and specification. The function or algorithm being modeled can be programmed in C, C++, or SystemC [5.26].
  - Hardware/software partitioning. Chip "hardware" separated into three categories—value added IP blocks, custom blocks and third-party IP. The fabless company should focus its resources on the development of the first two and leverage already available third party IP.
  - Verification of chip function and behavior against system models and specifications.
  - Micro Architecture definition [5.27].
  - RTL coding in languages such as VHDL or Verilog to capture behavior.
  - Creation of a verification environment and generation of test benches.
  - Verification of RTL of third party IP and custom blocks against their specification. Verification of analog IP blocks becomes a challenge and is sometimes accomplished through the use of Verilog-AMS [5.28].
  - Synthesizable RTL code generation.
  - Verification of logical equivalence of the gate level model against the RTL.
- **Implementation phase**
  - RTL synthesis into a netlist of logic gates.
  - Formal verification of RTL versus the gate level netlist.
  - STA of the gate level netlist.
  - Gate level regression tests.
  - Physical implementation.
  - Extraction of RC parasitics and timing sign off via STA.
  - Formal verification of pre- and post-layout netlists.
  - Test vector generation and simulation of the gate level netlist against the test vectors.
  - Physical verification checks and database preparation.
  - Sign off review and tapeout.

A simplified version of the design flow in the implementation phase of a (D/a) mostly digital design having some analog and other IP blocks and memories is shown in Fig. 5.19. Starting with an RTL

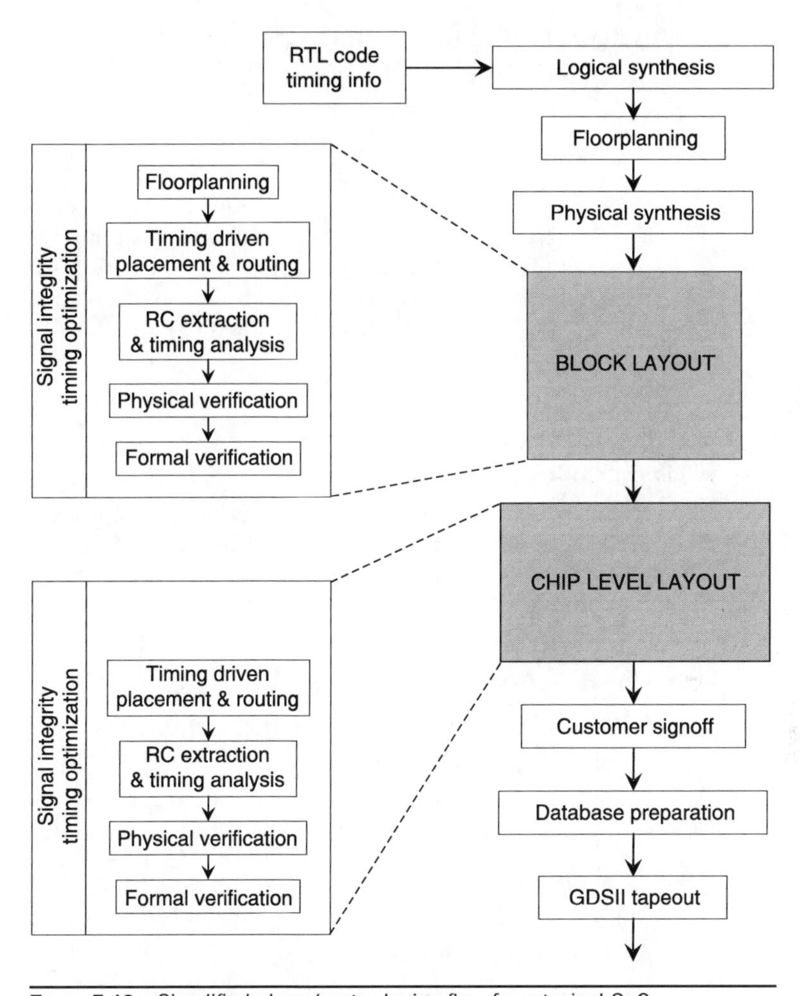

FIGURE 5.19 Simplified place/route design flow for a typical SoC.

description of the design behavior and timing, the first step involves a logic synthesis. A brief description of the major activities in each of the operations is as follows:

- **Logical synthesis and scan insertion**
    - Convert the behavior captured in RTL into a logic gate netlist.
    - Convert the standard flip-flops into scan flip-flops.
    - Partition the design into manageable physical blocks (e.g. 10–20) with "instances" (or placeable elements) of up to 500 K in each block.

  - Add pre-layout timing margin, e.g., greater than 10%.
  - Set timing constraints for paths as follows:
    - input to register (~25%);
    - register to output (~25%);
    - block to block (~50%).
  - Set up blocks for managing low power constraints as described in Sec. 5.9.2. Examples are clock gating, partitioning into voltage islands and identification of isolation gate control signals for power shutdown.
  - Verify that the functionality of the RTL and gate level netlist are equivalent.
  - Perform an initial STA.
- **Floorplanning**
  - Using estimated block sizes and pin-outs, perform an initial, global placement of hierarchical blocks. Example blocks are digital, analog IP, other IP, memory, and the I/O ring.
  - Low power management:
    - define power and voltage islands;
    - add I/O ring breaker cells for different power domains;
    - calculate power consumption.
  - Power and clock planning.
  - Virtual flat placement.
- **Physical synthesis**
  - Optimize the design for timing based on a global placement.
  - Apply timing constraints including:
    - clocks with insertion delay and skew;
    - input arrival and driving cell;
    - output delay and loading, false and multi-cycle paths.
  - Low power management:
    - insert isolation gates and/or level shifters;
    - prepare timing libraries for multiple voltages.
  - Perform STA:
    - setup, hold and transition time checks, for multiple corners PVT corner limits (or "corners");
    - create timing models for blocks from timing constraints;
    - validate timing with newly inserted isolation gates, level shifters, multiple voltages, or other low power management methods used;
    - estimate wire delays from the global placement.

  - Update block sizes, pin-outs and placements.
  - Update IP, memory and I/O placements.
  - Update power and clock distribution.
- **Block layout**
  - Digital blocks go through a timing driven placement and routing operation, as is discussed with reference to Fig. 5.19.
  - These blocks could include analog and other digital hard macros (IP blocks) as sub-block. The sub-blocks are input with their footprints, timing and power constraints.
  - Memory blocks are input with the footprints, timing and power constraints.
- **Chip level layout**
  - Once the blocks are complete they are integrated into the top level along with any top level hard macros, IP cores and memories, and the I/O ring. Timing driven place/route is performed to create the final GDSII.
- **Customer sign off**
- **Database preparation per the foundry requirements**
  - metal fill;
  - documentation structures;
  - final cleanup and verification;
  - formal foundry release documentation.
- **GDSII tapeout**
  - upload the design database;
  - foundry request form detailing information such as chip size, special process options, disposition when completed;
  - description of the mask layers, design rules, models and PDKs used and any special instructions.

5.9.2 Block and Chip Level Layout Detail

Now let us describe in a little more detail the steps during block and chip level layout. Upon completion of the physical synthesis, there is a gate level net-list available in a Verilog format. This net-list defines the gates used and how they are connected. For the blocks, also available are block pin locations, floorplan guidelines, power, and clock requirements and timing constraints. With these inputs, the major activities that follow are:

- **Floorplanning**
  - Generate power bus layout.
  - Placement of hard macros, IP cores, memory blocks.
- **Timing driven place and route:**
  - Placement of standard cells.

- Generate clock trees, which in some cases may include double wire spacing, wide wires and shielding.
- For low power management:
  - protect nets crossing voltage/power domains from being buffered or optimized in the wrong island;
  - separate power busing for voltage/power domains;
  - insert multi V_T cells for leakage reduction.
- Signal integrity adjustments:
  - wire spacing;
  - shielding;
  - upsize drivers;
  - add repeaters.
- Electromigration adjustments:
  - widen wires;
  - add redundant vias.
- Testability adjustments:
  - reorder scan chains.

- **RC extraction and timing analysis**
  - Extraction of best and worst case parasitics using accurate 3-D algorithms.
  - Extraction of cross-coupling capacitances on signal nets.
  - Signal integrity analysis:
    - glitch (noise);
    - delay;
    - signal/power electromigration;
    - IR drop.
  - Clock line analysis.
  - Switching noise and ground bounce analysis.
  - On chip variation.
  - Low power management:
    - multiple extractions;
    - library corners for multiple voltage islands.
- **Physical verification**
  - DRC (Design Rule Checks) for any violations of process design rules.
  - LVS (Layout Versus Schematic) for any deviations from the electrical schematics.
  - Antenna checks for violations of long wires that could aggravate charging.

- Density rule checks are related to uniformity in the patterning and polishing processes.
- Slotting rule checks were introduced for stress relief.
- DFM rule checks.
- Low power management:
  - verify proper power supply connections for voltage and power islands.

- Formal verification
  - Verification of place and route optimizations, additional clock tree gates.
  - Low power management checks for:
    - invalid buffering;
    - invalid optimization of isolation gates and level shifters.

Having completed the block level operations, a footprint and timing model are created for each block which will then be used in chip level layout operations.

The major activities in chip level layout are similar to those in the block level place/route section. The main difference is that instead of block pin-outs, the locations of the chip pads, chip I/O buffers, and the associated I/O ring information are required. Typically many of the analog macro blocks are placed at the chip level. Then the timing driven place/route is executed followed by RC extraction, timing analysis, physical verification and then formal verification.

5.9.3 Design Tools Selection

The design flow described here has been generic and independent of the specific EDA tools. The four major EDA tools suppliers are listed here for reference:

- Cadence Design Systems (http://www.cadence.com).
- Synopsys (http://www.synopsys.com).
- Mentor Graphics (http://www.mentor.com).
- Magma Design Systems (http://www.magma-da.com).

Each offers a variety of tools covering the entire design flow. Details of the offerings are beyond the scope of this book.

There are also many specialty EDA suppliers that offer specific solutions. Some examples are:

- Apache—low power and noise analysis.
- CoWare—Electronic System Level (ESL) design.
- Clearshape—DFM solutions (Company acquired by Cadence Design in 2007).

- Javelin—virtual prototyping.
- OEA—3D modeling solutions.
- Sequence Design—power management.

The design flow needs to be refined to suit the specific tools of choice. The leading commercial foundry TSMC now makes available "Reference Design Flows" adapted to each of the four major EDA suppliers.

For a start-up fabless IC company it is important to decide whether to outsource the RTL to GDSII design activity. Approximately 75% of EDA tool cost is in this portion of the design flow. In addition one has to staff-up the group. There are many design services providers that are proficient in managing the execution of designs through this phase. Unless you feel there is a compelling reason to build this group internally, the start-up is advised to outsource this activity. It may be appropriate to build this expertise in-house as the number of designs increases.

The start-up should definitely consider procuring the tools for the "front-end" design to the synthesizable RTL phase. It is beneficial to have a close link between the architects, marketing and software folks to allow quick decision making and execution of the chip's feature changes. It is also likely that the company has differentiating features it wants to keep "close to the vest."

Selection of the right tools, the suppliers and the methodology is very dependent on the type of design that is to be implemented, the experience of the company's design team, the schedule, and cost goals. Some suggestions for getting help are to hire experienced individuals, whether full time or as part time consultants, and to hire a key lead designer. However, even in this scenario, outside resources can be used to augment the internal team or to add specific domain or tool expertise.

5.9.4 IP Selection—Memory, Library, I/O, uP, etc

Selecting the IP required and its suppliers is a very important area that requires significant due diligence. The IP is really an extension of and a bridge between the process and the design. IP can be available as "soft IP" (RTL) or as a "hard IP" (GDS) block. Soft IP gets instantiated early in the design flow. Hard IP usually gets merged during the chip level layout, although a simulation model is required earlier.

The foundries generally have some IP available. This is a good place to start. However, at times this offering is not a sufficient set. They also have an ecosystem which includes third party IP suppliers who have developed IP blocks targeted at the foundry's process. There are also independent agencies that provide assistance in finding the right IP. Examples of such services are:

- Chip Estimate (http://www.ChipEstimate.com).
- Design and Reuse (http://www.design-reuse.com).

Differentiation

	Early Days	2007
Custom	Gates, Mux,... Memory	***Special Blocks*** high end ADC, DAC, PGA,...(perf, area, power optimized) Datapath eDRAM, NVM, OTP RF Tuner
Complex		***Optimized*** high end µP, Configurable µP DSP, Configurable DSP Analog Front End
Value Added		***possibly Standards-based, Re-usable or Hardened*** USB, PCI-x, HDMI, 10.100,.. standard ADC, DAC synthesizable µP optimized Library (for density, perf, power, leakage) special I/O (SSTL, HSTL, LVDS<...) Repairable multi-port SRAM
Commodity	Transistors	***Generic*** Standard Cell Library I/O Library ALUs, µC, PLL, DLL,.. Register File, ROM Compilable SRAM (1- or 2-port) Voltage Regulator Bandgap Reference

Figure 5.20 Increasing levels of IC differentiation can be achieved through the use of higher levels of IP blocks.

As mentioned earlier, the availability of IP has been an important key to the success of the fabless industry. Figure 5.20 is an attempt to illustrate how the IP industry is providing a support infrastructure that allows designers to implement their ICs. In the early days one had to start with the basic transistor and handcraft it to provide the required circuit functionality. Today there is a plethora of "generic" IP available from numerous sources on many of the commercial foundry technologies. Increasing differentiation of the IC can be achieved by using blocks with value-added IP, complex IP or customized IP, as depicted in Fig. 5.20. A similar theme has been outlined recently [5.29]. Although the boundaries between these categories could be fuzzy, the figure and the examples help partition the various types of IP and the differentiation opportunities they offer. The value-added IP blocks could be based on standards or could be higher level blocks that have been pre-designed and "hardened" through implementation and verification in specific technologies. Complex IP blocks are also value added blocks but are generally customizable for individual applications. Examples are embedded microprocessors, configurable microprocessors, and DSPs (Digital Signal Processors), and analog front ends. Then there are custom designed or modifiable IP blocks that allow the highest potential for differentiation.

5.9.4.1 Selection Criteria

Selecting the right block and the right supplier are critical factors to successful implementation of your design. In the end the weakest link will likely cause a failure in your IC design. Some guidelines and important criteria in selecting the IP and the supplier are [5.30]:

- **Functionality**
  - Does the IP meet functionality and performance, in the right process and with the right process options?
- **Maturity**
  - Has the IP been implemented in the specific technology under consideration?
  - How many customers have used the IP in the same process?
  - How many ICs and how much volume has been shipped with the IP embedded in them?
  - How long has the supplier been in the IP business?
  - How much experience does the supplier and its team have in developing similar IP?
- **Characterization/qualification**
  - Review the available characterization data and reports. How was the IP verified? Make sure the conditions are applicable in your application.
  - Has the IP been qualified using industry standard tests, e.g., high temperature life test, ESD, latch up?
- **Support**
  - Is the supplier set up to provide support during design integration?
  - Is the supplier set up to provide support in the test, characterization and manufacturing phases?

5.9.4.2 Standard Cell Library

Standard cell libraries are generally available from the foundries or from third-party suppliers via arrangements they have with the foundries. Leading third-party suppliers are:

- ARM (http://www.arm.com).
- Virage Logic (http://www.viragelogic.com).

The libraries provided are generally an excellent choice for the emerging company. For companies targeting their designs at the very leading edge process node and for the highest density and performance designs, it is worthwhile investing in libraries that are optimized for the specific application. This is because the "generic" library offering is optimized for general use. Customization may offer improvements in density, performance and manufacturability.

Some considerations of a good standard cell library are listed here [5.12]:

- Density.
- Speed.
- ECO (Engineering Change Order) capabilities:
    - ECO cells can be used to add functionality and fix chip errors through metal and via changes. Such capabilities can save significant cost and cycle time during debug of the IC. Such cells can be added in the place and route process as "spare cells." If not used, the extra gates can be added as decoupling capacitors.
- Ability to shrink may be appropriate for some applications.
- Design for manufacturability features:
    - Is the library compliant with the foundry's DFM rules and advisories?
    - Does the library have an OPC friendly layout?
- Low power management:
    - Multi V_T devices allow timing and leakage tradeoffs.
    - Back bias adjustability allows leakage reduction (reverse bias) and also performance improvement (forward bias). The library must have proper substrate contacts.
    - Multi voltage island support through isolation cells and level shifters. The library must be characterized at multiple voltages.
    - Power gating—high V_T footer cell and sleep control to allow reduction in leakage.
    - Retention flip flops allow retention of data during power down.
    - Dynamic voltage scaling.
- I/O cells
    - Configurations to support various wire bond schemes:
        - single row, in-line;
        - staggered row of pads;
        - circuit under pad (CUP), where the I/O circuitry is located under the I/O pad.
    - Proven ESD and latch up performance.
    - Support bump processing through a redistribution of peripheral pads using an RDL (Redistribution Layer).
    - Support area array bumps.

The common business models for the use of standard cell libraries is a no-up-front-charge model. The library suppliers usually get a royalty from the foundry as it ships production wafers. Libraries and I/Os with specialty requirements generally have an NRE charge associated with them.

5.9.4.3 Static and Read Only Memory (SRAM, ROM)

SRAMs have been a mainstay for use as embedded random access memory on chip. SRAM memory chips are also used by process development engineers to perform debug and yield enhancement when a new process node is being developed. The minimum memory cell area for a traditional, six transistor memory cell is used as a figure of merit (FOM) of the technology packing density. Intel has recently reported a SRAM memory cell below 0.35 square micron in 45 nm technology [5.31]. TSMC has reported a cell area less than 0.3 square micron [5.9]. Apples to apples comparisons should include the current carrying capabilities, power dissipation, stability, yield and manufacturability especially for parameters such as read current and minimum operating voltage, V_{min}. As a trade-off of yield and reliability, the process design rules are also "squeezed" by the process engineers in order to allow designers to increase packing density.

Compilers have been developed that enable generation of memory blocks for custom applications. They usually reside at the foundries, at third-party IP providers or at large IC houses that design many chips with many memory blocks per chip. The fabless IC company can acquire the memories from either the foundry or the third party supplier. Considerations for selecting the right memory block and the right supplier are similar to the ones outlined earlier. There could be subtle differences in the cell and the block density as well as performance offered by the different sources. Unless the start-up fabless company has the need to design many memories, it is not advisable for the company to acquire a memory compiler.

Single port SRAMs are the most commonly used, although some applications do use dual port or multi port SRAMs. Such SRAMs could have combinations of read and write ports. There are also some applications that require CAMs (Content Addressable Memories).

Through the use of compilers, hard-wired ROMs (Read Only Memories) are available in a manner similar to SRAMs. The ROMs could be programmed at the diffusion, contact, or via levels. The later in the process the program level, the shorter the lead time and cost to implement program modifications. Business models are also similar to those outlined earlier.

5.9.4.4 Embedded μP and DSP

The leading suppliers of embedded microprocessor cores and DSPs (Digital Signal Processors) are ARM (http://www.arm.com), MIPS (http://www.mips.com), ARC (http://www.arc.com), CEVA (http://www.ceva-dsp.com), and Tensilica (http://www.tensilica.com). Availability of such cores, some of which are user configurable, has been a key in expanding capabilities of the chips being designed.

These cores are generally licensed directly from the supplier. Some design service providers also provide easy access programs. The IP selection considerations discussed earlier apply. Business models can be NRE based, or a combination of NRE and royalty based.

5.9.4.5 High Speed Interfaces

With so many interface standards, some of the foundries and third party IP providers make available IP that is compliant with commonly used standards. Examples of available IP are USB, PCIx, DDR, XAUI, SATA, HDMI, and Ethernet. There are many suppliers—Analog Bits, ARM, Chipidea, Dolphin, Silicon Image, Snowbush, Synopsys, TSMC, etc. The best way to find one is to do a web search on the suggested sites and then apply the selection considerations discussed previously.

5.9.4.6 Analog / Mixed Signal

Many suppliers have developed commonly used IP blocks such as data converters (DAC, ADC), PLL, voltage regulator and bandgap reference. There are many suppliers—Analog Bits, Cadence, Chipidea, Dolphin,iWatt, S3, Snowbush, TSMC, etc. The best way to find one is to do a web search on the suggested sites and then apply the selection considerations discussed previously. Matching the functionality and performance requirements can be especially tricky and proper due diligence is recommended. AMS blocks with special requirements are usually a good candidate for customization.

5.9.4.7 Embedded DRAM

As mentioned earlier, SRAM's have been the mainstay of logic based SoCs. The SRAM integrates well with the baseline logic process. Dynamic RAMs, however, are the mainstay of mass memory storage in personal computers. Many companies have attempted to integrate the higher density DRAM into the logic processes. Unfortunately, the increased cost of wafer processing and reduced yield due to the added complexity have made such a marriage undesirable. The DRAM is fundamentally built on the basis of storing charge on a capacitor and refreshing it periodically. Recently, the use of a MiM (Metal insulator Metal) capacitor has made it feasible to replace the traditional trench capacitor and therefore integrate embedded DRAM blocks onto a logic process. While the process cost is higher than the cost of the logic process alone, trade-offs indicate that for some applications it may be feasible to consider such an eDRAM solution. The driving motivation for designers would be the larger blocks of memory on chip. A significant bandwidth increase occurs because there is no longer a restriction to go off-chip to access the memory.

A trade-off needs to be made carefully to assess the usability of eDRAM blocks; the higher performance and the higher process cost must be compared with an off-chip solution.

Recently there has been demonstration of a 1-T memory built with the capacitor in the metal layers on a logic process [5.32]. Although the capacitance is smaller than in a conventional DRAM and the MiM cap based eDRAM, functionality has been demonstrated in a 130 nm process. If proven successful, this technology could be revolutionary.

5.9.4.8 Embedded Flash

For many years suppliers have tried to integrate NVM into baseline logic processes. As in the case of eDRAM, the increased processing cost makes it undesirable to include NVM on chip. There are, however, some applications which must have the embedded flash capability. Some foundries, therefore, have offered add-on modules allowing fabrication of embedded flash blocks. The additional processing represents approximately 30–35% additional steps. Until recently the most advanced technology having the embedded flash capability was the 180 nm node. Blocks having 4–32 K bits of eFlash were available. As will be discussed later, customers requiring such NVM blocks should consider using a multi-chip package with an external NVM memory.

5.9.4.9 Embedded OTP/MTP/eFuse

Other forms of on chip NVMs are now available. While the eFlash technology requires two levels of polysilicon, there are new families of devices possible using a standard, single poly logic process. Their integration into the mainstream logic process is straightforward. This capability allows fabrication of either OTP or MTP blocks. The block sizes are generally small (less than 1 Mbit) and are ideal for use in applications such as analog trimming, gamma correction, program storage, HDMI decode, and feature selection. An extension of the technology is the use of an electrical fuse for applications such as memory repair and programming of ID tags. The blocks are available either from the foundry or third party suppliers. Especially since this is a relatively new technology, due diligence using the proposed selection criteria is advised.

5.9.5 IP Selection—Custom

Although the design ecosystem provides much available IP, certain applications may require customization. This is especially true if the new design has some distinguishing features. There are specialty IP developers who focus on providing such services. Examples are Cadence, S3, Simple Silicon/iWatt and Snowbush. Examples of custom blocks are: ultra low power PGA, high precision ADC, analog front end for a multi-band radio, RF tuner.

5.10 Process Alternatives

There are a few process technology alternatives to CMOS that the reader should be aware of, especially the experienced reader.

5.10.1 SOI (Silicon on Insulator) CMOS

Deposition of a thin layer of silicon on top of an insulating substrate to enable fabrication of SOI devices has been reported since the 1960s [5.33]. Sapphire was used as the substrate originally to build SOS devices (Silicon On Sapphire). Initial barriers to its widespread use were materials quality and cost, device design issues, and the rapid growth and use of bulk CMOS. Since the invention of the SIMOX (Separation by IMplantation of OXygen) process in the 1990s, SOI processing has gained much ground. In this process a thin layer of silicon dioxide forms the insulating layer underneath the active devices. IBM has invested heavily in the use of SOI technology and has announced its use in the fabrication of PowerPC chips and also ICs used in their servers [5.34]. More recent publications indicate the use of SOI by Microsoft in their Xbox chip set, by Sony in their Playstation3 and AMD in their Opteron and Athelon processors [5.35]. Notice that these companies have large, experienced teams that have leveraged IBMs and their own investments in SOI technology and design knowhow. The major benefits of SOI are as follows:

- Performance advantage of 25–30% caused by;
    - the elimination of area junction capacitance [5.34];
    - a lower dynamic V_T due to a floating body effect [5.35].
- Power reduction due to the lower capacitance and the lower dynamic V_T. For the same switching speed the V_{dd} can be reduced because of the lower V_T.
- Reliability improvement:
    - latch-up free because the device is surrounded by an insulator;
    - radiation hardness improved because of a reduction of exposed silicon area [5.36].

SOI technology has come a long way and has many benefits. It is also being considered as an important candidate for integrated RF ICs [5.37]. However, it is a technology that has limited availability for the emerging fabless IC company. It is available for ASIC design and implementation through IBM's ASIC group and will be made available at Chartered Semiconductor as a foundry. Another word of caution for the emerging fabless company is that there are circuit design issues and design implementation issues (passgate leakage and history effect) that need to be addressed [5.34]. Also, there is limited

IP available on SOI relative to CMOS. Unit cost is expected to be lower when the manufacturing learning curves are comparable. SOI consumes less silicon area and the process is indeed a little simpler than bulk CMOS.

5.10.2 SiGe BiCMOS (Silicon Germanium Bipolar CMOS)

Bipolar processes have been the technologies of choice for fabricating high speed ICs since the early days. The reduced logic signal swing of bipolar ECL (Emitter Coupled Logic) and the higher transconductance were keys to the fast switching speed. Bipolar transistors could therefore switch faster and also drive large currents. CMOS devices operating at 5 volts with rail-to-rail signal swings and higher series resistance (lower transconductance) were no match for bipolar devices. However, bipolar ICs had limited packing density and the complex process limited its cost reduction. CMOS transistors offered a simpler process, increased packing density, and low power. BiCMOS technologies combine the high speed characteristics of bipolar devices with the low power and high packing density of CMOS on the same chip. The use of germanium doped base regions of the bipolar transistor enhances flexibility to engineer improved bipolar transistors. SiGe-base BiCMOS processes are used in the fabrication of high speed and high power ICs, for example, RF front end ICs and power management ICs.

SiGe BiCMOS technologies are indeed available from some foundries—Jazz Semiconductor (http://www.jazzsemi.com), IBM and TSMC. However, more and more functions are being implemented in CMOS and RF CMOS processes. For the emerging fabless IC company it is advisable to consider CMOS as the starting point. If there is a compelling reason to use SiGe BiCMOS technology to meet the functionality and performance goals of the differentiated ICs, the technology is indeed available.

Gallium arsenide (GaAs) process technology has been used for fabricating very high speed devices. Figure 5.21 is a pictorial

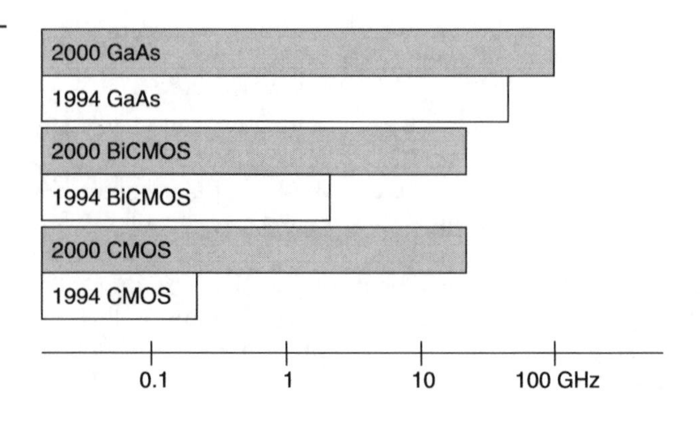

FIGURE 5.21 Comparison of maximum performance of CMOS, BiCMOS, and GaAs devices in 1994 and 2000 [5.38].

representation of the maximum performance of GaAs, BiCMOS, and CMOS devices in 1994 and 2000 [5.38]. Notice that CMOS device performance has caught up with the performance achieved using BiCMOS devices. GaAs devices still operate at higher frequencies, but the packing density is limited. It is indeed true that the designer's job is a little more challenging to achieve the 10+ GHz operation in CMOS than in BiCMOS. A word of caution about this chart—these results are likely to represent individual device and simulated results, and not chip level clock rates.

5.11 Nanotechnology Co-design Solutions

5.11.1 Process Considerations

Device scaling is continuing to provide a packing density increase from one technology node to the next. However, performance benefits are being limited by various fundamental limits. One such limit is the reduced rate of gate oxide scaling on process nodes below 130 nm, as shown in Fig. 5.22 [5.12]. A reason for this is the increased gate oxide tunneling current as the thicknesses are approaching atomistic levels. The relative gate delay of various process nodes compared to the 250 nm node are shown in Fig. 5.23 [5.12, 5.39]. Notice the 40% improvement between 250 and 180 nm nodes. Since then the gate delay is shown to reduce only by about 10% per node. Another observation is that in scaling devices from one generation to the next, one achieves only a fraction of the expected performance improvement, as shown in Fig. 5.24. One example of this is the increase in parasitic series resistance and strain relaxation, which do not

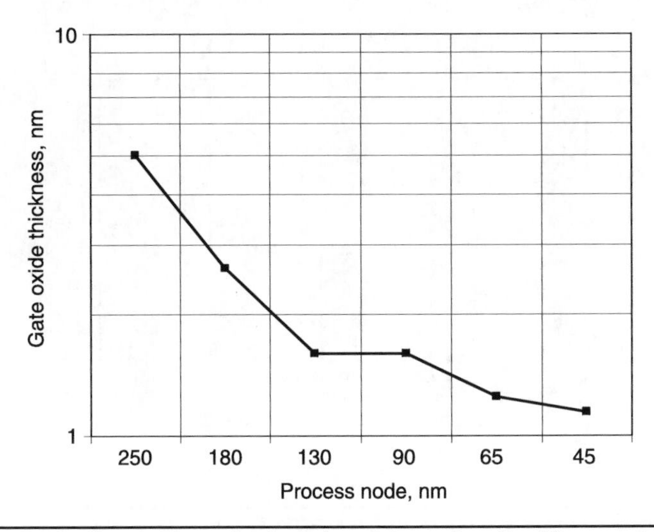

Figure 5.22 Gate oxide scaling has slowed down since the 90 nm node.

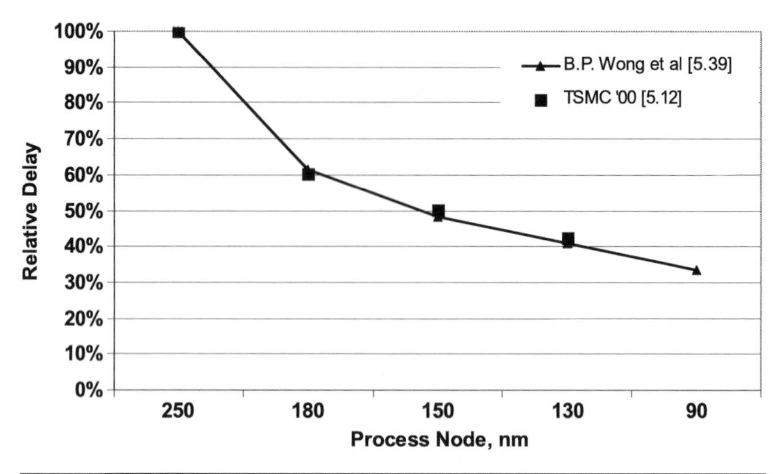

FIGURE 5.23 Relative gate delay for various process nodes [5.12, 5.39].

scale well with dimension reduction. Also shown in this figure is the increasing amount of innovation required to fill the performance improvement gap.

Some of the innovations are focused in areas represented by the transistor performance equation in Appendix B.1. Current innovation activities are focused on improving the gate work function, the dielectric constant (high-K material), increasing channel mobility (strained silicon), and reducing interconnect parasitics (low-K interconnect dielectrics). The following is a summary of some salient aspects of work in the important innovation areas. Development activities are ongoing at various semiconductor companies.

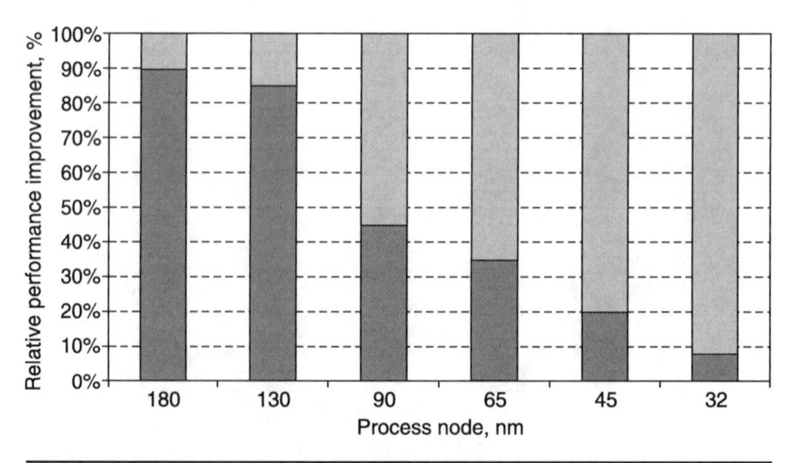

FIGURE 5.24 Increased amount of innovation is required to achieve expected performance gains at the new technology nodes [5.10].

- **Hi-K/metal gate**—There has been much discussion in the industry about when transistors with Hi-K gate dielectrics and metal gates will be introduced in manufacturing. Intel recently announced the use of Hi-K metal gate transistors in the 45 nm process technology node [5.5, 5.17, 5.31]. While other leading process development groups are considering introduction of Hi-K/metal gate structures, the Intel announcement is the first in the 45 nm technology node. A new materials set is used to form the metal gate which can be fabricated with a thicker, Hafnium-oxide based gate dielectric (3.0 vs. 1.2 nm). The transistor has been demonstrated to have a 60% increased capacitance and a 100 fold reduction in sub-threshold leakage [5.17]. In this example, the gate dielectric permittivity appears to be 4 times that of traditional SiON based dielectrics. One use of TiN as the metal gate has been reported [5.40].
- **Strained silicon**—Significant mobility enhancement (~35%) has been reported [5.10] for traditional, 45 nm devices through the use of a dual stress liner and selectively implanted source-drain junctions. The NMOS device gets tensile strain along the channel through the use of a tensile liner. A compressive liner is applied on top of the PMOS device, which also has selectively ion implanted source drain regions. A cross section of the reported device is shown in Fig. 5.25.
- **Low K dielectrics** (in interconnect layers)—Starting with the 130 nm node, interconnect delay became the dominant component of delay [5.39]. Global wiring delays on larger chips became a significant factor. Copper metallurgy was introduced as a standard offering at the 130 nm node to reduce the resistive component of delays. Dual damascene and CMP processes are now commonly used. Although these

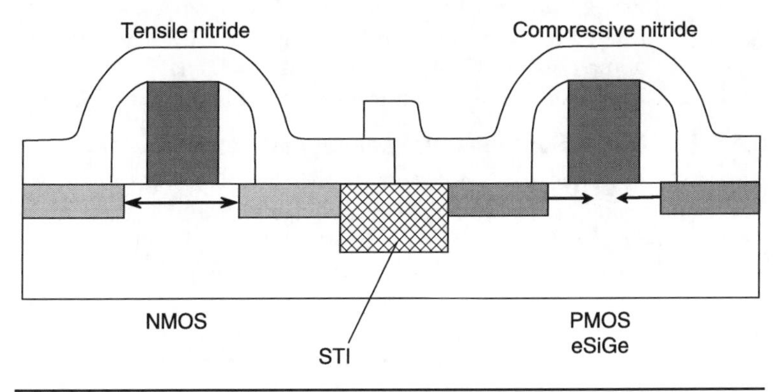

FIGURE 5.25 Cross section of NMOS and PMOS transistors with tensile and compressive strain engineering [5.10].

are complex processes, they are in mature production. Much of the development efforts are now focused on new, low K dielectric materials. There have been evolutionary enhancements to the dielectric material in order to lower the dielectric constant. The adjustment is based on variation of material properties to increase the porosity and has yielded dielectric constants around 2.5, compared with 3.9 for silicon dioxide (SiO_2). Careful optimization is required because increasing the porosity lowers the mechanical strength of the dielectric [5.39]. The jelly-like materials are susceptible to moisture absorption and contaminants. The material also is affected by erosion and yield loss in the CMP process. There have also been issues with wirebond assembly when using low K dielectrics.

- **Lithography**—Innovations in the lithography and patterning processes have been, and continue to be key to the advancement of ICs. One of the techniques for bridging the gap between the lithography wavelength and the minimum feature size (Fig. 5.18) is OPC. This operation, usually implemented in the mask making operation, makes adjustments to rectangular design features in order to print a more true image on the wafer. However, this adds to mask cost. An example is shown in Appendix B. Other ways to address lithography at reduced features sizes are:
  - High NA (Numerical Aperture) lens. As shown in the Rayleigh equation in Appendix B, a higher NA can help reduce the minimum feature size.
  - PSM (Phase Shift Masks). The mask making process adjusts the phase of the light on either side of critical features [5.41].
  - Immersion lithography. A liquid such as water is used instead of an air gap between the lens and the wafer in the lithography machine, called a step-and-repeat aligner. By doing so, the NA of the lithography system goes up by approximately 35% (from ~0.93 in air to ~1.3–1.35 with water). This results in a similar improvement in resolution [5.10].

5.11.2 Design Solutions for Leakage and Power

As discussed in Sec. 5.6.1, holistic solutions are required to manage active and leakage power, both in the active and the standby modes. These holistic solutions need to involve improvements in the process (e.g., HI-K/metal gates), the design process, the chip architecture and the software. Architectural considerations such as chip domain definitions, chip partitioning, and memory utilization can impact the chip power significantly. There is ongoing development activity at many institutions and the specific solutions are evolving. Here, let us study approaches from three different suppliers. There is some commonality, but the terminology and specific implementations vary.

The following is a list of factors reported by Texas Instruments that can affect chip power in the IC design phase [5.10]:

- adaptive voltage scaling; dynamic power switching; dynamic voltage/frequency scaling; static leakage management; multiple power, voltage and clock domains.

In the silicon process and standard cell library development, the following approaches are being used to manage leakage and power dissipation:

- low leakage process; retention memory and logic; multi-threshold CMOS cells; multi-domain support cells; on chip process and temperature sensors.

Chartered Semiconductor and IBM have outlined [5.12] an approach that uses voltage islands, which are defined as a chip region supplied by a separate, dedicated power feed. Chip regions are fed by a global V_{DD} source, but are independently controlled by a power gating circuit. The voltage island concept involves the following:

- Turn down the voltage to circuit blocks with non-critical performance in order to reduce dynamic power and leakage.
- Turn up the voltage to performance critical circuit blocks in order to reduce delay and improve timing.
- Turn off the voltage to some subset of system operations in order to drive the power to zero.

These are simple concepts that require innovative engineering of new methodologies and advanced design tools.

TSMC has organized their power management solutions into three categories, according to the task at hand—reduction of dynamic power, active leakage and standby leakage [5.9, 5.42].

- Dynamic power reduction
  - clock gating, where the clock connection to unused parts of the chip gets shut off;
  - power shutdown, sometimes called "sleep mode";
  - dynamic voltage frequency scaling, where an IEM (Intelligent Energy Management) circuit helps adjust the voltage up or down depending on the required performance need. The voltage gets raised only when full performance is required, and gets lowered when a slow response is required.
- Active leakage reduction
  - multi V_T devices are used. Low V_T devices are selected only for high speed devices—these will dissipate more power. When high V_T devices are used they will be slower but will have lower leakage;

- back bias is a scheme where back bias is applied to transistors to make them operate in the slow-slow corner. In this corner the leakage is the lowest, but so is the performance. The back bias is removed when high speed operation is required;
- Voltage scaling requires separation of operating power domains. Circuit regions that do not operate at high speed can then operate at a lower V_{DD}.
- Standby leakage reduction
  - fine-grain power gating: a high V_T (HVT) transistor is used as a "footer" switch that cuts off the flow of leakage current of a circuit block that has leaky low V_T devices;
  - power shutdown: there is a shut off to some subset of circuitry which is not required to operate at the time;
  - coarse-grain power gating: in this approach power multiple HVT devices are used as a "header" switch between a V_{DD} and a virtual V_{DD} for the particular block having leaky, low V_T devices.

In addition, adjustments must be made to SRAM cells and the standard cell library in order to accommodate the extra functionality required to manage implementation of these low power circuits.

5.11.3 Design Solutions for Variability/DFM

From the discussion throughout this chapter it is evident that solutions in the sub-45 nm technology nodes require very close cooperation between the wafer fabs, the IC designers and EDA tools developers. The reason is that many of the process problems that are limiting traditional scaling can be resolved, or certainly get assistance from design solutions. The last few years have seen the entities come together nicely. Now there is an ecosystem that fosters sharing of process variability information, models, and design kits to allow customers to bring their designs to fruition. TSMC, as the leader in the commercial foundry business, has done well in establishing relationships with EDA companies, design services companies, and IP companies. The "DFx" arena, where x has been used for manufacturability, yield, reliability, quality, silicon, and testability is in its nascent stages, although the term was coined first in the early 1990s by a team led by the author. The most common of these terms is DFM. There are many enhancements to tools and methodologies being made on a continuous basis. Every quarter brings something new in the space.

The industry has converged well on the need for "physical" DFM. This refers to rules that have been developed and are in place to check for physical layout violations that are known to cause process problems. Problem areas are identified as "hot spots." Examples are metal to active space, poly to active space, CMP constraints. There are now EDA tools available that check for such hot spots after the layout has been completed. The tools will then make fixes to the tight spot in the layout. Electrical DFM methodologies are still evolving. New tools are in development to allow simulation of the designed layout as it would appear on the silicon, and then doing a circuit simulation of electrical behavior—this is called S2E (Shape to Electrical). Similarly, a T2E (Thickness to Electrical) tool allows simulation of thickness variation of the metal layers and can estimate RC delay variations. EDA tools are also currently available for LPC (Layout Parameter Checks) and CAA (Critical Area Analysis).

5.11.4 Include Packaging and Test!

Accurate modeling of resistive (IR) drops, inductance and parasitic capacitance must include not only the chip, but also the package and the board, especially for performance critical applications. Chips operating at lower V_{DD} in the newer technology nodes will have higher current flow for the same power dissipation. This requires a more accurate determination of the resistive drops. Table 5.7 shows the breakdown of path delay in the chip core, from the chip I/O to the package pad, and from the package pad to the external ball as a percent. The chip is implemented in a 90 nm technology and is packaged in a typical Ball Grid Array (BGA) package. The delays in the package are 14–19% and cannot be ignored. Also, the parasitic voltage drops cannot be ignored, since a 10 mV drop represents 1.7% degradation in speed performance in this example [5.10].

5.12 Packaging Considerations

In the early days of integrated circuit development the major focus areas were silicon technology and circuit design considerations. Packaging and test were an "after thought". Mainframe computer companies were leaders in recognizing the value of packaging innovations in the development of their systems. There were two aspects to the importance of packaging. The first was an ability to provide packaging that met the electrical, mechanical, thermal, and reliability constraints on the implementation of ICs and systems. The second was a pro-active use of packaging technologies to gain further advantage in cost and performance at the system level. Some examples are described here. The use of Multi Chip Modules (MCM) allowed the

	Core	Chip I/O	Package
V_{DD} path	53%	28%	19%
V_{SS} path	40%	46%	14%

Table 5.7 Breakdown of Path Delays for a 90 nm Chip in a BGA Package [5.10]

packaging of multiple silicon die in the same package. Such an approach provided an opportunity to reduce the size of the systems, improve cost, and performance. However, the increased number of die in the same package caused an increase in the power dissipation in the module, and therefore an increase in individual device junction temperature. Reliability studies indicated a degradation of reliability as the power dissipation increased. Therefore it became necessary to cool the devices in order to maintain acceptable reliability. This led to careful management of air flow at the system level, the modeling, and minimization of the thermal resistance path from the die to the external world, the design of 'heat sink' devices attached to the package and in some special applications, the development of liquid cooled systems.

As mentioned in Chap. 1 there has recently been an industry focus on adopting "more than Moore" strategies—these include advantages at the system level from the use of innovative packaging techniques. This section provides an overview of the evolution of IC packaging considerations in selecting the right package and the right packaging technology. Some of the packaging level trade-offs are becoming extremely important in the context of optimizing ICs built using present nano-scale silicon and design technologies. Packaging of the IC includes the following:

- The package, its design and manufacturing. The package enables electrical connection between external pins and internal "landing" pads for making connection to the silicon die. The package also provides mechanical protection of the assembled die.
- The "assembly" process which is used to:
  - mount the silicon die in the package
  - make connections between the die's bonding pads (or connection points) and the "landing" pads internal to the package.
- The electrical, mechanical, thermal and reliability characterization of the die/package combination.
- Connections to an external "heat sink" device, if required.

Figure 5.26 shows an example of a die assembled in a package. This example is of a popular device using a copper leadframe. The device is called a QFN (Quad Flat No-lead) or an MLF (Micro Lead Frame) by the top two assembly houses: ASE and Amkor, respectively. The die is attached to a copper "paddle" using an epoxy adhesive. Wire connections ("wire bonds") from the die are made to a copper trace which leads to the external pad. The assembled package is molded using an epoxy based "plastic" molding material. Left exposed are the copper pads and the bottom of the die paddle for

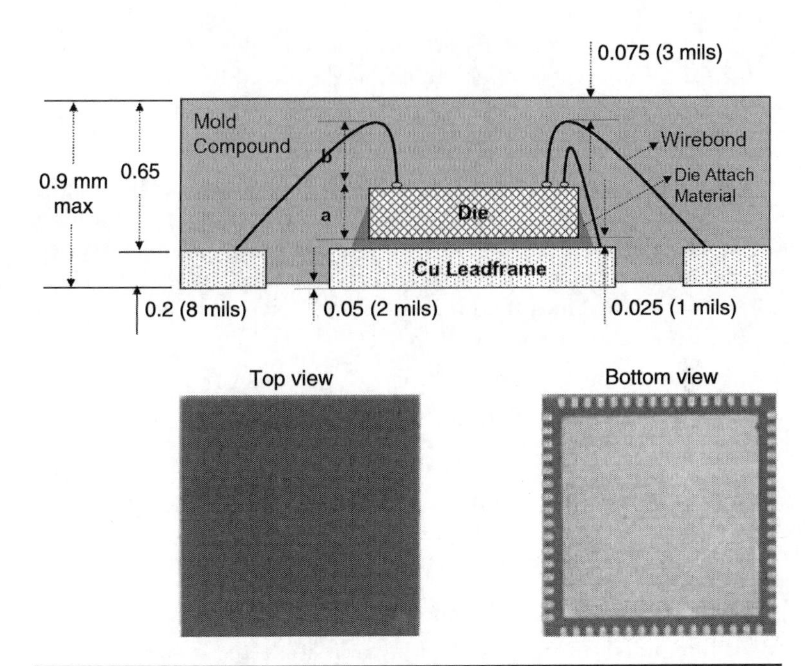

Figure 5.26 Cross section, top, and bottom views of a popular QFN package.

making external connections. The external copper is plated with a tin based material that facilitates attachment of the package on to the PCB. Exposing and attaching the die paddle to the PCB reduces the thermal resistance and assists in the management of the maximum junction temperature in the die. The dimensions show a total package height of 0.9 mm. The package body size varies depending on the number of leads. For example, a popular 48 lead QFN package could have body sizes that are 6 or 7 mm or larger.

Now let's discuss the various considerations when selecting a packaging approach. The list is applicable to the development of any IC. The fabless IC company must be aware of these considerations in offering an IC solution that meets the customer's requirement at the system level:

- **The application**—For example a desktop computer, a laptop, a cellular phone, an RFID tag, or a medical implanted device have vastly different power dissipation and package size constraints. This will dictate the use of single chip or multi-chip packaging, the cooling system required and other considerations as discussed here.
- **Cost per unit and package development cost**—These are both important. Note that package development cost for mature and mainstream packages is usually in the tens of

thousands of dollars and is much smaller than the silicon development costs. Chapter 7 is dedicated to a discussion of cost and will include some guidelines.

- **Number of I/Os**.
- **Body size**—In addition to the footprint, nowadays the height has become a major consideration, especially for slim-line cellular phones. One FOM the industry has used sometimes is the package footprint relative to the size of the chip. In recent times the ratio has come down from around 4–10 and is approaching one.
- **Power dissipation**.
- **Performance**—both electrical and thermal
- **Reliability**.
- **Environmental constraints**—Customer's requirement for hermetic, non-hermetic, or environmental constraints. Lead-free and "green" packaging options are becoming mainstream.

5.12.1 Packaging evolution

Most integrated circuits used to have a single die assembled in a single chip package. High end mainframe computers sometimes incorporated multiple die in a single package called an MCM. In the last 10–15 years packaging of multiple die in the same package has become popular. Multichip configurations could be achieved using Stacked Die (SD), or Side by Side (SxS) or Stacked Packages (SP). These 3-dimensional (3D) configurations have been called System in Package (SiP) or Package on Package (PoP). As mentioned in the context of SoCs, one is hard pressed to find an example of a complete system in a single package. To some extent, the term SiP has become popular as a manifestation of the "more than Moore" concept, and as a counter to SoCs.

Table 5.8 shows a summary of packages available for both single chip and multi-chip configurations. Note that there are many package types and many package families. The start-up fabless company is advised to select and use a package type and pin count within the mainstream envelope to reduce risk. There must be a compelling reason to select a packaging technology and package outside the mainstream. Ceramic packages were the mainstay through the mid 1990s. Since then plastic molded and laminate packages have become the high volume runners. Similarly there has been a transition from Through-Hole (TH) packages to Surface Mount (SM) packages. Here the reference is to the way the package is attached to the PCB. As the terminology indicates, surface mount devices do not require drilling of holes in the PCB fabrication process. Dual and quad packages have terminal connections (leads, pins or pads) on either two or all four sides of the package body, respectively. PGA (Pin Grid Array) and

		Surface mount or thru hole	Multichip
Plastic / Molded	3D/PoP	SM	SP, SD
	3D/SiP	SM	SD, SxS
	WLCSP	SM	SD
	CSP	SM	SD
	BGA	SM	SD, SxS
	Quad	SM	SD, SxS
	Dual	SM	SD, SxS
Ceramic	BGA	SM	
	PGA	TH	
	Quad	SM	
	Dual	TH	

SP: Stacked package
SD: Stacked die
SxS: Side by side dice

Table 5.8 Summary of Package Types

BGA packages have a matrix of either pins or solder balls pre-attached on the package body. CSPs (Chip Scale Packages) are BGAs that have a package to die area ratio approaching 1. This is usually achieved with a tighter placement of the ball array. WLCSPs (Wafer Level CSP) eliminate the package body—the die has the solder ball array attached to the die, which is mounted directly on the PCB.

There is a plethora of package types with a variety of pin counts, thermal, electrical and reliability characteristics. Figures 5.27 through 5.31

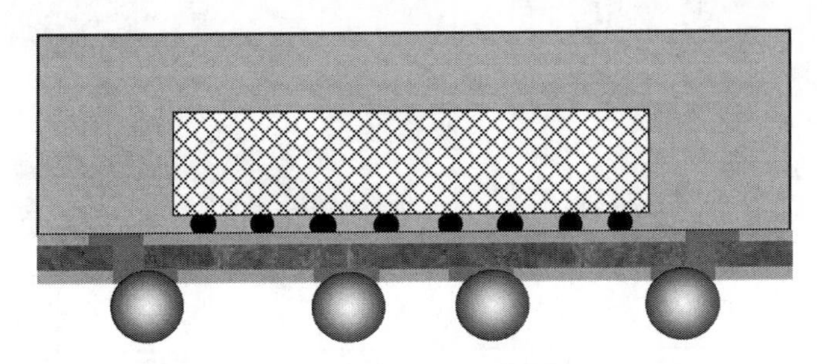

Figure 5.27 Single, flip chip–chip scale package (FC-CSP) cross section. (*Source: ASE.*)

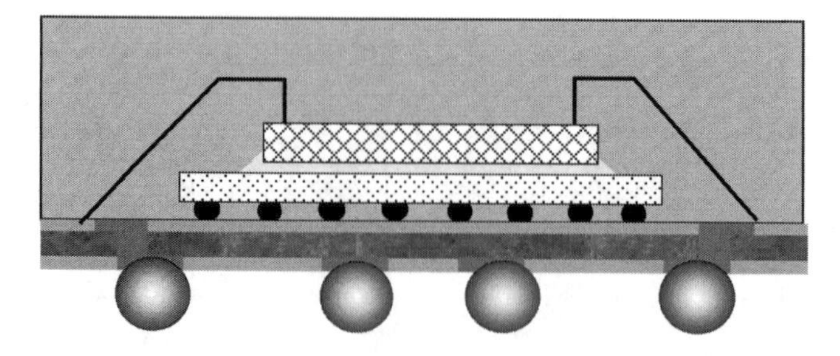

FIGURE 5.28 Two chip stacked die (SD CSP) package using flip chip and wirebond interconnections. (*Source: ASE.*)

are some examples of single chip and multichip packages. Some are conceptual drawings. Some enlargements are also shown. The best way to get further details of availability is through the web sites or the representatives of the assembly and package houses. The top four, in alphabetical order are:

- Amkor (http://www.amkor.com).
- ASE (http://www.aseglobal.com).
- SPIL (http://www.spil.com).
- STATS ChipPAC (http://www.statschippac.com).

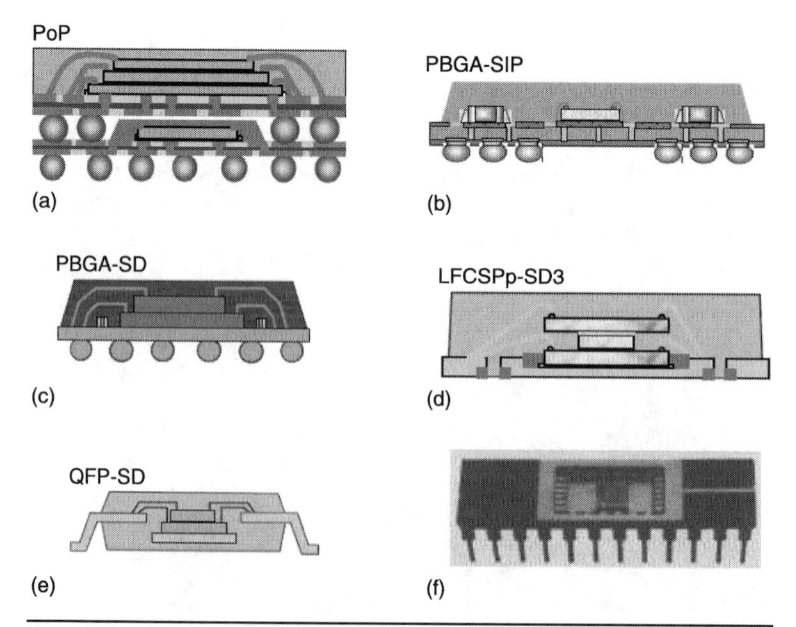

FIGURE 5.29 Examples of various package types: (a). 3D/PoP, (b). Plastic BGA–SiP, (c). PBGA–SD, (d). CSP–SD, (e). QFP–SD, (f). DIP. (*Source: STATS chipPAC.*)

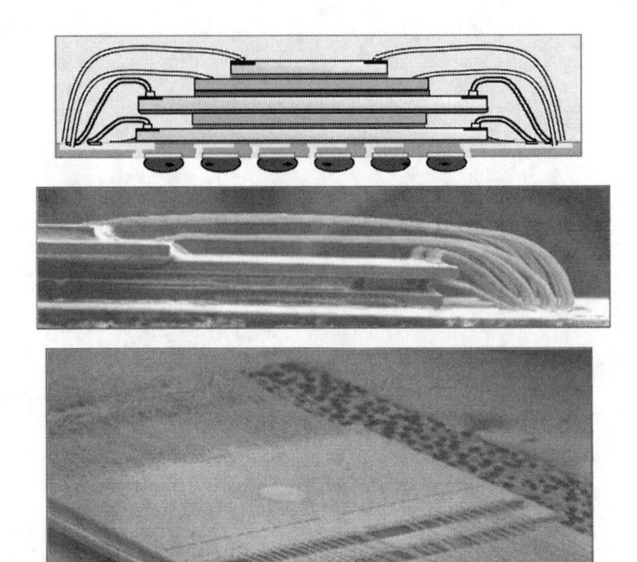

FIGURE 5.30 Example of a 4 die stacked, wire bonded CSP, or ball grid array (BGA) package. Shown are cross sections and a perspective SEM (Scanning Electron Microscope). (*Source: ASE*)

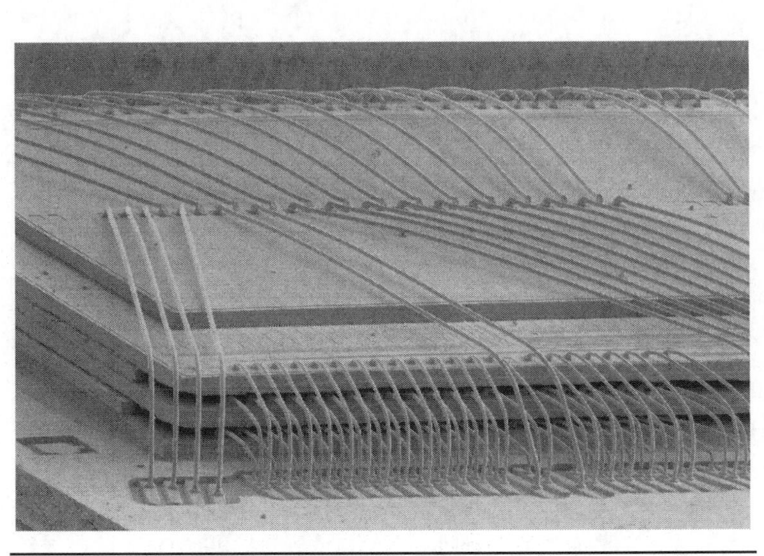

FIGURE 5.31 Examples of a 4-die stacked PBGA package. (*Source: STATS chipPAC.*)

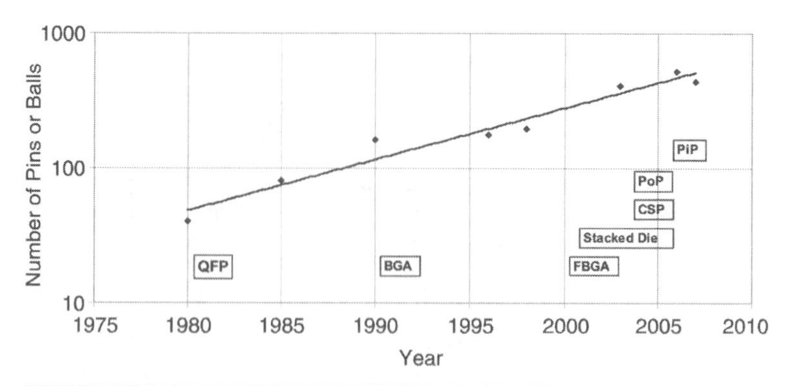

Figure 5.32 Evolution of terminal count and packages used in cellular handset applications [5.43].

The cellular phone and hand-set market has been a strong driver of assembly and packaging technologies since the mid 1990s. Figure 5.32 is an illustration of the evolution of the number of terminals (pins or balls) on packages used in the handset market place. In addition to the pin count escalation, there has been a continuous decrease in the package height and footprint. This is shown for a family of BGA packages in Fig. 5.33. Cellular handsets initially used QFP (Quad Flat Pack) packages followed by BGAs and FBGAs (Fine Pitch BGA). Further reductions in form factor resulted from the use of SD, CSPs and 3D configurations such as SiP, PoP, and PiP (Package in Package).

Figure 5.34 is an illustration of how key aspects of assembly and packaging technology for plastic molded ICs have evolved. There has been a continuous shrinkage of wire bond pitch. This has been required to allow shrinkage of the minimum die sizes constrained by peripheral wire bond pads. Peripheral leaded packages such as QFPs hit the minimum limit of 0.5mm lead pitch in the 1980s. This limit was determined primarily by PCB capability. This limit, when

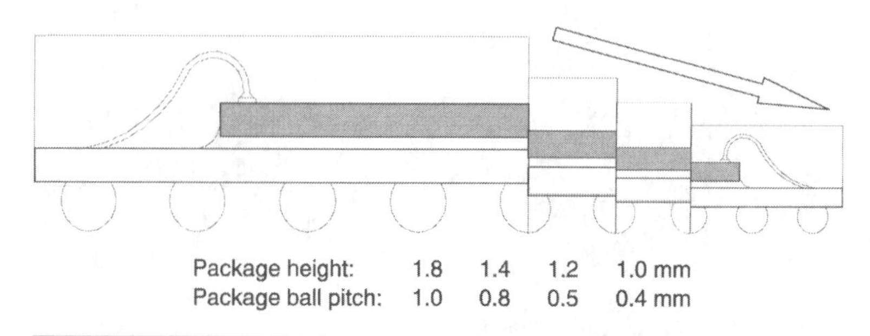

Figure 5.33 Package height reduction for BGA packages.

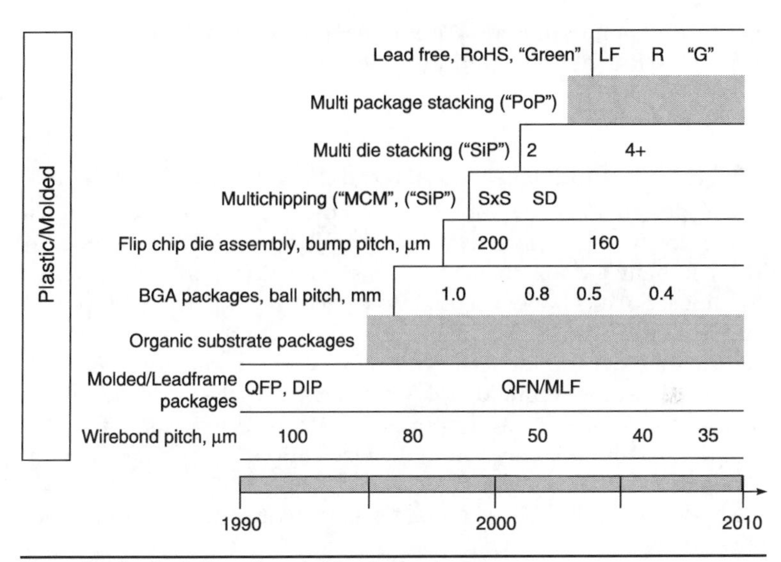

Figure 5.34 Evolution of packaging technologies.

coupled with the body size of a quad package, restricted the maximum number of package leads to around 300. This led to the use of packages having an array of connections, such as the BGA. A reduction in the package ball pitch has resulted in the reduction of package footprint for the same number of connections. Figure 5.35 is an illustration of the BGA package footprint reduction as a result of reduced ball pitch. In this example, there is an increase in the total number of connections in spite of the reduced footprint. There is also a reduction in package height from 1.8 to 1.0 mm, as shown in Fig. 5.33. The package substrate manufacturer needs to fabricate PCBs with tighter design rules in order to support these smaller packages. The use of flip chip die connections to the package will be discussed in the next section. Other evolutions shown in Fig. 5.34 are the use of SiPs with

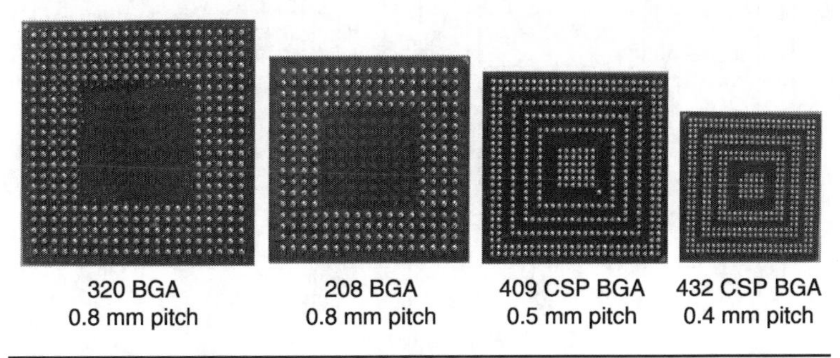

Figure 5.35 BGA footprints for decreasing ball pitch allows increased number of package connections in a smaller footprint.

SxS and SD configurations. Another major event in the industry has been the elimination of lead and the incorporation of environment friendly packages ("green").

5.12.2 Chip to Package Connections—Peripheral and Area

The IC designer usually knows a target die size, the number of signal connections to the chip, size constraints and a target performance metric. The packaging expert then has to make available possible package solutions. So far I have discussed the package families and types. In this section let us discuss two major options of making connections between the die and the package—wire bonding and bumping.

Wire bonding technology has progressed to where pitches under 35 μm are feasible. One important piece of information that the designer needs is the minimum die size supported by the minimum allowed pad pitch, called "pad-limited" die size. Figure 5.36 shows the pad-limited die size when using a single row of pads for various wire bond pitches.

It should be noted that the number of pads refers to the *total* number. The designer usually cares about the number of signal pins. In addition to the signal pins, there are power and ground pins that must be strategically distributed in order to minimize cross-talk between signal pins, ground bounce, and other signal integrity issues. For digital designs with high performance buses clocking at high speeds, a good rule of thumb is to add 30% additional pins to the signal pin count in order to estimate the total number of pads.

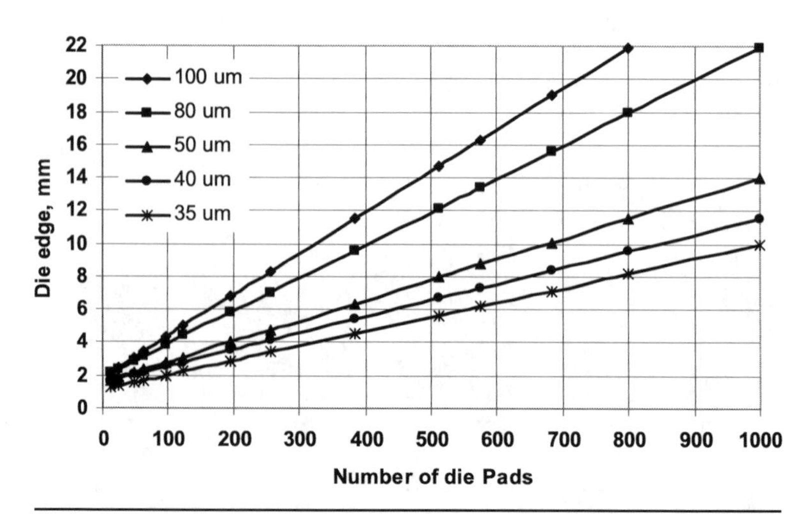

FIGURE 5.36 Pad-limited die size estimation for various pad pitches.

The minimum allowed pad pitch is determined by three major considerations:

1. Minimum wire bond pitch, which is determined by the wire bond equipment capability and the diameter of the wire used by the assembly house. Typical wire diameters have been reduced from about 35 μm to 20 μm. While the smaller diameter allows increased number of pads, the trade-off is reduced wire strength, increased resistance, and inductance.
2. The minimum allowable pitch for electrical wafer probe.
3. The I/O buffer pitch, determined by the circuit designer and the I/O library provider.

An alternative to wire bonding is the Controlled Collapse Chip Connect (C4) technique that was developed at IBM in the 1960s [5.44, 5.45]. The cornerstone of the technique was the formation of solder balls on the die. These solder balls or bumps could be arranged in an area array all over the die surface. Such placement required careful enhancement of the design automation tools. The die was flipped face down and solder bumps were coarsely aligned to pads on the package surface. During the solder reflow process the chip balls self-aligned themselves to the package pads. Replacement of wire bonds through this "flip chip" process allowed reduction of parasitic resistance and inductance. In the early days there were many concerns, including the additional cost of the solder bump and flip chip processes, and the reliability degradation of the solder joints. The degradation was due to a mismatch in the expansion coefficients of the silicon, the solder and the underlying PCB material. In the mid 90s the reliability concerns were eliminated as a serious risk when an underfill process was developed. The technology was licensed and has become commonly available at many wafer fabs and SATS. Subtle differences in the solder bump technology widely used these days are related to the use of low temperature solder. Also, ball grid array packages usually have pre-fabricated solder balls placed on landing pads; the balls get attached to the package during a reflow process.

The major benefit to using solder bumps is the ability to increase the number of die connections in a given area. This results in a reduction of the peripheral pad-limited die size and it eliminates the need for pushing down wire bond pitch. It also allows additional power and ground connections to reduce signal integrity issues.

Since design solutions were not available to allow placement of I/Os and solder bumps in the middle of the die, the industry adopted a methodology that allowed "re-distributing" connections from peripheral die pads to an array of bumps in three to four rows of pads along the die perimeter. Figure 5.37 is an illustration of the two

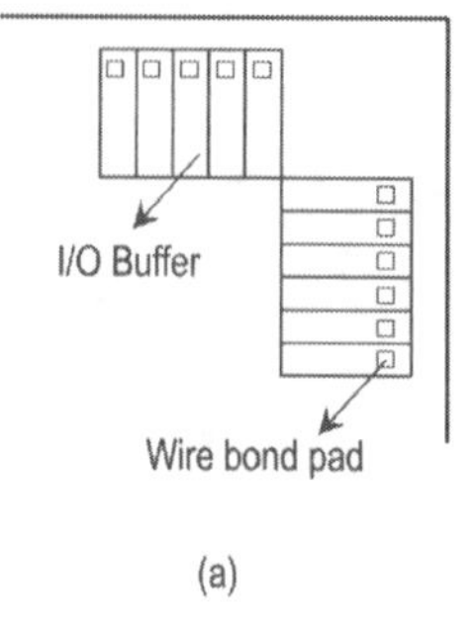

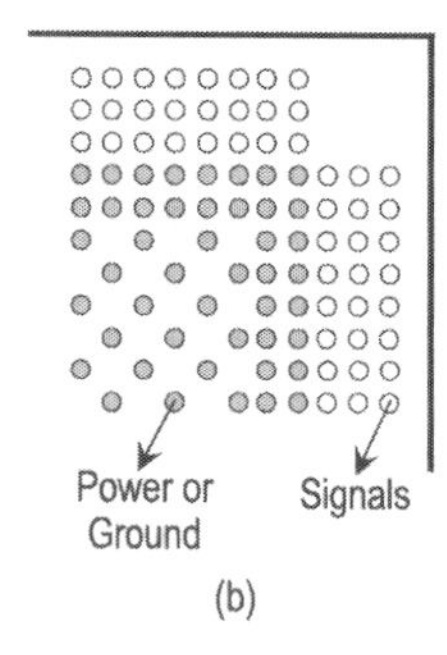

Figure 5.37 Pictorial representation of: (a) peripheral wirebond; and (b) area array solder bump patterns, in one corner of the die.

common approaches—wire bond pads and an array of four rows of bumps using a redistribution layer.

5.12.3 Package to Board Connections—Peripheral and Area

In keeping with the upstream value chain considerations discussed in Sec. 2.4, it is recommended that the fabless IC supplier design and deliver package solutions consistent with the customer's capabilities to assemble and manufacture the packaged ICs. For instance, when selecting the package peripheral lead or ball pitch, it is imperative that the customer has the ability to design and manufacture PCBs that mate to the fine pitch selected.

5.12.4 Example Packages and Products

- QFN 40 lead, 6x6mm body size, Broadcom, DVB Dual band Tuner, 0.18µm CMOS (http://www.broadcom.com/collateral/pb/2900-PB00-R.pdf) also available as a WLCSP 3.14 × 2.71mm, for multi-chip applications.
- 20 QFN, 3×3mm, Silicon Laboratories, FM Transceiver (http://www.silabs.com/tgwWebApp/public/web_content/products/Broadcast/Radio_Tuners/en/Si4720-21.htm).
- 389 FBGA, 12×12mm, 65nm CMOS, Broadcom, Single chip baseband+rf+multimedia processor (http://www.broadcom.com/collateral/pb/21331-PB00-R.pdf).
- 196 FBGA, 7.5×7.5mm, 65nm CMOS, Broadcom, Single chip 802.11 1/b/g MAC/BB/Radio with integrated Bluetooth and FM receiver for low power mobile handset and handheld applications(http://www.broadcom.com/collateral/pb/4325-PB00-R.pdf).
- 8 ceramic LCC, 5×5mm, Analog Devices, Dual axis, high-g iMEMS Accelerometer (http://www.analog.com/en/prod/0%2C2877%2CADXL278%2C00.html).

5.12.5 Cellular Handset Packaging

For any electronics system, the system and chip architects need to define the block level description of the various components of the system.

As an example, the following list shows that cellular handset blocks are optimally implemented in many different process technologies.

• Baseband processor	CMOS
• Application Memory	Flash or EEPROM
• Memory	DRAM
• Transceiver	RF-CMOS or SiGe BiCMOS
• LNA	SiGe BiCMOS
• RF Interfaces	Analog
• Audio/other Interfaces	Analog
• Filters	SAW
• Power Amplifier	Analog
• Power Management	Hi Voltage Analog

Customers would ideally like to have a single chip cellular handset for the lowest cost, the smallest area, and the highest reliability [5.43, 5.46]. Given the advancements in silicon CMOS technologies, the ubiquity of their availability from commercial foundries, and their cost effectiveness, it is prudent to capture as much of the functionality in a mainstream CMOS technology. Figure 5.38 shows the various functional blocks in a typical cellular handset. The mostly digital baseband processor, along with user interface functions, are best implemented as an SoC IC in CMOS technology. External memory chips are commonly stacked on top of such an SoC in

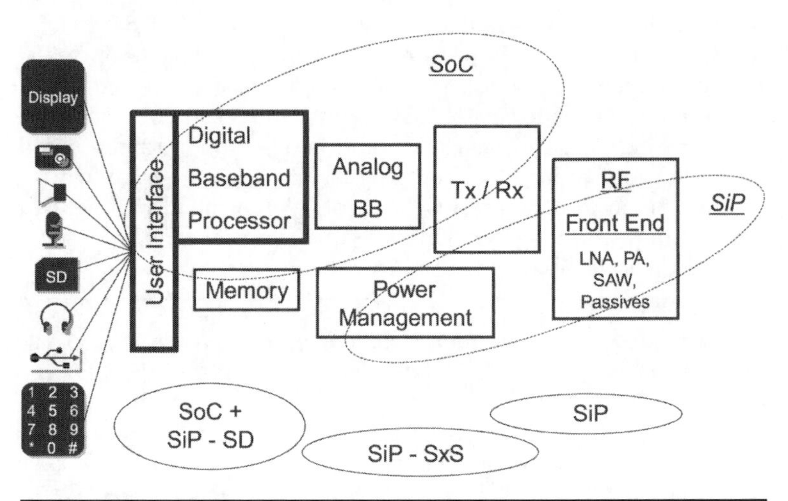

FIGURE 5.38 Functional blocks in a typical cellular handset. The shaded and dotted ellipses represent two possible ways to partition the cellular handset.

a single SiP. It is also common to package the analog baseband and the transceiver, and possibly the power management IC in a second SiP. Because of RF and power dissipation considerations a SiP that has the three die side by side may be a more desirable solution. Front end SiP modules generally include the power amplifier, duplexers, SAW (Surface Acoustic Wave) filters and passives that allow matching circuitry. Such a three SiP implementation is represented by the three shaded ellipses in Fig. 5.38. For some cellular handset applications such as GSM (Global System for Mobile communication), it has been demonstrated that the transceiver can be incorporated on the same SoC as the baseband [5.46]. Texas Instruments discussed a two or three module solution—an SoC with the baseband and the transceiver, a power module and a front end module [5.46]. This solution is represented by the dotted ellipses in Fig. 5.38. Both solutions have been proven viable since there are very high volumes being shipped in the industry today.

Judicious use of packaging technology can be used to offer solutions that minimize cost and board area without complicating the silicon processes. An example is the use of SiP technology to assemble an analog die, an RF die and a digital baseband die all in the same package. The three die could be fabricated in three different silicon technologies. The die must be thoroughly tested beforehand; the industry uses the term KGD (Known Good Die). A drawback is additional cost if one of the die is found to be non-functional after assembly, the throwaway cost is higher. Such yield losses can be factored into the cost model.

Why is all this important to the emerging fabless IC company? While some of these decisions could be made by the customer, it is important for the fabless company to be aware of some of the tradeoffs being made. The package that is selected could be a deciding factor for design-in of the IC into the customer product. For instance, the fabless company could lose a design-in opportunity if the customer would rather have the IC as a KGD, but the fabless IC company only offers it in a BGA package. Early and good communication could allow time to plan for a KGD offering and getting the design win!

Another important factor is that in developing an IC product that requires multichipping, the fabless company must be aware of some of the pitfalls, as discussed in the accompanying sidebar.

Multichip packaging–It's not just the package

Over the years, the SATS have made available the package design, assembly and test infrastructure to enable multichip packaging. The industry ships many billions of multichip packaged ICs annually. Many different options for multichip packaging are available, such as side-by-side die and stacked die configurations using flip chip or wirebonding, using laminate substrate-based as well as leadframe-based packages. In many situations, the packaging interconnect is the easy part. The harder part might be more the management of business challenges. Here is a short list of some of the major challenges:

- Die procurement

  Managing the availability of die from multiple suppliers, some of whom have their own products and whose core business is not die/wafer sales.

  Managing the additional complexities and risks to sustaining the yield, quality, reliability, short term and long term supply of your product.

- Memory die in a package

  Dynamic price variations that can lead to unpredictable cost and margins for the fabless company's IC product. The burden of price fluctuations when using external memory ICs is normally managed by the system house.

- Test

  For SiP's with digital, analog, and substantial memory content, final functional testing of a product is also a challenge. Often times the testing of the memory cannot be done as quickly and as low cost as commercially packaged memory which leads to excessive product cost.

The end customer would always enjoy enhanced integration, but you, the IC supplier, should consider all aspects to multichip packaging well beyond the packaging technology of the package to ensure it can be delivered in large volume and at the costs levels that are expected to maintain a lasting, profitable business.

Rich Rice
Vice President, ASE US, Inc.

5.12.6 Emerging 3D Packaging Technologies

As the industry is focused on leveraging packaging technologies to facilitate the concepts of "more than Moore" to continue the phenomenal growth in the semiconductor industry, there are some new technologies that have shown promise.

The first technology of interest is one where an IC or numerous passive devices are embedded in the PCB during the fabrication of the package substrate. This can facilitate the subsequent package assembly process.

The industry is also developing ways to stack memory chips to make available higher density and higher performance DRAMs and Flash memories. Another form of 3D stacking of chips requires the etching of holes throught the silicon die. This technology is called

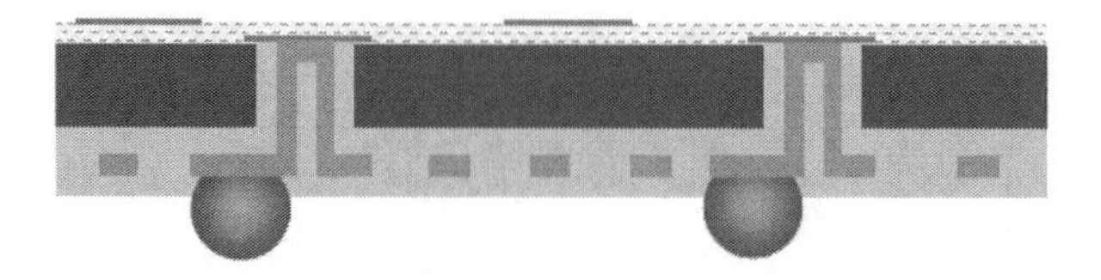

Figure 5.39 TSV process implemented after complete fabrication of the silicon IC where large vias are etched from the backside. (*Source: IMEC.*)

TSV (Through Silicon Via) technology in order to stack multiple silicon die. The technology for stacking multiple die using through silicon vias is called TSS (Through Silicon Stacking). Besides some IDM companies, lead research has been conducted at consortia—IMEC in Belgium and at Sematech in the US.

Figure 5.39 shows the cross section of a die with a via etched through the back side after the IC fabrication process has been completed. Solder balls are attached to the die backside, which can then be stacked.

A four die stack using this TSV technology is shown in Fig. 5.40.

The diameter of vias etched from the die back side are generally quite large—20 microns or more. An alternative method of forming small diameter (≤ 5 µm) vias involves a more complex process that is partially completed in the wafer fab, prior to the fabrication of the Back End Of Line (BEOL) metal interconnects. When the IC fabrication is completed the wafer back side is thinned to expose the small diameter vias. Copper pads are then fabricated on the back side and multiple die can be stacked, as illustrated in Fig. 5.41.

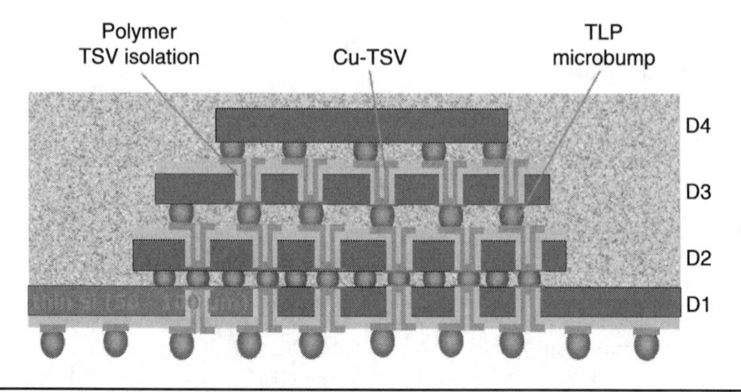

Figure 5.40 Four die stack using TSV process. (*Source IMEC.*)

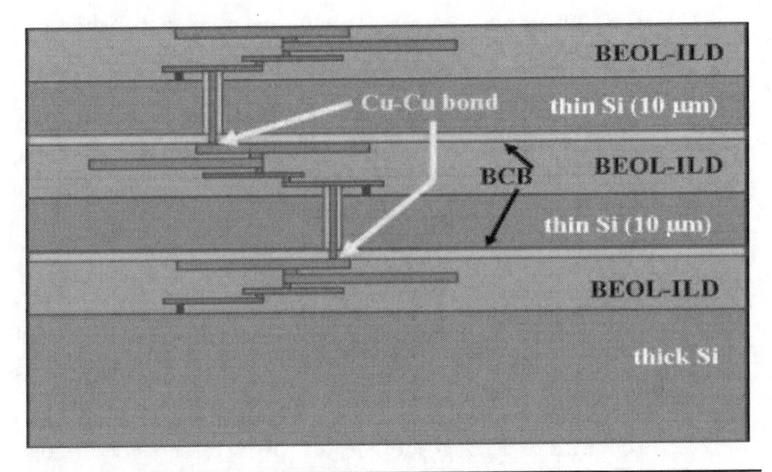

FIGURE 5.41 A 3-die stack using TSVs etched prior to completion of the BEOL (back end of line) processing in the silicon wafer fab. (*Source: IMEC.*)

5.12.7 SiP or SoC?

As pointed out in the context of Fig. 1.6, the industry has and will continue to drive towards pushing Moore's law—smaller lithography nodes, and more transistors per chip. The industry will continue to develop new SoCs. However, there is now also another alternative called "more than Moore." One consequence of this new thrust is the advent of SiPs. While multi chip modules have been around for many years, recent advances in this arena have been propelled by two key factors:

- 3D assembly and packaging of multiple die in the same package;
- availability of KGD.

A few helpful guidelines are presented here:

- If the fabless IC company is planning on designing an SoC, it should include *all* the functionality that can be implemented in the *same* process technology in that SoC. Make sure that this will not require the addition of too many I/Os. One must, of course, make sure that the new SoC will be cost effective and will meet the expected functionality, performance, and relaibility.
  - If adding features in the SoC involves new process technology options, careful assessments must be made of the technical and business feasibility. Examples are the addition of eDRAM, eNVM, complex analog or RF blocks.

- If the fabless company is developing a specialty chip that mates to an SoC, it is worthwhile considering and offering options to the customer to make the IC integratable into a SiP.
- If you want to evaluate the feasibility of creating a SiP solution from two or more separate chips, the following analysis should help. Let us assume there is a logic chip that costs L. The plan is to package this chip in the same package along with a memory chip costing M.

First you must make sure that the memory chip is available as a KGD. Generally, the cost of a KGD is approximately the same as the packaged part or may be within ±5%. While one saves on packaging cost, the supplier usually tests more thoroughly at wafer probe.

The cost of the SiP package design will be higher than the cost of the separate logic package because of increased layout complexity. The cost of assembling the two die in the SiP package will also be slightly higher. Another problem is that you have to design multiple packages to test and validate the individual die in addition to the SiP.

The net result is that the two chip SiP will cost more than (L+M). How much more is dependent on numerous factors. One must use a cost model to get an accurate estimate. A cost increase of 5-15% is a reasonable "ballpark" estimate.

So why consider a SiP? The key reason is reduction of PCB area, which allows the OEM to develop a system with a better form factor.

5.13 Key Points

- Start by assuming CMOS as the process technology of choice. There must be compelling reasons to deviate from this assumption.
  - Many technology options are available within the baseline CMOS process.
  - There are many challenges requiring close cooperation between process and design disciplines in implementing designs on leading edge process technology nodes.
- Leverages existing elements of the design ecosystem (design flow, library, IP, EDA tools, . . .) as much as possible.
- There are many packaging options, both single chip and multichip.
- Partitioning the IC content and leveraging the available technologies can be critical for success in meeting customer requirements

CHAPTER 6

Implementing the COT Approach

6.1 Introduction

Implementation of an ASIC design usually begins after the following key steps have been completed:

- feasibility assessment as outlined in Chap. 3;
- sourcing methodology selection, as discussed in Chap. 4;
- technology and supplier selection, as described in Chap. 5.

While these steps have been discussed in the context of a start-up fabless company, the process is also applicable for new ASIC products at existing fabless companies. Shortly after the architecture team has established a high-level specification and has made an initial assessment of the ASIC's required capabilities and features, it is important to go through the "global planning" stage described in Sec. 3.4.1. The global planning activity could be split into phases (e.g., phase one and phase two) if the chip definition is not very firm initially. The following list serves as a meaningful checklist:

- Implementation method—ASIC or COT or ? (Chap. 4).
- Design partition parameters and trade-offs (provided by the architecture team):
  - gate count;
  - memory—how much and what kind;
  - IP blocks required;
  - package type, pin count;
  - die (or chip) size;
  - performance metric(s).

- Technology selection and trade-offs (Chap. 5).
- Supply chain partner identification and early commitments of support (Chap. 5):
  - ASIC supplier;
  - foundry;
  - library, memory, IP;
  - package assembly, test.
- Cost estimation, both unit cost and development cost (Chap. 7).
- Qualification plan (Chap. 8).
- Overall program plan development, including resource estimates (Chap. 9).

In the early stages of a fabless IC start-up, design activity is of paramount importance. Design is a necessary, though not a sufficient, ingredient to the success of the emerging fabless IC company. As is described later in this chapter, there are many other activities necessary to make a successful fabless IC company. Over half of the company's time to significant revenue comes after design tapeout! In this chapter, let us focus first on implementation of the design. This will be followed by operations-related implementation activities, and a description of the support infrastructure required at a fabless IC company.

Assessments in the global planning phase require access to numerous skill sets—technical (design, silicon process, silicon device, packaging, assembly, test), business and operations. A just-funded startup usually does not have such a skill set readily available. This causes an "operations dilemma"—hiring operations people too early is a cash drain and getting them on board too late can be devastating when ramping up production [6.1, 6.2]. This and other aspects of operations infrastructure required at fabless IC companies are discussed later in the chapter.

Positioning of the global planning activity relative to the overall ASIC development is shown in Fig. 6.1. The next major set of tasks at the fabless IC company are IC design, IC prototyping, IC production, and establishment of an operations infrastructure. Discussion of IC design implementation follows in the next section. An implementation schedule is discussed in Sec. 6.3. IC prototyping and production is partitioned by discipline—wafer fab/mask making, packaging and testing in Secs 6.3, 6.4 and 6.5.

6.2 Design Flow and Supply Chain

For describing the design implementation, it is assumed that the design will be completed internally at the fabless IC company. By doing so, the following discussion will better prepare the reader to engage with a design services provider if the design implementation

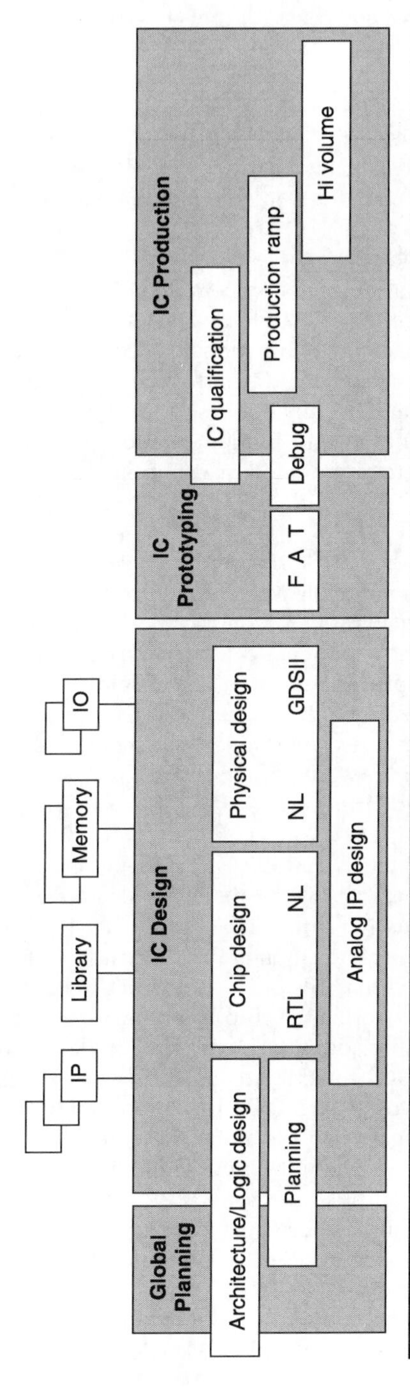

FIGURE 6.1 ASIC development activities partitioned into four phases.

is outsourced instead. It is also assumed that the technology, the suppliers, the design flow, and the EDA tools have been selected.

6.2.1 Design Flow Implementation

The three major phases in the implementation of present day SoC ASICs were discussed in Sec. 5.6 and are the concept phase, the definition phase and then the implementation phase.

6.2.1.1 Concept Phase

In the concept phase the major system level activities are focused on consolidating marketing inputs and a feasibility assessment of whether the ASIC can meet functionality, cost, and performance goals. The activities also include performance modeling and the generation of a preliminary functional specification for the IC.

The overall definition and implementation phase activities are discussed next and are represented in a simplified form in Fig. 6.2.

6.2.1.2 Definition Phase

System level designers work first on high level design concepts including the behavior, architecture, and chip structure. Transaction level modeling and generation of the specification follows. The function or algorithm being modeled can be programmed in C, C++, or SystemC [5.26]. All of this results in the generation of a chip level block diagram. An example chip level block diagram is shown in Fig. 2.10. A partitioning exercise determines which part of the system will be implemented in hardware and software. A micro-architecture is then defined which identifies some of the chip level hardware implementation choices, such as the arrangement of the execution pipeline, the size and width of the cache [5.27]. A chip level specification is also developed. Next is a translation of the behavior into an RTL description. RTL code is usually written in either the Verilog or VHDL language. There is much optimization of the logic and the state machines.

It should be noted that designers and managers at the fabless IC company need to categorize the chip level functions and blocks into three categories—value added IP blocks that are key to differentiating the company's IC, custom/analog blocks, and third-party or already available (re-used) IP blocks. It is important to align the company's precious design resources along these lines. Too many times decisions are made to design new blocks not critical to the chip's differentiation, causing additional cost and risk.

Design verification has become an extremely important task for complex ICs. Verification occurs throughout the design cycle as shown by the numerous "loop-back" arrows in Fig. 6.2. The first set of verification activities compares chip function and behavior against the system models and specification. Development of test benches to verify the design is an important and challenging task these days.

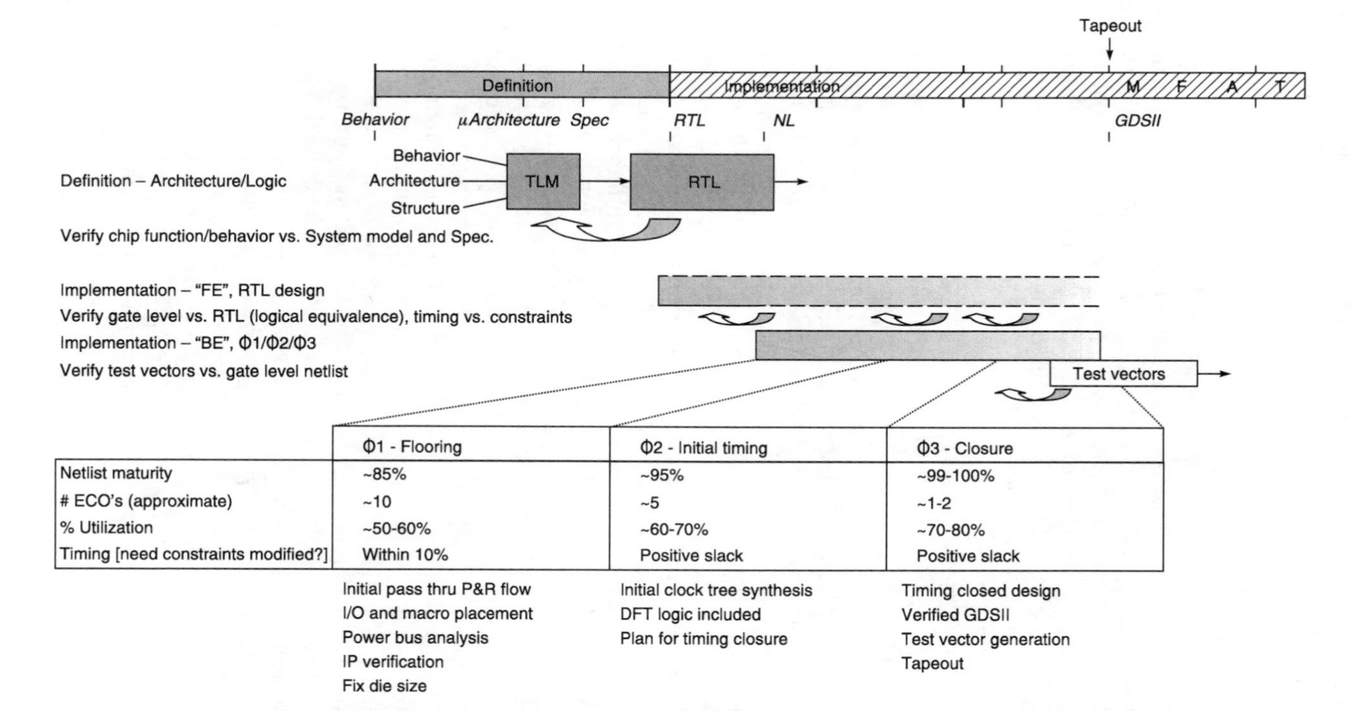

FIGURE 6.2 A simplified view of the overall ASIC design activities. FE refers to front end design whereas BE refers to back end design. ECO refers to engineering change order.

In addition, third-party IP and custom blocks must be verified against their specification. Verification of analog IP blocks becomes a challenge and is sometimes accomplished through the use of Verilog-AMS [5.28].

Behavior-level RTL code then leads to the development of synthesizable RTL [5.25]. The synthesized RTL instantiates clocking registers for the synchronization of the logical flow within the design. The clocking registers break the logical paths into segments that can be traversed within one clocking cycle. The logical length of these paths is set by the targeted clock frequency. Once the synthesizable RTL has been generated the process of RTL to gate level synthesis begins. The gate level design then requires logical equivalence checking of the gate level model against the RTL.

6.2.1.3 Implementation Phase

Availability of synthesizable RTL code generally signals the beginning of Front End (FE) design. At this point, the design behavior can be converted to logic gates and as the design matures, can be expected to meet timing. Meeting timing means that logical signals released from one clocking register can traverse the logic between the clock synchronization registers and be properly captured by the receiving clocking register with some level of margin. Generally the designers keep track of how firm the RTL code is and how much of it has been verified. When the maturity level reaches around 85%, it is generally sound practice to generate and transfer a "trial" RTL code to the physical design team, whether internal or external to the company. This signals the beginning of the Back End (BE) design phase. The reason to get started is to reduce overall lead time to get the design completed. There is also an opportunity to fix iteratively tight spots in timing by making adjustments at the RTL level. A structured approach to this handoff between FE and BE design teams is also important to getting convergence of the design in a timely manner. Many projects get into trouble for allowing too many functional and timing changes during this stage of the design, or for starting this stage too early—too much "churn" can cause inordinate and unacceptable delays! The gate level design is verified against the RTL by equivalence checking during each iteration.

In BE phase one the intent is to complete an initial floor plan, do an initial placement of IP blocks (or "macros"), make an initial pass through the timing driven place/route flow and do a power bus analysis. Timing constraints are verified to be achievable typically within 10% and the die gets set to an initial size. There is usually some flexibility for the RTL designers to make adjustments to fix any timing paths that are really tight. Note that the die size assumes a relatively low level of utilization level (~50–60%) at the start of phase one, but by the start of phase three the utilization will be greater (~70–80%). This allows for the initial die size to remain fixed in most cases, and yet allows room for the designers to add logic to their design as it matures. The utilization from the beginning to the end of phase three will also increase due to the addition of buffers, repeaters, upsizing of drivers, spreading of wires, etc. This allows closure of timing on the design and to meet all the required signal integrity rules. A number of ECOs (Engineering Change Orders) related to RTL changes are to be expected.

In BE phase two the team performs a more detailed pass through the timing driven flow, including clock tree synthesis. DFT logic for Memory BIST (Built In Self Test) and JTAG (Joint Test Action Group) are included in this RTL release. An action plan for timing closure in the final phase should be in place by the end of this phase. There should be a significant reduction in the number of ECOs related to the RTL in this phase.

In the final BE phase the RTL code is expected to be final with maybe one or two ECOs. Anything major that affects the die size or adds many gates could cause a major re-start of the BE phase. At the end of this phase the design is complete in terms of its required functionality, meets all timing requirements and has been verified to be compliant with the foundry design rules.

As the design progresses through BE phases, the RTL matures beyond the 85% initial "handoff" and also incorporates gate level ECO changes required in the design execution. There needs to be continuing verification that the implemented design will meet functionality and performance. These verification activities are represented in Fig. 6.2 by the "loop-back" arrows.

Another very important activity in phase three is the generation of test vectors that will be used to verify the chips functionality on the Automatic Test Equipment (ATE). In the design phase, these test vectors must be simulated against the gate level netlist to verify validity. These test vectors fall into three general categories:

- characterization vectors to verify the chip I/Os, voltage levels, currents, etc.;
- functional vectors to test basic functionality; and
- structural vectors automatically generated to check each circuit node for manufacturing defects. These are generally referred to as ATPG (Automatic Test Program Generation) vectors.

6.2.2 Mixed Signal Designs

Mixed signal designs require careful placement of the analog macros and incorporation of guard rings, and other techniques to minimize unintended effects from nearby circuits.

Additional work is also required to separate the power buses for these macros versus those for the standard digital power domains, using structures such as breaker cells within the I/O ring.

Often the physical verification process of Layout Versus Schematic (LVS) is more complicated, as extraction of the analog devices requires special recognition layers and special constructs that are not part of the standard foundry-released LVS decks.

Connections between these macros sometimes require differential signals whose routing lengths must be matched. Often signals must be shielded or other special care must be taken to reduce interference from other signals.

Analog verification is becoming a very important aspect of successful implementation of ICs that have an increasing analog and RF content [6.3, 6.4]. There needs to be focus on verifying that the IC will meet both functionality and performance goals.

To meet today's consumer, wireless, and wired system requirements, analog design has become *functionally* complex. Circuits operate in different power modes. They implement multiple standards. Components are tunable to allow calibration. Algorithmic architectures are employed to meet ever tougher performance requirements on manufacturing processes that are increasingly unfriendly to pure analog design techniques.

With functional complexity, functional errors arise. Examples of functional errors are chips that do not power-up properly, control bits that are inverted, buses that are misaligned, and bias lines that are incorrectly connected. Often the errors are "human errors," such as from miscommunication between members of the design team. Unlike missing a performance specification, functional errors can often be fatal, meaning that they lead to a chip that cannot be used to test the system or write the firmware. Delayed product shipment leads to delayed sales, missed market windows and a possible loss of customers

Analog verification is the separate and systematic process of catching these functional errors. Individual analog block designers usually choose to focus on designing to meet the performance specifications. It is often the job of the lead designer to look for these functional errors. Unfortunately, the lead designer usually does not have the skills or the time to focus on this task. Analog verification requires the creation of regression tests, testbenches, and behavioral models – these tasks can be assigned to a verification engineer. The design is fully checked against the specifications, the models are checked for functional equivalence to the transistor level implementation.

Any analog, mixed-signal, RF fabless semiconductor company designing complex chips needs to put in place an analog verification team alongside their digital verification team. Putting this team in place from the very beginning establishes best practices allowing for the consistent design and delivery of chips free of these errors.

Henry Chang and Ken Kundert
Designers Guide Consulting
http://www.designers-guide.com

6.2.3 DFT

Memory BIST, JTAG, scan insertion and ATPG are all part of a process that supports DFT.

Memory BIST incorporates one or more memory controllers that allow memories to be tested with only minimal control and interaction from outside the chip. These controllers are created as RTL structures and integrated into the functional RTL, to be later synthesized.

JTAG, or Boundary Scan, implements logic around the I/O cells that allows the board level interconnect to be easily tested and failures to be quickly diagnosed. As with the Memory BIST controller(s), this logic is created as RTL and integrated into the functional RTL.

Scan design requires the conversion of standard flip-flops into scannable flip-flops that are connected into scan chains [5.24, 6.5]. These flip-flops are then used as control and observation points for ATPG software, which reads in a Verilog netlist of the complete design and generates scan vectors. These ATPG scan vectors are typically stored in special memory on the tester and shifted into the chip via the scan chains. The scan chains can be re-ordered during place/route to maximize their connectivity based on the actual placement of the flip-flops.

6.2.4 ECO (Engineering Change Order) Capability

Spare gates allow for metal-only revisions of the design. They can be implemented post-synthesis by distributing a set of predefined logic cells, typically including the major cell types, around the core of the design. When a change is required, the nearest spare cell of the correct type is used, with only routing changes, on as few layers as possible, to effect the logic change.

Another technique is to use special cells that can be customized by a metal mask only, similar in nature to the gate array cells of the past. Any flip-flops that are included in either scheme must be pre-wired to the clock tree, so there will not be any unexpected changes to the clock skew or insertion delay after an ECO is implemented.

ECO features of the place/route software are used late in the design process (e.g. phase three of the physical design). With these features, only the layout of those portions of the design which are changed are modified by the place/route system. There should be little or no effect on the areas of the design that are already meeting their constraints.

There are three general categories of design contingencies: **firmware fix**; **targeted contingency**; and **spare resources.** Each has their own scope of applicability, costs, and implementation options. Some of the techniques can be used to fix almost any bug on the chip, while others are much more limited. The costs can be monetary, schedule, or new risks.

The implementation of these fixes varies from external solutions to mask changes. External implementations include changes to an external memory that gets loaded into the chip, or external circuitry to correct for the problem. Some fixes can be implemented by changing the chip bonding, which only affects the chip assembly operation; or by changing the contents of on-chip non-volatile memory, which can be done during chip testing. Other fixes require changes to the mask set, but may be limited to the top few layers (metal). If a few wafers are held back before metal on the prototype or engineering lot in fab, the impact of a metal-only change can be significantly less than a full-mask change.

If the chip has a processor, a firmware fix may be an option, but these solutions are limited to problems that the firmware has visibility into. Even if the firmware has access to the information to correct for the bug, it may have undesirable or unacceptable consequences, such as increased processing delays. While this option may appear free, it adds complexity to the firmware development, which translates into increased firmware development schedule, verification effort, and firmware risk. The firmware fix may be implemented in an off-chip memory, resulting in no changes to the chip; or changes to on-chip writable non-volatile memory, resulting in only a change to the test program; or changes to on-chip ROM, requiring at least a metal mask change, if the ROM is programmed in metal.

Targeted contingencies are most appropriate if you can anticipate where you are likely to have a problem. For example, if you are uncertain how your algorithm will perform in the real world, and you are not able to adequately model that behavior before producing silicon, you can provide several alternative solutions in silicon, and then select the best performing alternative after evaluations in the lab. These contingencies can even be provided for analog circuits by having extra capacitors or resistors that can be switched in or out of the circuit. Targeted contingencies require identifying the potential sources of problems up front, and increasing the design complexity to provide multiple solutions for a single problem. There has to be a mechanism to select between them. This adds silicon area, some of which is guaranteed not to be used. A targeted contingency can be designed to be exercised externally (by having a mechanism to load a selection register), as a pin or bond option, as a nonvolatile memory option, or a metal option, depending upon what resources are available.

Spare resources provide the most generality. The best bug fix often involves adding extra gates or circuitry. If the appropriate resources are already available on-chip, then it is simply a matter of utilizing those resources. The biggest problems with this are the limited spare resources available, the types of resources, and gaining access to those resources. Digital designers typically provide spare NAND and NOR gates and flip-flops, sprinkling them across the chip, and wiring them together to create spare routing channels. Utilizing spare resources requires at least a metal mask change, but providing these spare resources reduces the chances of requiring a full mask set change.

None of these techniques is guaranteed to be the proper answer for any particular bug that you encounter. But, if provisions are not allowed for any of these, then you are guaranteed to require a full mask spin for any bug found.

Kurt Stoll
Director, ASIC and Hardware Development
Echelon Corporation

6.3 Implementation Time Line

In addition to the design activities, the fabless company needs to proactively engage in other "operations" activity prior to the design tapeout. The major activities are illustrated in a simplified way in Fig. 6.3 and will be discussed throughout the rest of this chapter. The time line includes packaging, test and qualification activities that must proceed in parallel with completion of the design.

Design and fabrication of the package, test program development, ATE load board and other hardware, the qualification plan and fabrication of the qualification life test boards are examples of activities that must be started in parallel with back end phase two design.

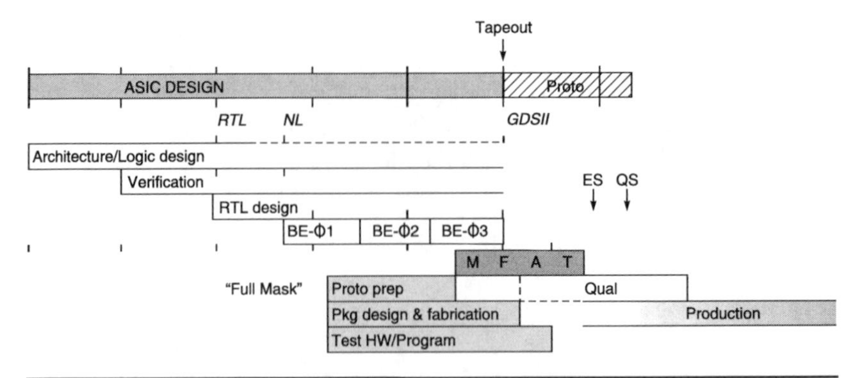

FIGURE 6.3 Simplified time line of activities in a one-step, "full mask" prototyping approach. Some packaging, test and qualification activities precede tapeout.

The proper way to manage all these activities is to develop a project plan. Since Microsoft ® Project charts do not reproduce well in a book format, I have chosen to show this relative relationship between the various important activities in a pictorial format. Also shown in Fig. 6.3 are the mask making, fabrication, assembly and test operations. The arrows show delivery of ES and QS parts. The latter are usually shipped upon completion of the first 500 hours of life test during the qualification. Once the fabless company's customer has approved the QS parts they will likely place orders for production IC units (refer to discussion in Chap. 2). The production ramp usually begins in parallel with the qual. The number of units that the fabless company wants to build "at risk", before real purchase orders have been received, is a delicate balance that must be managed carefully. Large investments have been known to cause fabless companies serious financial pain.

6.4 Silicon Prototyping and Production

OK, so the design is completed! This is usually a major benchmark. Certainly from the designer's perspective, the rest just happens! Well, not quite. As you will see in the next few sections, completing the design is only about the half-way point to first production—if everything goes well on the rest of the execution! If time to money is an important benchmark, there is much to be done.

6.4.1 MPW vs. Full Mask Prototyping

What one does with the design database is dependent on a key strategic decision. There are two approaches to getting first silicon—a "one-step" or a "two-step" scenario. Let us refer to Table 6.1 for this discussion.

	Initial Silicon	
	One step	Two step
	"Full mask"	"Shuttle" + "Full mask"
Design confidence	High	Low/medium
Design "maturity"	1st design in mainstream node	1st design in new node
Design "envelope"	Mainstream	Leading edge
Design content	Mostly digital (D/a)	Mostly analog (A/d)
Process node	Mainstream	Leading edge
Confidence in metals-only fix	High	Low/medium
Cash flow	Solid	Tight
Initial volume of units required	Several K	Up to ~1–2K
Time to 100K production units	1–3 months	6+ months
Time to 1M production units	~6 months	12+

TABLE 6.1 Comparison of Two Approaches for Getting to Initial Silicon. The Numbers and Lead Times Are Typical and Are Shown for Illustration Purposes

In the one-step approach the design gets taped out to the foundry for fabrication of an "engineering lot" using a full mask-set. This mask-set is dedicated by the foundry for the exclusive fabrication of the single customer's design. The cost of the mask-set is borne by the customer, the fabless IC company. As shown in Fig. 1.11, such a mask-set can cost \$50K–3M, depending on the process technology node. This approach is appropriate for use by an experienced team for a mainstream complexity design in a mainstream node, when there is a high level of confidence that the design will be functional and may require only metal level fixes. On the business side, the company must have cash available. Another set of considerations is related to parts delivery. This approach is appropriate if many engineering samples must be provided to many customers and if a rapid ramp to high volume production is anticipated. The time to 100K and 1M units is shown for comparing the two approaches.

The alternative approach offers a lower cost by sharing the initial mask and silicon cost among many customers. Such an approach has become commonplace and is offered by the foundries as well as a few other agencies. This MPW ("multi project water") approach combines designs from numerous customers on a single mask "reticel" which is printed on silicon wafers in an engineering lot called a "shuttle." The foundries run shuttles on the most popular technologies at regular intervals. The intervals can vary based on demand, but typically there is one every month or two. Figs 6.4 and 6.5 illustrate examples of two MPW reticles with 40 and 9 different designs, respectively. Two agencies that consolidate designs for incorporation into shuttle runs are:

- MOSIS (http://www.mosis.org), located in southern California.
- EuroPractice (http://www.europractice.com), which is a European Union initiative to support the use of advanced technologies in industry and universities.

There are guidelines for allowable "drop-in" area. Shuttles are not available on all process technologies and do not offer all the process options. Usually the customer receives 20–40 parts, either in die form or as packaged units. When the foundry uses 300 mm wafers, expect the number of prototype units to be 50–100. While the foundry ensures that the wafers meet process parameters, the parts are delivered without any functional testing and are referred to as being "blind-built." The cost of using this approach is usually around 10% of a full mask-set cost. Using such a ratio applied to the mask costs listed in Table 7.3 is a good starting estimate. Sometimes a cost per square mm is used as a guideline. However, one must be careful about the allowable die sizes. Some fabs restrict each drop in to be 5 × 5 mm. If your die size is less, you still pay the fixed amount. If the die size goes over this restriction, you may have to pay for two or four

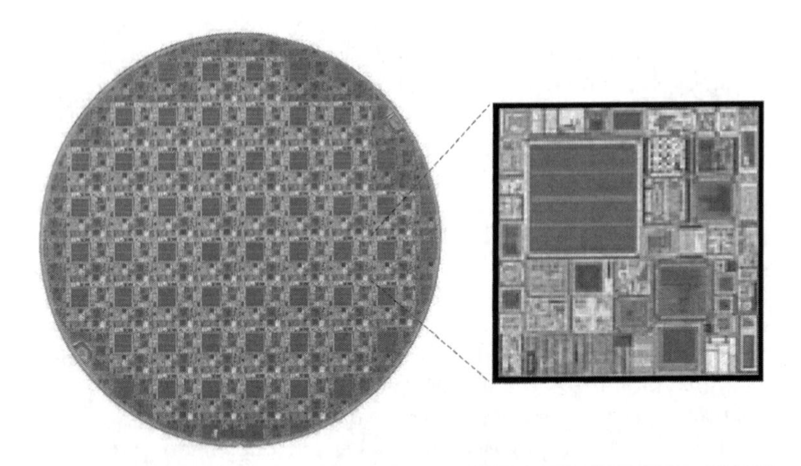

FIGURE 6.4 Photograph of an MPW wafer along with an enlarged image of one reticle containing approximately 40 designs. (*Source: MOSIS.*)

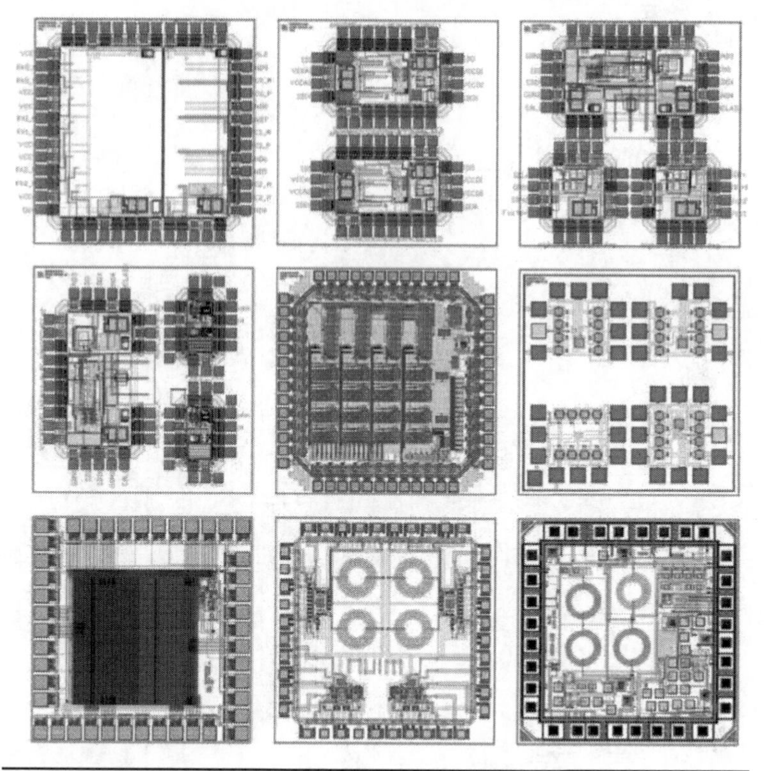

Figure 6.5 Plot of nine different designs assembled as one MPW reticle. (*Source: EuroPractice.*)

drop-in slots, or some pro-rated charge. It is prudent to check with the foundry or the agency under consideration. The drop in size has been reduced to 4 × 4 sq mm for the 90 nm node, 3.5 × 3.5 for the 65 nm node and 3 × 3 sq for the 45 nm node. Also, it is now possible to purchase additional units up to about 1–2K to meet initial customer demand.

The two-step approach uses a shuttle to get some initial prototypes. This approach is appropriate if design changes are anticipated and it is expected that the design fixes will need changes in the premetal (or "diffusion") layers. When the design has been debugged and there is a high level of confidence in the design implementation, the company commits the funds to subscribe to a "full-mask" and an engineering lot of wafers. The two-step approach is also appropriate when a slower production volume ramp is anticipated, relative to the single-step approach.

The time lines in Fig. 6.6 show a comparison of the single and the two-step approaches. Clearly, the "operations" development time is longer for the two-step approach. A slower cash flow results in the two-step approach. Note also that complex analog and RF designs sometimes use more than one shuttle run.

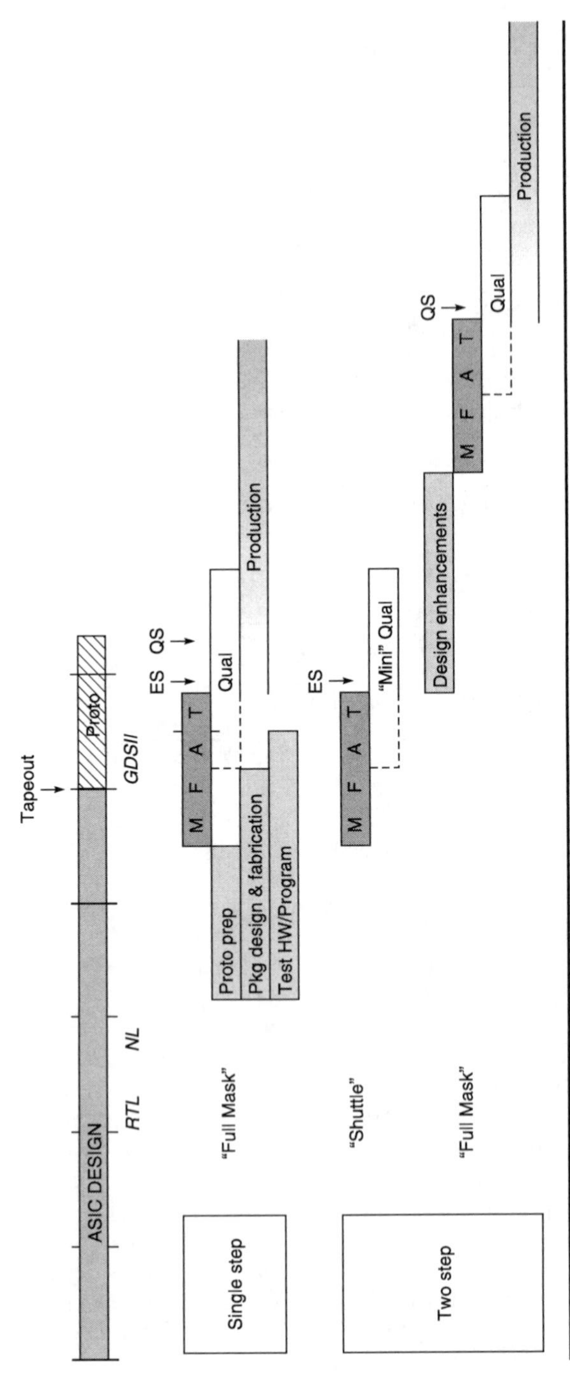

FIGURE 6.6 Comparison of development time lines for the single-step and the two-step approaches.

6.4.2 Tapeout Process

The foundries usually have carefully documented processes for tapeout of designs to enable a smooth transfer of information and an accurate description of the actions required. In addition to a purchase order, there are usually important forms that need to be completed by the customer. There will also be a need to establish an FTP (File Transfer Protocol) account for the design transfer [6.6]. The information requested on these forms is summarized here.

- **A service request form**
    - Customer's product information.
    - Process technology information—node, number of poly and metal layers, voltage levels, special features and options, etc.
    - Service(s) required—wafer, wafer probe, assembly, packaging, test. The customer can usually buy:
        - Wafers that have been "e-tested" to insure compliance with electrical process parameters.
        - Wafers that have been "circuit probed" (CP) in wafer form. Sometimes CP is also called "wafer-sort" or "wafer-probe."
        - "Turn-Key" parts that have been assembled in packages and have been tested. The foundry usually acts as a contractor to coordinate execution of the assembly, packaging and test operations.
        - Bumped die, where the wafers have gone through an additional set of operations that form bumps for flip-chip assembly.
    - Specifications used in the design, including revision numbers—design rules, mask layers, SPICE models, PDK.
    - Lot handling.
    - Any special instructions.
- **A mask tooling form**
    - CAD database details:
        - design and mask layer description;
        - GDSII formats;
        - chip size;
        - naming conventions used;
        - dummy fills.
    - Scribe lane and seal ring description.

- Data transfer details.
- Mask making instructions:
  - pattern density and OPC information;
  - mask quality levels. Not all mask layers require the same line width and spacing tolerances. There are usually three grades of masks that are used.

6.4.3 Mask Making

The mask tooling operations can begin upon transfer of the design database, the service request form, the mask tooling form, and any other documentation required by the foundry. By the way, not all foundries have their own mask making operations. If such is the case, or if the fabless IC company chooses to have masks made by a third party mask supplier, additional care must be taken to make sure there are no "glitches" in requirements, communications, etc. You must make sure that the mask supplier has been "certified" by the foundry for the specific process node. It is best to follow the foundry procedures for documentation forms and checks anyway.

The mask shop will usually go through three major operations:

- Frame layout.
- Data processing required for converting the GDSII database to a format compatible with the MEBES mask making machines. This involves a translation of the physical, rectangle based information into information that allows the MEBES machine to raster the pattern onto a mask.
- Job view—this operation provides an opportunity to verify and compare the converted database with the original. The customer usually has to sign off the completion of this operation.

Making of a typical mask set usually takes about one week. Sometimes, in a time crunch, the foundry will start wafers upon completion of the first three mask layers.

6.4.4 Engineering Lots for Prototyping

In the full-mask approach the foundry will usually start a 12 wafer engineering lot. The term "engineering lot" implies that there is usually a foundry engineer assigned to watch the progress of the lot through the fabrication process. Wafers in the engineering lot will typically have a premium of 20–25% above the production wafer price. This is reflected in the "lot charge" for the silicon.

There are usually three types of cycle times offered on the lead engineering lot:

- Standard, usually one calendar day per mask layer. Can vary from foundry to foundry and fab to fab within the same foundry.

- "Hot Lot", usually 20–30% faster than standard cycle time. There is a premium on the lot charge.
- "Super Hot Lot" (SHL), usually 20% faster than the hot lot cycle time. There are usually a very limited number of these available because each fab may only allow one or two in the fab line at any time. There is also a waiting list and a significant premium charge. Usually the customer's lot starts as a hot lot, with a request to boost the status as a SHL becomes available.

Although 12 wafers get started, it is normal practice to hold six wafers at the polysilicon level in order to save cycle time in the event of a design change requiring a new polysilicon (or subsequent) mask change. The remaining six wafers get processed through contact mask and first metal deposition. At that step, three more wafers are held in order to allow a "quick turn" fix in the metal layers. The rest of the three wafers are then processed through to completion. These are delivered to the customer or to a SATS partner.

In order to estimate variations in manufacturing parameters, the foundries usually run at least one 12 wafer "skew lot" or "corner lot." During the fabrication of these wafers, the nominal values of the following parameters are skewed to create typical, slow-slow and fast-fast variants:

- Gate oxide thickness.
- Polysilicon CD (Critical Dimension).
- Threshold voltages for both n and p transistors.

While there is usually an extra charge for fabrication of these wafers, it is a highly desirable investment and is strongly recommended.

6.4.5 Production Ramp

Production shipments of the ASICs usually require the following conditions to be fulfilled.

- Product qualification completed:
    - As discussed in Chap. 8, this assumes that the foundry process and the assembly process have been qualified by the respective supplier.
- Customer's validation and qualification of the part.
- Customer purchase orders.

As pointed out in Chap. 2, the fabless company needs to work closely with their customer to manage the ramp up, especially for products with a short life cycle that require a rapid ramp. In these situations, the 12–16 week production cycle time to assemble and test ICs can become a problem. The IC company needs to start silicon at risk and "build-ahead." This does introduce financial risk. Of course for many companies, this is a good problem to have!

6.5 Packaging Considerations

6.5.1 Package Design

The package is usually selected in the global planning phase from the choices described in Chap. 5 and supplier information. The package usually requires customization to meet requirements. The amount of customization effort could vary significantly. Here are some examples of package design work that must be completed in parallel with the IC design activity. The reason for the parallelism is that the lead time for the design work and fabrication of sample packages could be long.

- PBGA package design. Even if one selects an industry (or JEDEC) standard package body size and footprint, the package needs to have traces that connect I/O pads from the new ASIC to the package pins. At a minimum this involves routing of traces and vias in the package substrate. For performance critical IC's there must be an electrical and thermal analysis of the new package and die combination.
- Packages with leadframes. It is preferable to select a leadframe that has been designed and is available for general use ("open tooled"). While this may cut down on the lead time, the package performance may not be optimum. There must be a careful assessment of the cost, performance and lead time trade-offs.
- Complex packages using 3D stacking of silicon die need to be modeled and assessments made for availability, testability and yield of all the components.

6.5.2 Prototype Packaging

The foundries and the shuttle service providers do offer a limited number of package choices for the initial debug of the prototype parts. Sometimes a problem develops if the production package is not avialble as a choice. The suggestion is to find people with a breadth of knowledge in the availability of packages at the various suppliers. It helps to be connected. This kind of a problem may also require revisiting some of the assumptions and decisions made in the global planning phase.

In this cyclical industry, access to low volume assembly packaging has been problematic for the small companies. Some of the smaller

Figure 6.7 Examples of die assembled in open cavity prototype packages. (*Source: Quick-pak.*)

SATS have been more amenable to accepting start-up business. Once again, it helps to be connected.

One option for prototyping in plastic packages is the capability to create open-cavity versions of plastic packages. The plastic is removed from mechanical samples of finished packages using a patented process at a company in San Diego (http://www.icproto.com). The new ASIC die is then attached and wire bonded in the prototype package. For instance, if the design needs a 28TSSOP or a 48QFN package for which mechanical samples are available, this company can create a cavity by removing the plastic encapsulant and assembling the diein. A "glob top" protectant can be used to finish the assembly. For the class of packages they can work with, this approach offers an interesting alternative. Examples are shown in Fig. 6.7.

> Our patented process for creating open cavity plastic packages has proven very useful for prototyping of new designs. An integrated assembly capability has allowed flexibility and quick turn cycles times for numerous applications – MEMS, RF, chip-on-board to a few.
>
> Steve Swendrowski
> General Manager, Quik-Pak

6.5.3 Production Packaging

High volume production packaging of parts for the emerging fabless IC company must leverage the learning curve at the SATS. This can be achieved by defining the package, assembly and test requirements to fit within the mainstream envelope (Fig. 4.16). Having special, leading edge requirements with a limited volume forecast can cause serious disconnects and support issues at the SATS. So, with a judicious choice of package and assembly requirements for the new ASIC, the fabless company can expect to leverage the high volume learning curve, the quality and reliability capabilities already demonstrated at the SATS. A plan must be developed and executed for the characterization of the assembly process, the package and test for each new ASIC/package combination. High volume fabless IC companies pushing the limits in any of these technology areas need to have plans in place to investigate and characterize the processes for each new technology.

6.6 Test Considerations

6.6.1 Pre-Tapeout

As in the case of package design, there are also some long lead time activities in the test arena. Some examples of activities necessary to get ready for bench and ATE testing are:

- Hardware and test suites for bench validation of the ASIC. This validation is usually conducted under the guidance of the ASIC design team. They need to validate functionality and performance of the ASIC.
- Load boards and other hardware for the ATE tester.
- Test program development and debug for use on the ATE tester.
- Proactive engagement with the DFT team to make sure that the design can be tested cost effectively.

6.6.2 Prototype Testing

First silicon parts coming back from the foundry are usually siphoned off to the bench validation team. The product engineering team that manages ATE testing also gets some parts. Once functionality and the ATE test program have been validated, all the parts are usually tested on the ATE tester and a yield calculated from the results.

Following the functional testing of the parts, a detailed characterization of the ASIC is highly recommended [6.7]. This characterization must include the following:

- Functional test at a low frequency (e.g., 10MHz).
- Functional test at a high frequency (e.g., 50MHz).
- D.C. parametric tests, such as continuity, currents (I_{ih}, I_{il}, I_{DDQ}) and voltages (V_{ih}, V_{oh}, V_{il}, V_{ol}).
- A.C. parametric tests such as set up times, delay times, rise times, etc.

- Measurements and analyses at:
  - various temperatures from –55°C to 125°C;
  - various power supply voltages set at nominal, and at nominal +/– 20%;
  - various process corners—slow slow, slow fast, fast slow and fast fast. These variants come from the skew lot described earlier. The SS wafer has high poly CD and high V_T's. The FF wafer has low poly CD values and a low V_T.

6.6.3 Production Testing

ATE test programs in the engineering mode include detailed testing to allow debugging of functionality and performance issues. Once the ASIC has been released to production, the test program can be streamlined to reduce test time. In production testing of high volume parts, reducing even a fraction of a second of test time can affect unit cost. Testing at low and high temperatures can also be eliminated if there is proper "guard-banding" of parameter limits. The detailed characterization report comes in handy in developing these limits.

Another important factor is the cost of the tester that must be used. The more sophisticated the tester required to test the IC, the higher the test cost. Some of the factors that affect tester cost are:

- Number of pins.
- Performance.
- Number of channels.
- Features such as RF and analog.
- Temperature testing.
- Automatic handlers.

Some of these factors are affected by the design and hence, close cooperation between the DFT team and the manufacturing team can be invaluable.

6.7 Quality and Reliability Considerations

Elements of the quality manual and the qualification plan are discussed in Chap. 8. In order to execute the qual plan in a timely manner, it is essential to make arrangements with a services provider. It is also essential to design and build life test boards. ASIC parts are mounted in sockets which are wired on boards and the whole assembly is subjected to an operating life test at temperatures as high as 125°C. Therefore the sockets and boards must be rugged. Lead times for special sockets can be quite long, especially if you require a new socket or a modification to an existing socket; hence the recommendation to start early.

In the two-step approach it is always a good practice to implement a "mini qual" (or "early look qual") with the first batch of parts. The purpose is to look for any glaring reliability problems with the silicon/assembly/package processes and their interaction with the design. As an example, if in the qual the high temperature operating life population is planned at 77 using three boards, the early look qual could be set up with one board. Other tests such as ESD and latch up could also be implemented on the initial silicon.

6.8 Supply Chain Considerations

Operations execution and the selection and management of the supply chain are intricately linked to the success of fabless IC companies. A great idea and a great design go nowhere unless the company is able to deliver prototypes and production as planned, on schedule, and cost effectively. This requires careful planning of operations activities during the lifecycle of the company. Supplier selection and management, operations activities, skill sets, and resource requirements are addressed in the next few sections.

6.8.1 Supplier Selection Checklist

Having discussed the many activities in the development and implementation of a fabless IC, Table 6.2 provides a summary checklist of possible suppliers. The reader should review the list and make sure they customize the checklist for their individual needs.

Fabless company requirements		Who supplies?		
Process				
Design rules, etc.		Foundry	Mosis, EuroP	
PDK (process design kit)		Foundry	EDA supplier	
MPW proto		Foundry	Mosis, EuroP	ASIC/TK
Proto engineering lot		Foundry		ASIC/TK
High volume		Foundry		ASIC/TK
Design IP				
SC library		Foundry	3rd parties	ASIC/TK
SRAM		Foundry	3rd parties	ASIC/TK
ROM		Foundry	3rd parties	ASIC/TK
Input/output		Foundry	3rd parties	ASIC/TK
Specialty interfaces	USB, DDR, HDMI,...	Foundry	3rd parties	ASIC/TK

Fabless company requirements		Who supplies?		
Analog/mixed signal	PLL, DAC, ADC,..	Foundry	3rd parties	ASIC/TK
Embedded processor	ARM, MIPS, Configurable,..		3rd parties	ASIC/TK
Embedded DSP			3rd parties	ASIC/TK
Embedded DRAM	1T,..	Foundry	3rd parties	ASIC/TK
Flash memory		Foundry		ASIC/TK
OTP (one time program)			3rd parties	ASIC/TK
Specialty applications	Bluetooth, Mobile,...		3rd parties	ASIC/TK
EDA tools				
Front end		EDA supplier(s)		
Physical design		EDA supplier(s)		
Verification		EDA supplier(s)		
Design services				
Front end		EDA supplier(s)	3rd Parties	ASIC/TK
Physical design		EDA supplier(s)	3rd Parties	ASIC/TK
Assembly & test				
Package design		SATS	3rd parties	ASIC/TK
Prototyping		SATS	3rd parties	ASIC/TK
High volume		SATS		ASIC/TK
Test characterization		SATS	3rd parties	ASIC/TK
Other services				
Qualification, reliability			3rd Parties	ASIC/TK
Debug, failure analysis			3rd Parties	ASIC/TK

Note: TK: turn key supplier; EuroP: EuroPractice.

TABLE 6.2 Summary Checklist of Possible Suppliers for the Various IC Development Activities

6.8.2 Supplier Selection

In this section let us focus on supply chain activities in the four development phases. Supply chain activities such as partner selection and contract development vary in the four development phases of the company's lifecycle. The use of a concurrent engineering approach using the four development phases is recommended here. An alternative approach has been recommended elsewhere [1.24]. Table 6.3 is a summary of an approximate time line for supplier selection. The recommended approach is to have an initial list of suppliers in the global planning phase. The table also illustrates approximate time frames when final supplier selection must occur.

6.8.2.1 Global Planning Phase

In the global planning phase, and shortly thereafter, the major partner selection activities are:

- Assessment of various available partners and the determination of a technical match:
  - Does the foundry have the right process technology with the right process options?
  - Are the Library, IP and PDKs available in the right technology node at the foundry and at the EDA supplier of choice?
  - Do the SATS have the right package, the right tester, the quality and reliability information?
  - Is there a linkage between low volume sourcing options and the high volume production capabilities?
- Assessment of a business match.
  - Will your business be acceptable to the supplier?
  - Is the partner willing to provide the support required to meet your company's technical and business requirements?
  - Cost match—unit cost and development cost?
  - Quality match?
  - Legal match?
  - Personal links, both at a technical and a management level
- There must be positive indications of early commitments from the suppliers to support the business. Some suppliers have formalized processes to assess and commit to new business. Others may have ad-hoc processes:
  - Get a "handshake" with key executives.

	Global planning	IC design			Prototyping	Production
Process	Initial		Final (Proto)		Final (Initial Prod)	Final (Hi Vol)
Design IP	Initial		Final			
EDA tools	Initial	Final (FE)	Final (BE)			
Design services	Initial		Final			
Assembly & test	Initial			Final (proto)	Final (Initial Prod)	Final (Hi Vol)
Qual	Initial			Final		
Debug/FA	Initial				Final	

TABLE 6.3 Summary Time Line of Initial and Final Supplier Selection during the Four Phases of IC Development at a Fabless IC Company

- Don't count only on informal technical connections and informal "commitments" made by unauthorized supplier personnel.
- Getting budgetary quotes is usually a right step. This should include estimates for development, wafer and/or unit pricing. Some foundries have methodologies for validating the fabless company's estimates of product cost and schedule in the selected market segment.
- Engage the authorized supplier personnel that can help facilitate internal decision making for acceptance of your business. Initiate processes for due diligence, if appropriate.

For a startup fabless IC company, the ability to get their business accepted by the supply chain partners of choice can vary depending on factors such as overall industry health, the capacity/demand fluctuations and the individual supplier. In general, though, every supplier likes customers that are potential winners. So, if you really have a compelling argument for the worthiness of high volumes for your IC, you are half-way there. The other half is finding the right contact links and convincing the suppliers of your company's potential. As a start-up, expect to do some selling with the major suppliers and support their due diligence processes.

Beware also that from time to time, suppliers do have shifts in their acceptance strategies for emerging company business. A recent example is the cutback of emerging customers at the major SATS. I am a witness to how this created serious issues at a number of fabless companies.

How to do all this can also be tricky for the start-up company. Managing all this requires experienced personnel that are not on board. In general people resort to past experiences and ad-hoc inputs from associates about technical and business feasibilities of the various suppliers. Another possible approach requires recruitment of a "foundry manager" [1.24]. Key ingredients for such an individual are a breadth of knowledge and experience as well as business acumen and a technical and management network of connections. Recommendations for the proper "virtual operations" team have been made previously [6.1, 6.2].

There must be a technical and a business match when selecting the suppliers. By the way, expect that perfect matches will be difficult to find. Hence the need for compromises, negotiations, and adjustments to plan. Like in the development and fostering of any relationship, the fabless IC company must count on investing efforts to make the supplier relationships work.

There are a number of mantras we undertook from the start. First, we ensured everyone in the value chain from design to manufacturing were working with the same information and completely aligned to the same goal. Second, we wanted to instill a strong sense of ownership amongst our suppliers right down to the individual contributor level. One technique we employed was the occasional management visit to supplier locations in order to explain product strategy and confidential plans to the entire supplier team. It seems that semiconductor engineers rarely get such info on the end product. Our communication approach seemed to provide great motivation and resulted in a strong sense of ownership. Lastly, we had a strong Program Manager embedded at the supplier development facilities.

By our own estimations, we were the 4th largest fabless semiconductor company in the world on a procurement basis in our first year of production. For making that happen, credit must be given to our extensive supplier base. The credit goes not only to the silicon suppliers but the substrate assembly & test providers for believing in the product and managing our ramp and their capacity exceedingly well.

Bill Adamec
Senior Director
Semiconductor Technology Group
Microsoft XBox

6.8.2.2 IC Design Phase

As the company transitions into the IC design phase, it is important to focus on the following:

- Obtain the foundry's design rules, PDK's and any other design related information. If the foundry has indicated an acceptance of your business, it is relatively painless to obtain this information. There may be a need to execute Non-Disclodure Agreements (NDAs) and possibly other agreements related to the use of foundry supported IP.
- Obtain the "front end" models for the library, the memory blocks and the relevant IP blocks to get the front end design launched. Analog models may be required if custom analog blocks are included in the design. Some of this information comes from the IP supplier(s). If it comes directly from the foundry, the foundry engagement paperwork may be sufficient.

When using third party IP suppliers, there needs to be the normal selection due diligence, agreements on business terms, an NDA and possibly a contract with each supplier. It may be possible to get the designers started once a Letter Of Intent (LOI) or a Memorandum Of Understanding (MOU) has been executed with the supplier(s) of choice.

As the time line progresses in the design phase the agreements need to be finalized and executed. These must be in place before work begins on the synthesized RTL netlist. If the physical design is to be outsourced, the Design Services supplier selection requires due diligence, NDA, and agreements to be put in place.

In the second half of the design phase there must also be finalized arrangements for the package design and the prototyping wafer fabrication, assembly, and test. The qualification services provider must also be engaged.

6.8.2.3 Prototyping

During this phase there must be finalization of the failure analysis and debug suppliers, especially if there is a need for them! Final arrangements for support of initial production at the foundry and the assembly and test suppliers must be complete.

6.8.2.4 Production

During this phase the fabless IC company must have final agreements with the foundry and the SATS.

6.8.3 Contracts vs Purchased Services

Established suppliers have standard processes for engagement with customers. Ask to review their proposed agreements and make sure your company is protected. Initial agreements must include clauses such as an agreement to be a supplier, confidentiality protection, IP rights, and indemnification. High volume manufacturing agreements must include clauses on warranty, cost, default, jurisdiction, and the like. It is possible in some cases to get started with purchase order based services. Which approach to use is dependent on the individual circumstances.

6.8.4 Supplier Relationships

It should be clear by now that establishing and maintaining relationships with suppliers is an important key to a fabless IC company's success. Some will require more investment and nurturing than others. Executive level contact and selling could be the key to getting your business accepted. Assigning key interface points at the supplier and the fabless company is a good practice once the business relationship has been established. Table 6.4 captures typical interactions at the executive, management, technical, and business levels between the supplier and the fabless company. For a start-up implementing the first design, the interactions could be limited to an executive and a program manager. Note also that many of the interactions can be managed remotely through conference calls and web meetings. However, the value of face to face meetings in establishing and maintaining relationships cannot be emphasized enough.

6.9 Operations Best Practices

It must be clear by now that an emerging fabless IC company has many technical, marketing, and sales challenges. Operations issues are generally not on the minds of the principals. An "operations dilemma" facing such companies has been discussed previously [6.1, 6.2]. A systematic, concurrent engineering approach to addressing the operations activities was described. In this chapter let us describe and summarize the operations processes and practices that must be addressed at emerging fabless companies in order to position themselves as a leading or best-in-class supplier of integrated circuits. Similar discussions have been reported by the author previously [6.8, 6.9] and more recently in reference [1.24]. In established companies the concept of best practices has usually been explored and implemented in some or all of these business process/practice areas. The four key steps to determining a best practice are:

- current assessment;
- benchmarking;
- cost–benefit analysis;
- revising the process.

Establishing best practices involves integrating business processes and allows the fabless companies to measure and improve their efficiencies, improve effectiveness of working relationships, and environment within the organization as well as with customers and their supply chain partners. However, establishing a best practice is

Contact	Organization role	Reviews	Interactions
Executive	CEO, COO, VP	Annual	As needed
Management	VP, Director, Program Manager	Quarterly	As needed
Technical	Program Manager, Engineer	Quarterly	Weekly, daily
Business	Procurement, Order Fulfillment	Quarterly	Weekly, daily

TABLE 6.4 Typical Contact Levels, Frequency of Contact and Who in the Fabless Organization Makes the Contacts with the Supplier

also time consuming and expensive. Judicious business decisions need to be made whether to elevate established practices into best practices.

Since managing cash flow is a very important aspect of life at a start-up, a miniset of operations processes and practices that must be in place during the various stages of development and production ramp are discussed here. An understanding of these processes and practices will be of assistance to senior managers at emerging companies. Practice areas include engineering, quality, customer support, production control, and finance.

In the following section there is discussion of the operations effort required in the four development phases. Section 6.10 provides a more detailed discussion of the practice areas in the production phase.

6.10 Operations Effort and Resources

Operations tasks and practices vary as the fabless company grows. A conceptual graph showing the operation tasks and resources in development and production is shown in Fig. 6.8. The phases shown correspond to the four phases discussed in this book.

A simplified model is presented here to illustrate an order of magnitude estimate of required operations investment at fabless companies. For simplicity, let us assume a single product company, a first time right design approach and three revenue "plateaus"—$1M, $10M, and $100M annual revenue. Outsourcing of production wafer fabrication, assembly, and test is assumed. It is also assumed that the fabless company will manage the supply chain and will have ownership for the WIP.

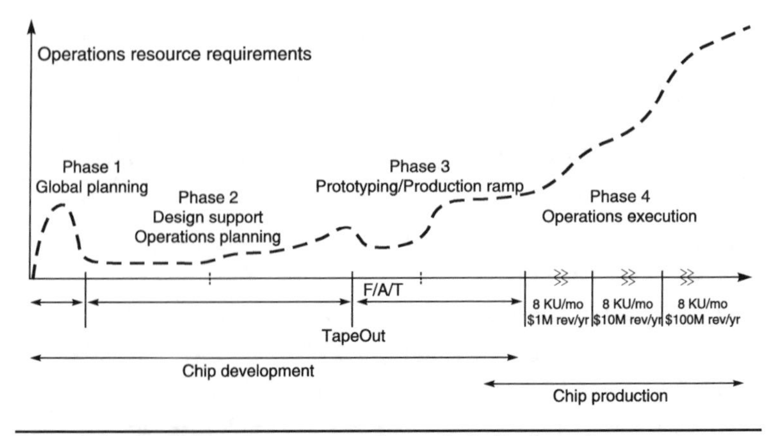

FIGURE 6.8 Operations resource requirements in the four phases of a fabless IC company. (*Source: GSA [6.8, 6.9, 1.24].*)

Table 6.5 is a summary of the annual revenue, the number of units shipped and the number of wafer lots that have to be managed for the three revenue plateaus. Also shown is an estimate of the number of operations resources required to support the three revenue streams. Operational expenses, both manpower and other infrastructure, are shown as a fraction of company revenue. These estimates provide an order of magnitude estimate to help the fabless company executive measure themselves against. There are, of course, many variants of the assumptions and different circumstances that can affect the model. The assumptions here are that the chip size is 5 mm × 5 mm, 130 nm 1P6M CMOS logic technology, 200 mm silicon wafers, a $5 ASP (Average Selling Price), $120–150K per man year. It should be recognized that these are typical/guideline numbers that could vary from company to company due to individual company specifics.

6.11 Resource Skill Sets

Now let us discuss the resource skill sets required in the various development phases. Figure 6.9 provides a pictorial representation. A detail of the skill sets and the numbers required is illustrated in Table 7.5.

Design engineering effort starts at the very beginning and is the major focus at the startup through the prototyping phase. Specific design resources and skill sets required are dependent on the technology node, the type and complexity of the design and on how much of the design is executed in house. Figure 6.10 provides a conceptual estimate of FE and BE design resources required. This assumes that

Revenue/Year, $M	1	10	100
Units shipped/year @ $5 ASP, M	0.2	2	20
200 mm wafer starts/month	18	180	1800
200 mm wafer lot outs/month	0.8	8	80
300 mm wafer starts/month	8	80	800
300 mm wafer lot outs/month	0.3	3.3	33
Operations resources required:	1–3	3–5	10–17
Operations cost to revenue ratio	15–40%	5–10%	2–4%

Assumptions:
Die size: 5 mm × 5 mm
Technology: 130 nm CMOS, 1P6M
Manpower cost: $120–150K/year

TABLE 6.5 Summary of Model to Estimate Operations Resources and Cost

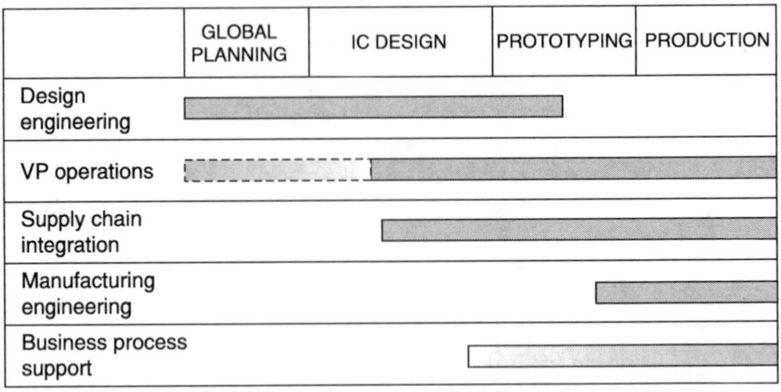

GLOBAL PLANNING
IC DESIGN
PROTOTYPING
PRODUCTION
Design engineering
VP operations
Supply chain integration
Manufacturing engineering
Business process support

Figure 6.9 Pictorial representation of operations resource requirements in the various development phases.

BE design is implemented in house. Models to estimate the design effort and resources have been developed and reported previously and are discussed further in Chap. 9. There is also further discussion of the phasing of design resources between FE and BE design. Peeling off resources from one project to the next also becomes a management challenge.

Major design activities in the FE and BE phases are:

- FE design:
  - architecture development;
  - RTL design;
  - verification.

Design resource requirements
Front end design
Back end design
F/A/T
8 KU/mo
$1M rev/yr
8 KU/mo
$10M rev/yr
8 KU/mo
$100M rev/yr
TapeOut
Chip development
Chip production

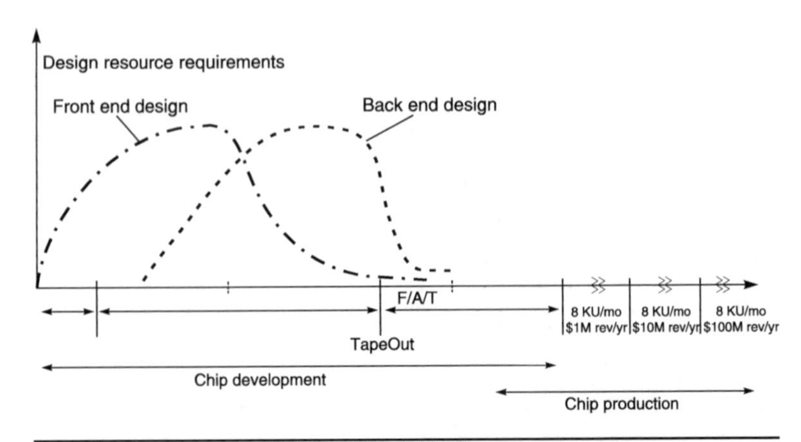

Figure 6.10 Design resource requirements in the various development phases.

- BE design:
  - physical design:
    - floorplanning;
    - place and route;
    - verification.
  - Infrastructure:
    - design flow—tool selection integration, verification;
    - library, memory, and other IP integration;
    - tapeout.

Operations activities in the global planning phase require a senior level individual with current knowledge of available technologies, relevant technical and business trade-offs, and industry contacts. This skill-set requires a Vice-President or director level person. As discussed previously, the need for this senior level person goes into hibernation for some quarters while the design is being executed. Hence the recommendation for a "virtual operations" team [6.1, 6.2].

Operations activities pick up momentum in the second half of the IC design phase and grow through the prototyping phase, except for a slight slowdown while the design is in fab, as shown in Fig. 6.8. The skill-sets required for this "supply chain integration" activity are listed below. Estimates of resources required are provided in Sec. 7.4. Specific needs at companies will vary depending on factors such as maturity of the technology node, design complexity, and features and design classification (leading edge, mainstream, or mature). The breadth of knowledge and experience of the team will also affect the numbers. These skill-sets are:

- Program and supply-chain management.
- Silicon engineering:
  - foundry relationship management;
  - silicon process expertise (e-test, models, yield, etc.);
  - acceptance criteria, test, disposition.
- Packaging/assembly engineering:
  - assembly relationship management;
  - packaging expertise;
  - management of bond diagram, assembly build instructions, etc.
- Product/test engineering:
  - test relationship management;
  - test program and hardware development;
  - design validation, characterization;

- failure analysis, FIB (focused ion beam) repair coordination;
- yield engineering.
- Quality and reliability engineering:
  - qual plan development and execution.

Operations activities are scaled up in the production phase as the product gets qualified and production ramp begins. Details of activities are described in the following section. Note also that in a small fabless IC company the boundaries between supply chain integration and manufacturing engineering are transparent. Manufacturing engineering activities can be classified in a manner similar to the previous phase:

- Silicon engineering.
- Packaging/assembly engineering.
- Product/test engineering.
- Quality/reliability engineering.

Business process support is also discussed in the next section.

6.12 Production Operations Activities and Processes

Within the fabless company, the operations department usually has ownership to procure and deliver product to customers on schedule and at a favorable cost. The following is a simplified view of operations tasks and activities. These were described initially in [6.8, 6.9] and are also discussed in Chap. 11 in reference [1.24]. Within the fabless company the operations group should also be the focal point for coordination of cross-functional issues with customers and suppliers. Some of these roles and responsibilities can be shared with other parts of the organization. For simplicity, I have assumed that operations is the "center of the universe" within the company and have used a fairly broad definition of the operations activity. How the total company manpower can be optimized through sharing with the finance, sales, engineering and other organization is not included in this discussion.

Operations activities at the fabless company can be categorized into three major practice areas:

1. Manufacturing engineering.
2. Quality, reliability, documentation and rrocedures.
3. Business processes.

In an emerging company with a $1M revenue stream, the major focus in operations should be on manufacturing engineering. The quality and business processes can be accomplished with an "ad-hoc" set of processes. A "systematic" focus for establishing the quality and business processes and practices as the company approaches a $10M revenue stream is recommended. A company must have established practices in all the areas as its revenue stream approaches $100M. The company may choose to invest in establishing a best practice only in some of the areas—areas that are critical to its core strengths and to its success.

6.12.1 Manufacturing Engineering

This function is fulfilled by a group that usually has a broad set of technical responsibilities, including product release to manufacturing, product yield, and quality maintenance, supply chain interface and technical support to the various business groups. The following is a list of typical responsibilities. Formalizing and documenting the processes associated with each of these responsibilities is the first step towards establishing a best practice in this area.

- Definition and documentation of the product, procedures and reports to monitor and control product yield, cost and performance.
- Supplier interface to address product, design, process, and test related issues.
- Yield monitoring and tracking:
  - lot-to-lot variances;
  - wafer-to-wafer variances;
  - design sensitivities.
- Yield enhancement projects, if any.
- Yield issue(s) resolution.
- Quality and reliability issue(s) resolution.
- Quality and reliability maintenance coordination and execution.
- Engineering support of customer returns.
- Failure analysis, debug, and reports.
- Engineering support of production planning.
- Cost reduction including test time reduction, price negotiations, technology migrations and alternate supplier strategies.

6.12.2 Quality and Reliability

This function plays a key role towards the graduation of a fledgling start-up into a best-in-class supplier of integrated circuits. Setting-up, monitoring, and maintaining the quality processes within the fabless IC company and its supply chain are the major responsibilities. The following is a list of items around which processes can be set up as a first step towards establishing a best practice. One must look to integrate these with the discussion in Chap. 8.

- Quality manual documentation:
  - documents the company quality policy, quality system, organizational responsibilities, control mechanisms.
- Formalize document control, and related procedures:
  - verify product design, manufacturing, and qualification documentation. An example of commercially avialable software that enables documentation processes can be found at http://www.scalarsoft.com
- Product qualification documentation:
  - qual plan, specification, and report;
  - supplier data—foundry, assembly, test;
  - incoming and outgoing QA (Quality Assurance) specifications;
  - reliability and quality maintenance program.
- Process certification procedures and maintenance:
  - SPC (Statistical Process Control);
  - yield.
- Continuous improvement program:
  - certification and reliability maintenance;
  - supplier audits;
  - change control procedures;
  - incoming and outgoing quality level improvement.

6.12.3 Business Processes

These processes seldom get the priority they deserve. Since most companies need to establish the basic financial processes early in their existence, the perception usually is that extending these processes to include operations and supply chain issues will be straightforward. While the operations activities can and should leverage existing business infrastructure at the company, a comprehensive look at developing the proper processes required to become a leading fabless IC company is recommended. The processes have been categorized into three major areas as outlined here. Formalizing and documenting the processes in each of these areas is a first step towards establishing best practices.

- Legal:
  - confidentiality;
  - supply agreements;
  - IP agreements;
  - IP portfolio.

- Production Planning and Control (PP&C):
  - demand and judged forecast coordination;
  - order management and fulfillment;
  - supply-chain forecast and tracking.
- Financial processes:
  - Cash flow for WIP and inventory;
  - material and cost tracking;
  - product cost—actual vs. plan.
- Customer communication and support:
  - response to customer requests;
  - RMA (Return Material Authorization).

6.12.3.1 Legal

Early on in the company's lifecycle there must be legal agreement documents that allow engagement with suppliers, customers, consultants, contractors, and similar entities.

Supply agreements must be executed with entities on both sides of the value chain. This means that the fabless IC company must have supply agreements with the suppliers such as foundries, SATS, service providers, and EDA tool suppliers. Since the fabless company is a supplier of ICs to its customers, it must execute supply chain agreements with its customer.

Agreements with IP suppliers can have many variants from a legal and business perspective. The fabless company must get a license to use the IP block, either a single use or a multi-use license. The business arrangement could be an NRE (Non-Recurring Engineering) based or a royalty based model, or a combination of the two.

It is of paramount importance for the fabless company to build its patent portfolio from the beginning. While this may not seem so important at the beginning, the significance of having a large portfolio will be apparent as the company grows.

6.12.3.2 PP&C

a. Demand Forecast Generation Processes

Operations is the focal point for pulling the demand forecast together. This is clearly a multi-disciplinary function and can be coordinated by different pieces of any organization—sales, marketing, finance, or program management. Since operations is the group that has to communicate the build forecast to the suppliers and to manage the order fulfillment process, it is an excellent candidate for this very critical function. Operations' role involves the following:

- Coordinate cross functional processes.
- Receive sales/marketing input.

- Receive "do-ability" from engineering and suppliers—quantity, schedule, quality.
- Coordinate "judged" forecast generation. This is usually a forecast agreed to by the company management and the various stakeholder groups.
- Keeper of up-dated demand forecast.
- Communicate forecasts to suppliers.
- Capacity allocation management.

b. PP&C—Order Management and Fulfillment Process

The following is a list of processes that must be followed and documented. In a best-in-class operation these activities are usually linked together in an ERP/MRP system connected to the company's financial system to control inventories, cost, COGS (Cost Of Goods Sold), and lead time.

- From/to sales:
  - verify availability and pricing;
  - receive/acknowledge PO (Purchase Order).
- Order entry:
  - enter order into WIP/inventory management system;
  - trigger shipment from inventory or WIP or new starts;
  - notify production control;
  - acknowledge PO and advise delivery date to customer.
- Parts shipment and logistics:
  - monitor ship readiness, advise sales 48 hours prior to shipment;
  - generate shipping/freight forwarding/import/export documents;
  - generate drop ship forms, if required;
  - pack;
  - follow thru with sales and customer.
- Invoice and payment:
  - generate and send invoice;
  - receive payment.

c. PP&C—Supply Chain Management

- Relationship management:
  - management level;
  - working level.

- Logistics management.
- Coordination and reporting of the following:
  - order placement, forecast communication/discussion;
  - WIP, yield, quality information flow, and dissemination;
  - invoicing, royalty payments (if any), cash flow management.

6.12.3.3 Financial Processes

In executing these activities it is assumed that the operations group works in conjunction with the company's finance group but is held accountable for achieving the operations goals.

- To/from customer(s):
  - PO receipt;
  - order fulfillment;
  - invoice issuance;
  - payment receipt.
- To/from suppliers:
  - forecast;
  - committed forecast;
  - order placement;
  - WIP ownership and liability;
  - inventory ownership and liability;
  - special materials order liability, if any.
- Management reporting:
  - COGS—plan vs. actual;
  - cash flow;
  - demand vs. build forecast reconciliation.

6.12.3.4 Customer Support Processes

Having an efficient and responsive customer support organization and associated practices is another key element of a leading fabless IC company. These processes have been categorized into three areas—Quality Assurance (QA), Customer Service (CS), and Product Planning (PP). The interface to the customer needs to be clearly identified—it could be either a sales person or a Program Manager.

Figure 6.11 is a pictorial view of the business processes that have been discussed here. A general perception in the industry is that the operations infrastructure required at the fabless company is eliminated when using an ASIC supplier. While requirements for

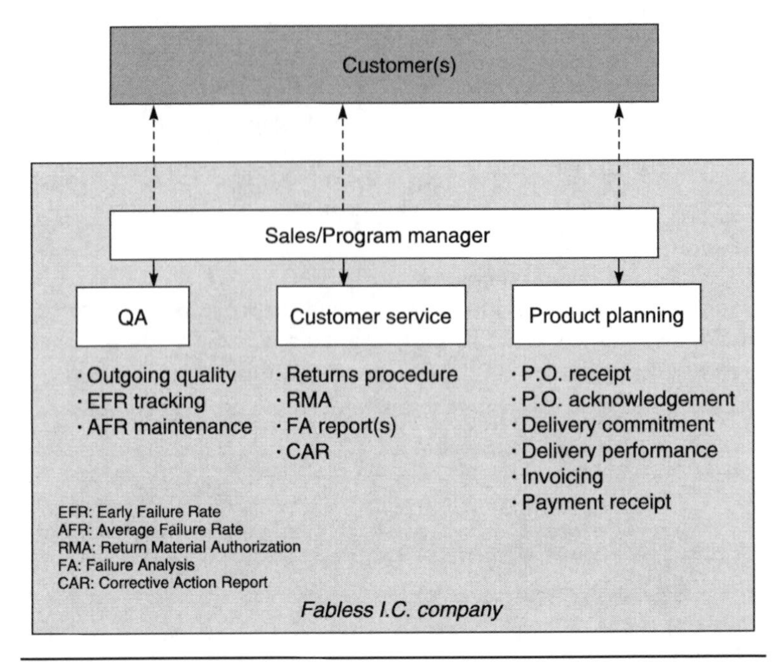

FIGURE 6.11 A pictorial view of business processes at a fabless IC company. (*Source: GSA [6.8, 6.9, 1.24].*)

operations best practices at the fabless company get reduced by 30–50%, they are not eliminated by using an ASIC supplier.This is because the fabless company maintains responsibility for the product quality, delivery and order fulfillment for their customer(s). Therefore, it is very important still to focus on establishing all the other practices discussed. While there is indeed some cost reduction associated with the reduced operations infrastructure, this must be traded off against higher unit price usually charged by the ASIC supplier/aggregator. The core best-practices must be in place regardless of the sourcing model used.

6.13 Key Points

- Design verification is critical.
- There are numerous tradeoff in selecting an MPW or a full-mask approach for silicon prototyping.
- Many operations related best practices need to be implemented at fabless IC companies.
- Supply chain partner selection and management is crucial for IC implementation.

CHAPTER 7
Managing Cost

There are many components of cost related to the development and manufacturing of an IC. The three major components of cost considered in this chapter are:

- Unit cost.
- Development cost:
  - mask;
  - prototype silicon;
  - design;
  - test;
  - package design;
  - qualification.
- Operations cost:
  - infrastructure and logistics;
  - returns management;
  - manufacturing support—engineering, quality, reliability, customer support, etc.

Other costs such as SG&A (Sales general and administration), facilities and depreciation are not discussed

7.1 Unit Cost Estimation

7.1.1 Cost Breakdown and GDPW

As discussed in previous chapters, a fabless company can buy wafers, or it can buy tested die, or assembled, packaged, and tested "finished" units. In this chapter there is discussion of how one can estimate die cost and the finished unit cost. Being able to calculate cost in the IC planning stages is important for making knowledgeable development, operations, and product pricing decisions, and trade-offs.

It is also important to understand the various components of cost in the manufacturing phase in order to set cost expectations and targets, sometimes called "standards."

The cost adders, yield factors, and calculations associated with the major operations steps are shown in Fig. 7.1. In this simplified version of the cost model, the wafer purchase price becomes a cost adder WC. Knowing the die area one can calculate the GDPW, the total possible candidates for the IC die on the wafer. A word of caution—colloquially the term "die size" is used to describe either the die area or the linear dimension, which is the square root of the die area.

Gross Die Per Wafer (GDPW) = total usable area on the wafer/ die area

Once the design is completed, the die area becomes a known quantity. In the planning phase, estimating the die area can be a challenge. Usually, experienced IC designers and managers create spreadsheets to estimate areas associated with the various components of the design. An example of the elements to be included in the spreadsheet is provided here as a "starter" list:

- Logic area calculated from the total number of gates in the design, the maximum packing density in the technology and an expected gate utilization factor. The last two factors are estimates provided by the foundry and experienced individuals.
- Memory area based on estimates from the foundry or IP providers or based on a scaling from a previous technology.
- IP block area based on estimates from the foundry or IP providers or based on a scaling from a previous technology.

	Cost adder	*Yield*	*Calculation*
Water purchase	WC		GDPW
Water sort	WS	SY	
NDPW			SY * GDPW
1st optical		FOY	
Die cost, DC			(WC + WS)/(NDPW * FOY)
Package	PC		
Assembly	AC	AY	
Final test	TC	TY	
2nd optical		SOY	
Finished unit cost			(DC/(AY * TY * SOY)) + ((PC + AC)/(AY * TY * SOY)) + (TC/(TY * SOY))

FIGURE 7.1 Summary of IC operations, cost adders, yield factors, and calculations at various steps.

The area of a "soft" IP block instantiated as a logic block can use the algorithm used for estimating logic density.

- Analog block area has to be estimated by the designer, or by the IP or design services provider.
- Input/Output (I/O) ring and scribe area based on estimates from the foundry or IP providers or based on a scaling from a previous technology.

It should be noted that there are numerous die size estimators available commercially. One example is from http://www.chipestimate.com. There are definite advantages of leveraging an industry tool that has been calibrated with the foundries and IP providers.

The other piece of the puzzle in estimating GDPW is the total usable area on the wafer. While this sounds simple, predicting the actual number of whole die "stepped" on the wafer during the lithography process is not that straightforward. Equations have been used for years to approximate the number of usable die. More recently computer programs have been developed to generate a physical layout and then counting the number of die within a circle. The wafer radius used in the calculation must exclude a perimeter area that is unusable, typically 3–5mm from the wafer edge. In generating the physical layout it has been observed that the GDPW varies depending on how the die stepping pattern is centered on the wafer. In the data shown in Fig. 7.2, the stepping pattern centering has been optimized to yield the maximum GDPW possible. Figure 7.2 shows a comparison

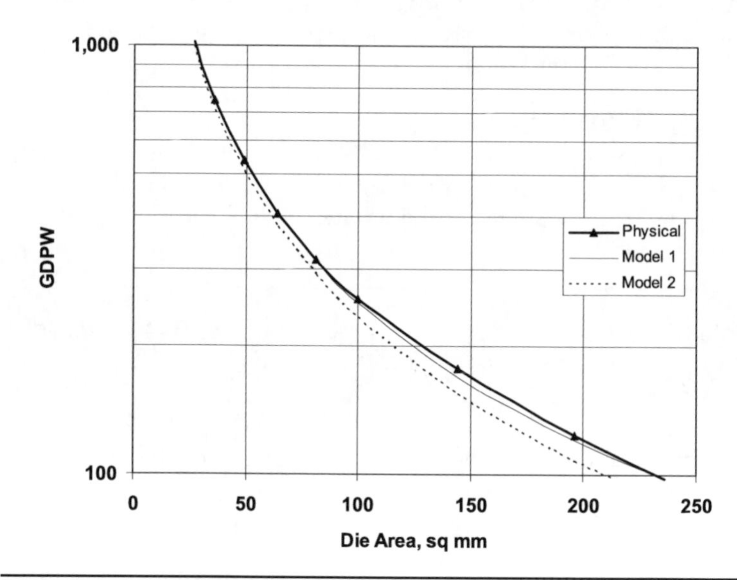

FIGURE 7.2 Comparison of two different models vs. a physical count of gross die on a 200 mm diameter wafer.

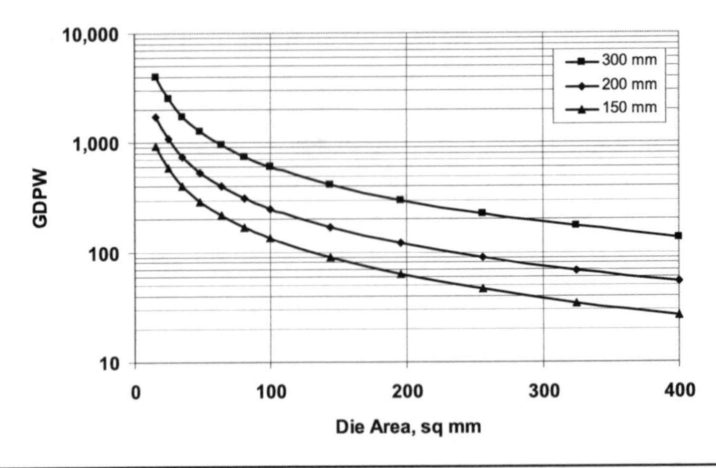

Figure 7.3 Gross die per wafer for 300, 200, and 150 mm diameter wafers.

of the physical gross die count on a 200 mm wafer with those predicted by two different user models. Model 1 approximates the actual count very closely. The second model has errors as high as 20%. Notice the large deviations at larger die sizes.

Model 1 has been used to calculate the GDPW on 300, 200, and 150 mm wafer for various die sizes, as shown in Fig. 7.3. This chart confirms the prediction in Chap. 5 that there are approximately 1.8 times as many gross die on a 200 mm wafer as on a 150 mm wafer, and approximately 2.2 times as many gross die on a 300 mm wafer relative to a 200 mm wafer.

7.1.2 Silicon Die Yield

The next step in the cost model is to calculate the number of good, usable die on the wafer. This is called Net Die Per Wafer (NDPW). This number is calculated as the product of the wafer Sort Yield (SY) and the GDPW.

Net Die Per Wafer (NDPW) = Sort Yield (SY) × Gross Die Per Wafer (GDPW)

The factors affecting overall sort yield can be broadly categorized into two areas [1.14]. These are:

- **Systematic yield:**
  - Systematic defects and yield loss are usually addressed by fixing the root cause. An example is the yield loss caused by an uneven metal thickness during the CMP operation;

this is fixed by proper installation of the pad, a re-adjustment of the process operating parameters, and by requiring "metal fills" in the design phase (Chap. 5).

- Design errors may be found.
- Predicted estimates are usually targeted at 90% [1.14].

- **Random defect-limited yield:**
  - This has been the most commonly addressed yield component. The major factors affecting this yield components are:
  - **Defectivity** (or defect density). This represents an average number of electrical defects per unit area and is denoted D_o. It is usually derived heuristically from the observed sort yield of a known product. The defectivity is then applied to predict yield on future products based on the same yield model but a different critical area.
  - **Critical area**. In the simplest approach it is assumed that the critical area, to be used in the models, is equal to the die area. However, there are at least two refinements that can be made. First by adding up the active area of the logic, memory, I/O and "white space," and assigning different defectivities to each of these components. This approach can provide some sensitivity to yield based on the content of the die. This refinement becomes meaningful if different defectivity numbers can be assigned to each of the components. Assuming that the wafer fab provides a single, average D_o, I have used a simple approach that assigns a 30% adder to D_o for memory blocks, a 30% reduction to D_o for active I/O areas, and a 50% reduction to D_o for white space (on a plot layout). The proper way to do this is to get yield information from chips with logic only, memory only and I/O only blocks and then calculating defectivities for the different components. As sophisticated EDA tools become available to derive detailed process related information and critical areas embedded in the design database, more sophisticated models are being developed and used. Examples are the number of vias and contacts in a design, the number of metal layer cross-overs and the like.
  - **The Yield model.** Many different yield models have been used in the industry for estimating yield affected by random defects. The fundamental difference in the models is the assumed distribution of the random defects. Simple models, such as the Poisson and the Murphy models using the die area as the critical area, were prevalent in the early days. The Bose-Einstein model using die area and identifying a defectivity per critical layer has been used extensively in recent years [7.1, 7.2]. IDMs generally have internally calibrated, custom models. Equations for the models are shown in Appendix B.2.

A comparison of the yield prediction from the Murphy, Bose-Einstein and the Poisson models for three different defect densities (0.2, 1, 2 per square centimeter) as a function of die area is shown in Fig. 7.4. For small defectivities (0.2/sq cm), all the models predict similar yield results. At a defectivity of 1/sq cm, there is significant deviation between the models. The Murphy model is the most optimistic (highest yield) and the Poisson model is the most pessimistic (lowest yield). Deviations of Murphy and Poisson yield predictions from the Bose-Einstein model are plotted in Fig. 7.5. It is observed that for large die sizes the deviations can be very significant. Even for a 100 sqm die the deviations can be around 5%. Commonly used defectivities for modeling yield on mature process technologies in world class fab facilities are 0.3–0.5 per sq cm. Notice that for relatively small die, under 100 sq mm, any one of the three models provides a fairly accurate prediction of yield.

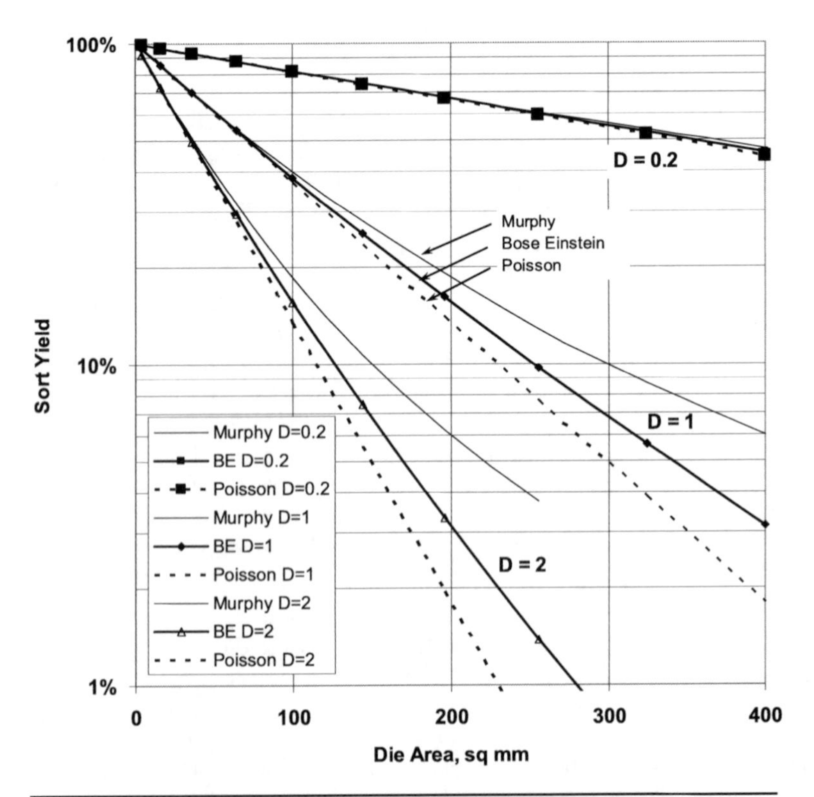

FIGURE 7.4 Yield predictions from the Murphy, Bose-Einstein, and Poisson models as a function of die size, for defectivities of 0.2, 1, and 2/sq cm.

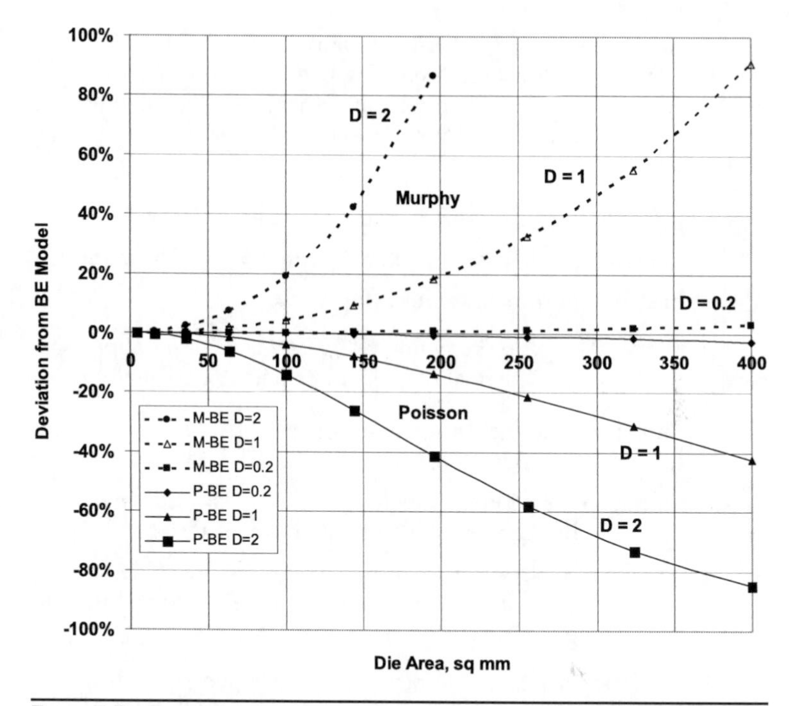

FIGURE 7.5 Comparison of Murphy and Poisson yield deviations from the Bose-Einstein model as a function of die size for defectivities of 0.2, 1, and 2/sq cm.

Most commercial foundries use the Bose-Einstein model—see Appendix B.2. The two key parameters provided by the foundry are:

- D_o, defectivity per mask layer per square inch;
- n, complexity factor.

If the foundry does not provide it, the fabless company will need to estimate a time dependence of D_o. See Fig. 5.2 for a trend of defectivity learning at Intel for four generations of process technologies. Notice that at the leading edge of a new technology, the learning curve is steep. The fabless company also needs to estimate the die area or the critical area for predicting sort yield over time.

Having calculated the GDPW and predicted the sort yield, the NDPW can be calculated.

7.1.3 Die Cost

Knowing NDPW, the number of Net Good Die Per sorted wafer, it is possible to calculate the die cost, since the basic equation for predicting the cost of an IC die is:

Die Cost = (Wafer Cost + Wafer Sort Cost) / NDPW

This presumes that the fabless company knows the cost of a wafer that the foundry will charge them. Predicting the wafer cost before availability of a budgetary quote or a negotiated deal with the foundry can be a challenge. Here are some ideas for sources of wafer cost information:

- GSA wafer pricing survey conducted quarterly;
- knowledgeable industry sources with current information;
- budgetary estimate from the foundry.

Another challenge in predicting wafer cost is its variability, especially in recent years. The major factors that affect this variability are:

- industry cycles of supply and demand affect factory utilization and available capacity;
- within any one foundry there is usually a variation in pricing favoring the high volume and long-term customers;
- competition between foundries affects pricing. The smaller foundries anxious to acquire a piece of high volume business tend to use low cost as one possible differentiator.

As mentioned in Chap. 6, a start-up should expect some difficulty in securing an account at the suppliers, and especially in securing favorable wafer pricing. Overall, it helps to have a network of connections and reliable sources.

A few words are appropriate here about how foundries set wafer pricing. The foundry needs to cover its wafer cost, which is determined by the cost of facilities, equipment depreciation, materials, labor and processing cost. IC Knowledge [1.15] has developed a model that can be used to estimate the manufacturing cost of a wafer. In additions the foundry needs to cover its overhead, sales, G and A and other miscellaneous costs. The wafer price then is calculated using the target profit margin.

While detailed modeling of wafer fab costs is a complex task, a simplified model provides approximate breakdown of a foundry's wafer selling price to a fabless IC customer into the following components. Note that different depreciation schedules are used for amortizing facility and equipment cost. Also, not all equipment is procured at the same time, as equipment is upgraded every few years. Therefore, the precision of the numbers varies during the lifecycle of the fab. From this information, the most important takeaway for the fabless company is that a significant portion of wafer cost is capital depreciation. Depreciation becomes a smaller fraction of wafer cost as the fab and the equipment ages.

- Depreciation ~25 %
- Cost of Goods Sold (COGS)
  - Materials ~20 %
  - Labor – direct and indirect ~ 8 %
  - Utilities and Maintenance ~ 8 %
- Gross Margin **~39 %**
  - SG and A ~12 %
  - Other costs (start up, IP, etc.) ~ 7 %
- Operating Profit **~20 %**

Glen Possley, PhD
Semiconductor Industry Veteran

Another component of DC is the Wafer Sort (WS) cost. Estimation of WS requires an assumption of the cost of ATE being used for the project. Components of WS are capital depreciation cost, operating cost and manpower cost. Typical numbers can be $70–400 per hour. The numbers are highly variable since they depend on the tester being used and the location of the test operation. The ATE depreciation and operating cost is used to calculate to a per second cost of tester time. In addition to the actual test time, the fabless company is also charged for wafer set up time and the "index time." The index time is the dead time between probing one die and the next.

Note that in this model, the cost adder for the visual, "1st optical" inspection is being ignored for the sake of simplicity. There is indeed a cost associated with this inspection; however, it is a small number.

7.1.4 Package, Assembly and Test Cost

Many factors determine the cost of packaging, assembly, and test. Examples of these factors are:

- package type—single chip, multi chip, SiP (System in Package), plastic or ceramic, BGA, QFN or MLF, DIL (Dual In Line)—there are a myriad;
- number of pins or balls;
- wirebonded or flip chip;
- ATE;
- clock speed of the IC;
- testing at multiple temperatures;
- unit volume;
- prototyping or production test;

- stress testing, if applicable;
- geographical location of the operation.

Predicting the numbers for package, assembly and test cost can have a challenges similar those in predicting wafer cost. Approaches outlined in the previous section are applicable for packaging, assembly and test. A general rule of thumb for package and assembly cost is approximately 0.5 to 1cent/pin (or ball) for plastic, leadframe, or BGA packages in high volume production. Test costs for ATE test are calculated in a manner similar to the wafer sort cost.

As illustrated in the next section, package and assembly, cost is 10% or 38% of unit cost in two examples. The more significant the package cost, the more due diligence is recommended in refining cost estimates.

7.1.5 Sample Yield and Cost Model

Now, let us illustrate yield and cost calculations with some examples. Table 7.1 shows calculations for a 130 nm CMOS process technology

	130 nm 1P8LM	130 nm 1P8LM	130 nm 1P8LM	130 nm 1P8LM
X mm (chip edge)	3	6	10	14
Y mm (chip edge)	3	6	10	14
Die Area (sq mm)	9	36	100	196
Wafer diameter (mm)	200	200	200	200
Edge ring (mm)	4	4	4	4
Gross die per wafer	2889	715	246	118
Logic D_0/layer, /sq inch	0.20	0.20	0.20	0.20
n	12.5	12.5	12.5	12.5
Sort yield bose-e	96.2%	85.7%	65.5%	44.3%
Net die per wafer	2780	613	161	52
Wafer cost (8LM)	$1600	$1600	$1600	$1600
Die cost	**$0.61**	**$2.75**	**$10.45**	**$32.09**
Package pins	44	362	400	896
Pkg cost/pin, c/pin	0.50	0.50	0.60	1.20
Package & ass'y cost	$0.22	$1.81	$2.40	$10.75
Test cost	$0.08	$0.12	$0.12	$0.25
Unit cost	**$0.97**	**$5.03**	**$14.24**	**$47.75**

TABLE 7.1 Comparison of Yield, Die Cost, and Unit Cost of ICs with Various Typical Die Sizes and Packages for 130 nm CMOS Fabricated on 200 mm Wafers

fabricated using 200 mm wafers and die sizes of 3 × 3, 6 × 6, 10 × 10 and 14 × 14 square mm. The GDPW is calculated as 2889, 715, 246, and 118 for the four die sizes respectively. The yield is calculated using the Bose-Einstein yield model with defect density and complexity factors typical for a mature process. The sort yield is a strong function of the die size—96.2%, 85.7%, 65.5%, and 44.3%, respectively. Using the average wafer price ($1600) from GSA's Q4 2006 wafer pricing survey [7.3], the unprobed die cost is calculated to be $0.61, $2.75, $10.45, and $32.09 for the four die sizes. Note that the cost of the die increases as more value is added (wafer sort, packaging, final test) and as there is yield loss at each step, albeit small. The finished unit cost is $0.97, $5.03, $14.24, and $47.75 for the four ICs. Different package types and pin counts have been assumed in these four examples. The parameters chosen are typical of real ICs evaluated recently. The four packages used are a 44 QFN (Quad Flat pack No lead), a 362 PBGA (Plastic Ball Grid Array), a 400 PBGA and an 896 FBGA, respectively.

The yield model results in Table 7.1 have been reproduced for 300 mm wafers, and are shown in Table 7.2. The wafer price used is the average price from [7.3]. The finished unit cost is now $0.94, $4.86, $13.38, and $44.29 for the four ICs. This data shows some reduction in die and unit cost when using the larger diameter wafers. This is consistent with the expectations outlined in the scaling section in Chap. 5. The reduction in die cost and unit cost is more significant for the larger die sizes. This is to be expected because the larger wafers allow a better optimization of die steppings on the wafer; there is less fallout of large edge die on the larger diameter wafer.

Next let us illustrate the impact of package pin count and package type on unit cost, as shown in Fig. 7.6. For example, let us consider a 6 × 6 sq mm die fabricated using 130 nm CMOS process on a 200 mm diameter wafer. When the units are packaged in a 362 PBGA, the die cost is 59% and the packaging cost is 38% of the unit cost. Instead, if the die is packaged in a 10 × 10 mm, 64 QFN package the unit cost drops from $5.03 to $3.44. In the QFN package the die cost is 85% of the unit cost while packaging cost drops down to 10% of the unit cost. The significance of this illustration is that the reader must carefully evaluate options when selecting the package pin count and package type. Packaging cost could make a significant contribution to unit cost. This is unlike the old days where the IC cost was mostly dominated by silicon cost.

The yield and cost model can also be used to predict the die cost and unit cost as one considers more advanced technologies. Figure 7.7 shows a graph of how the die cost component of finished unit cost increases as one fabricates a 6 × 6 sq mm die on 300 mm diameter wafers in 130, 90, 65, and 45 nm process technologies, relative to the 180 nm node. This data makes it appear that the more advance technologies are more expensive, and therefore unusable. It is indeed true

	130 nm 1P8LM	130 nm 1P8LM	130 nm 1P8LM	130 nm 1P8LM
X mm (chip edge)	3	6	10	14
Y mm (chip edge)	3	6	10	14
Die area (sq mm)	9	36	100	196
Wafer diameter (mm)	300	300	300	300
Edge ring (mm)	4	4	4	4
Gross die per wafer	6780	1704	599	295
Logic D_0/layer, /sq inch	0.20	0.20	0.20	0.20
n	12.5	12.5	12.5	12.5
Sort yield bose-e	96.2%	85.7%	65.5%	44.3%
Net die per wafer	6524	1461	392	131
Wafer cost (8LM)	$3600	$3600	$3600	$3600
Die cost	**$0.58**	**$2.59**	**$9.66**	**$28.94**
Package pins	44	362	400	896
Pkg cost/pin, c/pin	0.50	0.50	0.60	1.20
Package & ass'y cost	$0.22	$1.81	$2.40	$10.75
Test cost	$0.08	$0.12	$0.12	$0.25
Unit cost	**$0.94**	**$4.86**	**$13.38**	**$44.29**

TABLE 7.2 Comparison of Yield, Die Cost, and Unit Cost of ICs with Various Typical Die Sizes and Packages for 130 nm CMOS Fabricated on 300 mm Wafers

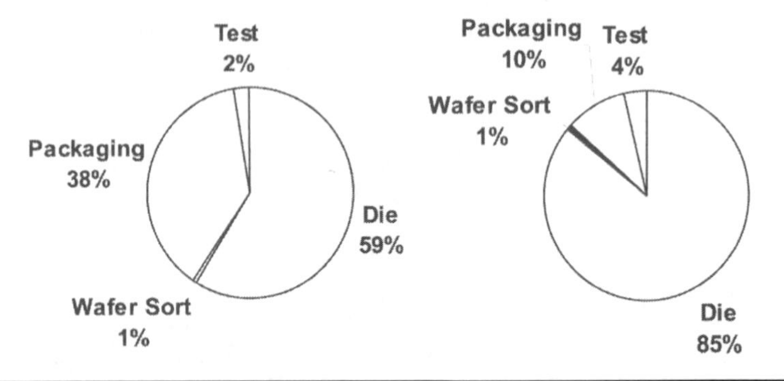

FIGURE 7.6 Packaging cost is a larger fraction of unit cost when using a 362 PBGA package versus a 64 QFN package. Assumptions are a 6 × 6 sq mm silicon die in 130 nm CMOS process.

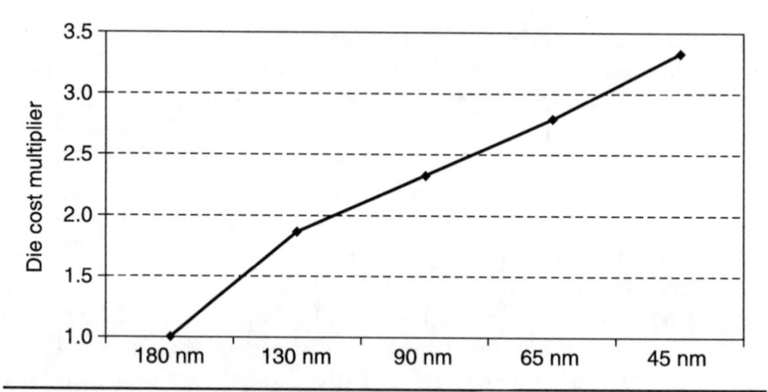

FIGURE 7.7 Die cost multiplier for 300 mm wafers, various CMOS process technologies. Same packaging and defectivities are assumed.

that a 6 × 6 sq mm piece of silicon in 45 nm technology will be 3.3 times more expensive. However, as discussed earlier in the book, there can be over 10 times as many transistors on the same size silicon in 45 nm technology, when compared with 180 nm. So, the cost per transistor decreases in the more advanced technologies. A more detailed discussion of how to optimize cost per gate in the various technologies will follow later in this chapter.

7.1.6 Some Simple Guidelines

The purpose of illustrating the yield and cost model information in this section has been to provide a flavor of the kinds of cost trade-offs that are possible and important. There are too many degrees of freedom in the selection of the parameters to be used in the model. Wafer prices, for instance, change frequently due to market fluctuations, they vary by supplier, and volume and by the "clout" of the customer. Defect densities generally improve as the technologies mature. Package prices have similar dependencies on market fluctuations, competition, the manufacturing learning curve, and the like. In order to have an accurate estimate the fabless company must have access to knowledgeable personnel with current information. It is important to get the model right. It is even more important to input the right parameters in order to get meaningful results.

7.2 Optimizing the Die Size and Packing Density per Chip

Selecting the optimum packing density and the die size becomes a challenge in this dynamic industry. Some of the die size trade-offs have to be made early in the development cycle, maybe one to three years before the high volume demand! Keeping in mind that a new technology node is introduced every two years, and wafer prices fluctuate, making predictions and trade-offs is a little tricky. The models presented here can be a start. Studying the sensitivities can help in making knowledgable decisions.

Models to predict the optimum die size and functions per chip have been developed and illustrated here. In Figs. 7.8 and 7.9 are shown examples of the cost/gate, in 2008, for 180, 130, 90, 65, and 45 nm nodes as a function of die size and packing density in millions of gates per chip. The curves illustrate a U-shape. If the die size is too small the cost is dominated by the overhead of the input/output structures, the scribe lane, etc. If the die size gets too large, the cost per gate increases due to the increased complexity. The shape of the curve is affected by parameters such as wafer cost, defect density, physical and electrical design rules, design tools' packing efficiency. Notice that the decreasing cost per function for newer process nodes is consistent with the analysis presented in Chap. 5.

For simplicity, gate count is assumed here to be an equivalent two-input NAND gate count. Each equivalent gate therefore uses four transistors. The optimum gate density and cost per gate can be converted to transistor density and cost per transistor. The actual transistor count per chip increases rapidly as large amounts of memory is included on the die. For reference, one of Intel®'s Pentium® processors is reported with 55M transistors (14M equivalent gates) in a 90 nm technology [1.15]. Referring to Fig. 7.9, this data point will be considered reasonably well optimized in this analysis. This is because

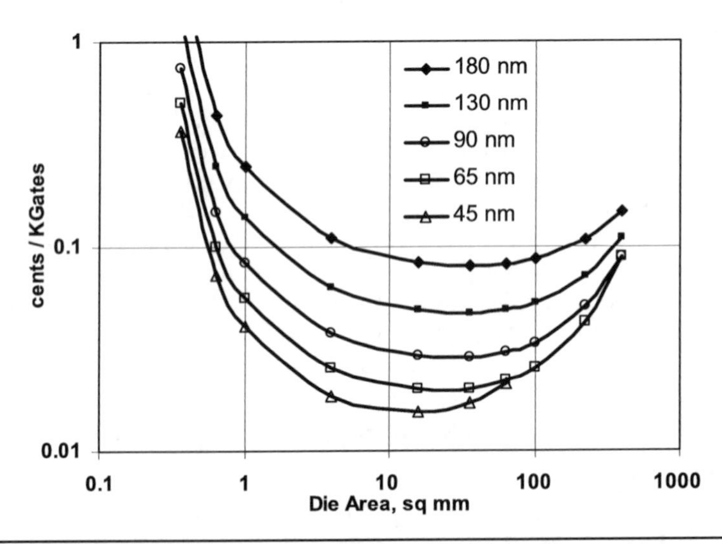

Figure 7.8 Cost per gate as a function of die size for various process technology nodes in 2008.

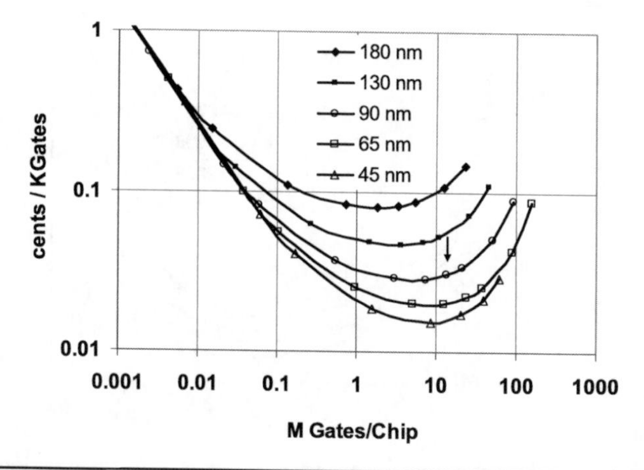

FIGURE 7.9 Cost per gate as a function of packing density (gates per chip) for various process technology nodes in 2008.

it is located near the minimum, just at the cusp of the steep slope, and is marked by the arrow. Notice also that if Intel's actual defect density and wafer cost are lower than the assumptions in this analysis, the curve will flatten out and then the arrow could point to the very bottom of the curve.

Figure 7.8 shows that the cost/gate decreases progressively for newer process nodes—the 45 nm has the lowest minimum cost/gate in this set of technology nodes. However, note that the width of the "sweet spot" region is narrower for 45 nm than 65 nm and larger nodes. The sweet spot is defined as the region bounding the minimum point of each curve. As the new technology matures, the wafer prices and the defect densities drop. This causes the curve to flatten out. This effect can be observed in Figs. 7.10 and 7.11 for the 65 nm node and Fig. 7.12 for the 45 nm node, where cents per gate is plotted for years 2007–2011. In these figures projections have been made of expected wafer cost and defect densities for the out years.

In each of the Figs. 7.9–7.12 it can be observed that the minimum point moves lower and to the right. The implication is that it is more cost effective to fabricate larger die with more gates per chip in more advanced nodes, and the cost gets lower over time. The ability to achieve this has been a cornerstone of the semiconductor industry and has been a driver for designers to move their designs to more advanced nodes.

As the minimum point in Figs. 7.9, 7.11, and 7.12 moves lower and to the right, significantly more gates can be fabricated cost effectively over time in the same technology. As the technology node matures, the wafer cost drops, and so does the defectivity due to manufacturing learning curve. Compare these curves with the conceptual curves

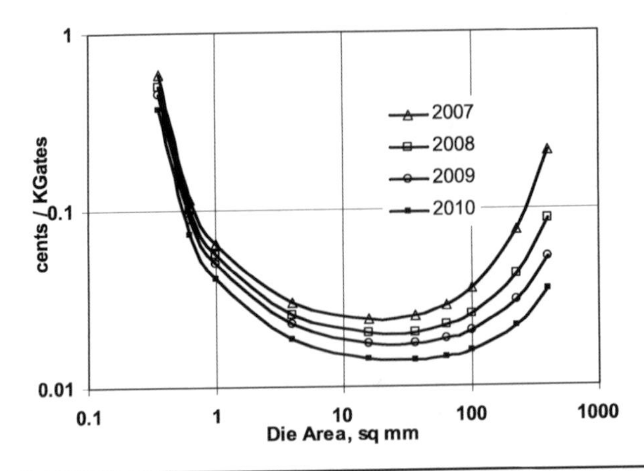

FIGURE 7.10 Cost per gate as a function of die size for 65 nm, showing a decrease over time. Larger chips can be fabricated more cost effectively over time.

shown in Fig. 5.1. Notice that the cost per gate in 45 nm drops below 0.01 cents/Kgate in 2011, and it is lower than any other node shown.

7.3 Development Cost

The analysis in the previous section illustrates the motivation to move ICs to more advanced process nodes in order to achieve the lowest cost per gate or cost per function. The analysis so far has been based on silicon die and unit cost. A complete financial analysis of unit cost must

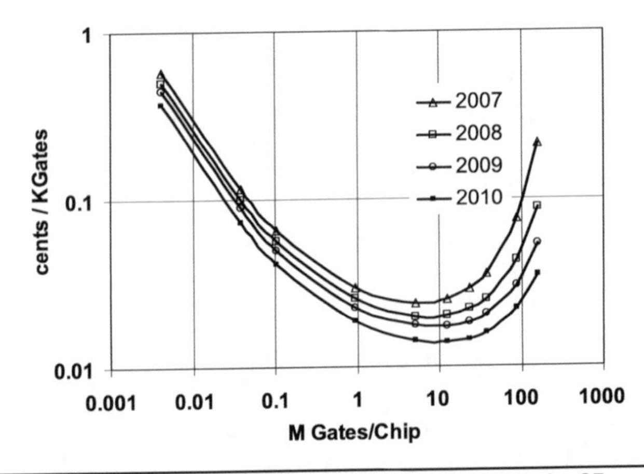

FIGURE 7.11 Cost per gate as a function of packing density for 65 nm.

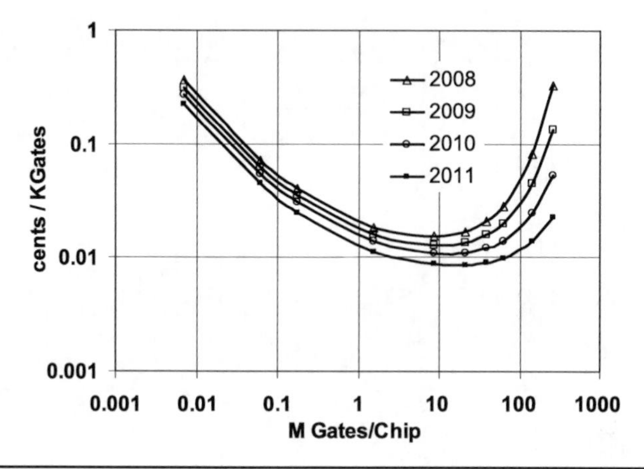

Figure 7.12 Cost per gate as a function of packing density for 45 nm.

also include an amortization of all the development costs per unit volume. It was shown in Figs. 1.9 and 1.11 that the cost of masks has been increasing steadily. As is discussed later in this section, the cost of design development and verification is also escalating rapidly. The following discussion will provide the reader some "order of magnitude" estimates of the various components of development cost. Refinements will need to be made for specific IC development environments.

7.3.1 Mask Costs

Mask set costs vary over time and are a function of the "critical" (defined later) and total number of masks used. As a new technology is introduced the mask making facility is usually pushed into the next generation of resolution and defectivity requirements. Therefore, the initial mask cost in a new technology node is very high, and the number increases for every node. The good news is that within the first year the mask set cost decreases rapidly due to the manufacturing learning curve and due to competitive pricing pressures. In Table 7.3 are shown average mask set costs for 130 nm and older process nodes based on GSA's quarterly surveys. Also shown is my estimate of cost and variance, to be used for planning purposes.

Some factors that affect mask set cost are:

- the number of process options used affects the total number of masks;
- the negotiating leverage, which is higher for long standing customers that purchase a high volume of wafers from the foundries;

$ K nm:	500	350	250	180	130	90	65	45
Q1 2005	$31	$49	$90	$206	$425			
Q4 2005	$32	$49	$75	$190	$396			
Q4 2006	$37	$40	$75	$198	$206			
Average estimate	$35	$40	$75	$200	$350	$700	$1,300	$2,000
Variance ±	10%	10%	10%	25%	30%	30%	30%	30%

TABLE 7.3 Mask Set Cost for Various Process Technology Nodes. Data Shown Is from GSA. Also Shown is the Author's Best Estimate of Cost and Variance.

- for more advanced nodes there are variations in grades of masks. The grades are determined by the minimum resolution and quality requirements. "Loose" tolerance protect masks can be fabricated less expensively. "Critical" mask layers require the tightest control on minimum resolution, quality and tolerance. Masks on critical layers in process technologies below 130 nm also require special operations such as OPC.

7.3.2 Prototyping Cost

As discussed in section 6.3.1, there are two options to silicon prototyping—MPW and full mask.

In the MPW approach the fabless company pays $15–150K to get 20–40 prototype parts of their design. When the foundry uses 300 mm wafers for the fabrication, expect the number of available parts to be 50–100. The price varies depending on the technology node. An approximate rule of thumb is 10% of the mask set NRE for mainstream technology nodes (250, 180, 130, 90 nm). Check with the foundry or MOSIS or EuroPractice for the specific options, restrictions, and cost. The parts are usually available in die form or as packaged units.

In the full mask approach, the fabless company makes arrangements with the foundry directly to run an engineering lot. The charges for this will be detailed in a quotation and are usually a fixed charge, which has the following components.

- Mask cost, as discussed in the previous section.
- Silicon cost for the 12 wafers (200 mm), or 6-12 wafers (300 mm). A good starting estimate of the fixed lot charge is to apply a 25% premium on the low volume wafer price that the foundry quoted you for production pricing.

In addition one must estimate the cost of assembly and test. Assembly cost should be detailed in a quote from one of the SATS. There is usually a fixed lot handling charge in addition to a per unit charge. Depending on the package type, budget a few thousand dollars. Test cost is discussed in Sec. 7.3.4.

7.3.3 Design Costs

As the complexity of ICs has escalated, so has the cost of designing them. And because of the high cost of masks and silicon on the leading edge process nodes, it is imperative that the design be thoroughly verified before releasing it to the wafer fab.

The industry is usually focused on and enamored by leading edge design, design tools, process nodes, process equipment, etc. This has been a good motivator over the last 40 years, since it drove the competitive spirit, sparked innovations, and led to electronics components and products with more features, better performance, and lower cost. As the semiconductor industry is maturing there has been much gloomy talk, especially related to escalating costs of IC design, wafer fabs and the fabrication process. As pointed out in Chaps. 1, 5 and 7 the industry has learned to deal with increasing cost of silicon/sq mm because the cost per function is indeed better when using the newer technology nodes. A similar concept needs to be applied to design development costs.

Recent estimates of IC design costs indicate that they are going "sky high" [7.5, 7.6, 7.7]. Information from these references is summarized in Fig. 7.13. The publications indicate that a 45 nm design costs approximately $45M, and that approximately 70% of this cost is expected to be for the design, verification, and layout. Cost for software, test/product engineering, and masks/silicon are all increasing. No further details or assumptions are provided.

A plausible explanation for this increase in cost is as follows. Today's nanometer, leading edge SoC ICs have complex functionality

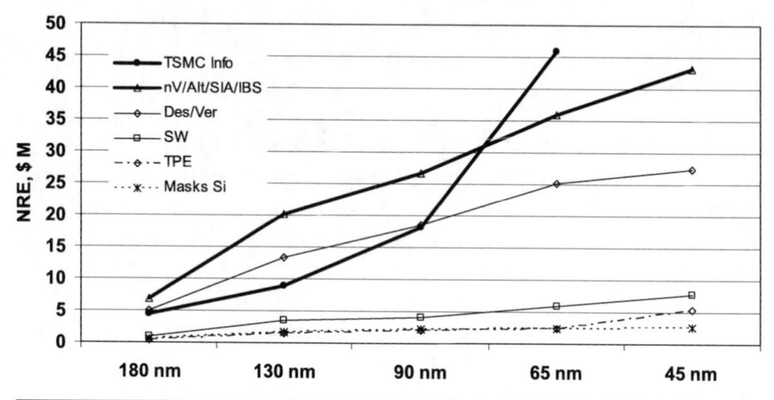

FIGURE 7.13 Reported information about increasing design development cost.

Design investment	$50M
Expected ROI	$500M
Product lifecycle	2 years
Annual revenue	$250M
Unit price	$5
Required annual unit volume	50M

Table 7.4 A Simplified Illustration of the Need to Ship at Least 50M Units Annually to Support a $50M Design Investment

and pack hundreds and thousands of times more power and performance relative to mainframe computers and other systems developed 10–20 years back. The cost of developing these SoCs should be compared to the cost of developing whole systems of yester year. The system level and logic level verification that is necessary in the development of today's SoCs, is work that was traditionally done by system designers. What IC designers did in previous generations of design was really the physical design portion of the design of present ICs.

The problem with the information in Fig 7.13 is that it creates a perception in the industry that designing on the leading edge process nodes is only for the large companies. It is implied that only companies with large resources and the ability to make large investments can design products on the leading edge nodes.

It is indeed true that products requiring such large investments must have very high unit volumes in order to justify the investments. Let us illustrate this point with a simple example, as shown in Table 7.4. If one assumes a design investment of $50M and the need for a 10 fold return on investment during two years of product shipments, the annual return will need to be $250M. Assuming a unit selling price of $5, the company must ship 50M units each year. Unfortunately there are only a few markets that require IC shipments this large. It is also difficult for the start-up fabless company to be a contender in such an exclusive market.

7.3.4 Is There Hope for the Small Fabless Company?

The following discussion will attempt to dispel the notion that the startup or small fabless company can never design in 45 nm or 65 nm process nodes in the early years of the technology lifecycle (2007–08). To facilitate this discussion, a new approach is suggested here. Consider NRE cost/gate as a parameter in making design and technology decisions. In keeping with the silicon concepts related to a cost per function in the manufacturing context, using the NRE/gate as a development cost metric can be enlightening. The importance of this is illustrated in Fig. 7.14 shown is a plot of NRE/gate based on

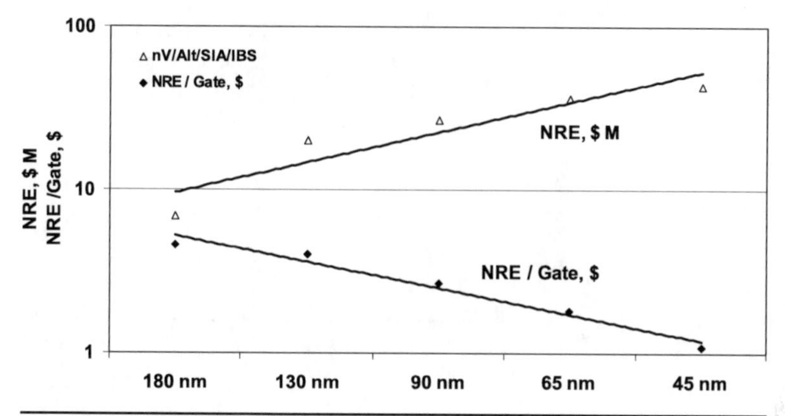

FIGURE 7.14 Published NRE data [7.5–7.7] indicates increasing total design cost. However, per gate NRE is shown to decrease.

the maximum expected gate packing density. What can be observed is an exponential decline in the NRE/gate from one process node to the next. This is good news! It is true, however, that the fabless company still needs to be able to make the large investment. Note that, on the silicon side, the foundries make the capital investment and the fabless IC companies are the beneficiaries.

The following analyses will show two very key opportunities to lower design development cost:

- starting the design effort one to two years after the very first design start;
- backing off slightly on the packing density and performance requirements.

Of course one must make these choices judiciously. Make sure you still have a viable product. The product must have differentiating features other than the use of leading edge process technology node.

To illustrate the impact of designing at the leading edge, consider the following. The cost to design the first chip in a new technology node is indeed very high. Design of subsequent designs using the same design infrastructure and the same architecture will cost less. This is because of many new elements that must be included in developing the required design infrastructure concurrently with the first design:

- a new architecture, implemented possibly as proprietary IP blocks;
- new software;
- new process models and the associated refinements normal in the early stages;

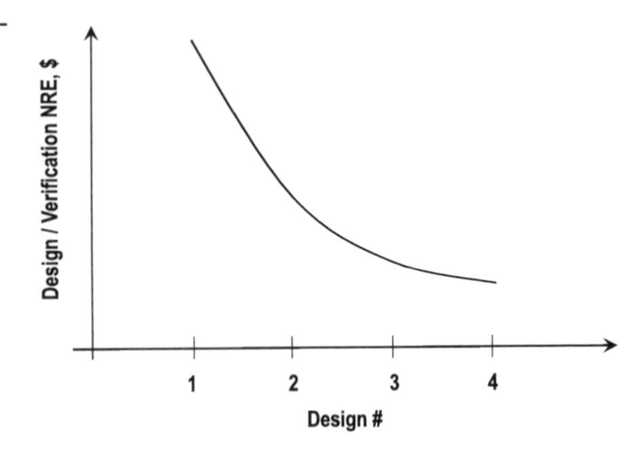

FIGURE 7.15 Design learning curve influences the cost of design.

- a new library;
- new IP blocks, both off the shelf and custom;
- new EDA tools, a new design flow and maybe new design methodologies.

It is postulated here that within the same fabless company, when using the same architecture, the second time those blocks are implemented, the design and verification effort will drop significantly. Figure 7.15 is a conceptual chart that captures this effect.

Several approaches can minimize the design verification cost for complex system chips:

The more pre-verified blocks are used in a design, the less the effort required to verify the total design. Derivatives of complex designs with limited modifications at the block level will require significantly reduced verification costs. This is because verification will need to be focused mainly on inter-block relationships and system-wide operation.

- Embedding assertions into the design code (System Verilog, Verilog, VHDL, etc.) makes the design verification more complete. The assertions stay with the block and continue to provide verification coverage without additional design effort. It makes the IP block self-verifying when it is reused in another design or when the design is tested as part of a larger system.
- Just as design blocks can be re-used, so can test benches. Unit tests for reusable or other design blocks are part of the total design database and could be incorporated in the system test bench. A system test bench, in turn, should be reusable at multiple levels of design abstraction, from architectural to RTL.

- Most simulation (or emulation) effort is spent re-testing functions that have already been verified. Reductions of verification cost can result from the use of methodologies and tools that specifically treat interface verification as distinct from functional verification. These can be used to create test benches that provide orthogonal testing, i.e. simulations that exercise logic that has not already been tested. Such "algorithmic test benches" are becoming available from many sources and promise to increase the effectiveness of verification and greatly reduce its cost.

Walden Rhines, Ph.D
Chairman and CEO
Mentor Graphics

Therefore, design cost can be lowered significantly by designing with a validated design infrastructure, one or more years after the release of the very first design in a new technology node. We will illustrate this through an example, shown in Fig. 7.16. It is important to focus on the trends in this analysis, and not the precision of the numbers. Each company embarking on such analyses must determine the right numbers for their own situation. Consider the 65 nm process node. The first commercial designs were taped out in 2005. At that time the lead designers had to develop their own library, memories, and IP blocks. A large fabless company designing at the leading

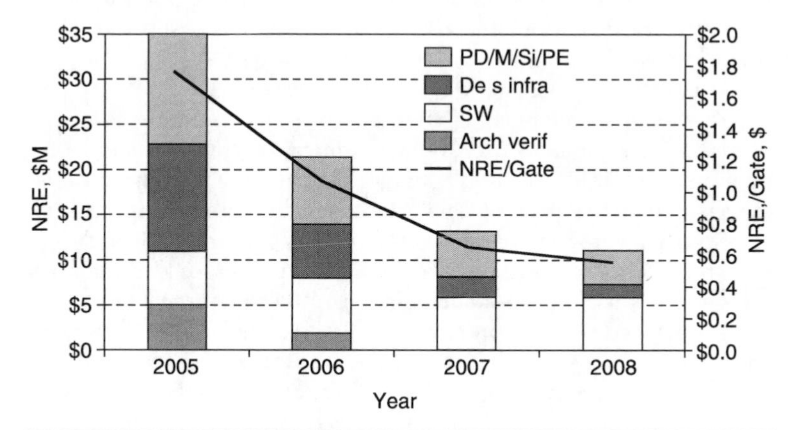

FIGURE 7.16 Simplified estimates of Design NRE cost and NRE/Gate for the 65 nm node. Design in 2005 would represent an industry leading design; Design costs decrease due to maturity and learning. A fixed, 20M gate complexity is assumed.

edge with its proprietary architecture and proprietary IP blocks had to invest much resource verifying the architecture and the design. In this model the various design costs have been adjusted to match the total, $35M published number [7.5–7.7]. The software development cost was also adjusted to be at the published $6M. The architectural verification cost drops in the second year for a follow on design. The rest of the cost—the physical design (RTL and Netlist to GDSII), masks, silicon and product engineering and tests were also assumed to be high because changes to the models, the libraries, design flows and tools are all to be expected. Multiple spins of the masks and silicon were assumed.

As a new technology node is announced to the general user community, the foundries and third-party IP providers invest in developing IP blocks in that technology. EDA companies make investments in design tools and flows. Design services providers gain some know-how and experience with design tools. So design implementation one year after the initial designs had some learning and maturity and therefore costs a little less. In 2007, about a year and a half after the initial design, validated IP, models, tools, and flows were available. In this model, a 20M gate design could be implemented for $13M, including the $6M software development cost. And the number gets slightly lower in 2008. The line on this graph indicates a reduction in NRE cost/gate from $1.70 to about $0.60. Note that this is a very impressive metric, considering that in 1986 it cost $50K for a 50K gate array design to do just the netlist to GDS physical design ($1/gate)!

Now let us look at the reduction of design cost by backing-off on chip complexity. By implementing a 65 nm design at the low end of 5–20M gate "sweet spot" shown in Fig. 7.9 and 7.11, one can lower the NRE cost, as shown in Fig. 7.17. In this analysis, the 20M gate design in 2007 matches the corresponding data point in Fig. 7.16. Keeping the software development cost fixed, this analysis shows a $2M saving in NRE if the design had only 5M gates. As expected, the NRE/gate increases as the complexity decreases.

This analysis offers some alternatives to the emerging company in making decisions to use the latest process technology nodes for their designs. Unlike the perceptions created by the referenced publications, the picture is not that bleak for the emerging companies. By judiciously leveraging off-the-shelf IP blocks, mature models, design flow, tools, libraries, and IP, choosing mainstream (not leading edge) design complexity, it should be possible for the emerging company to design in a leading edge process node. For example, limiting the gate packing to approximately 70% of the full technology capability, it may be possible to develop a 65 nm IC for $5M plus the cost of software development. This data has been verified against some design under way in mid 2007.

The following are some example budgetary development cost ranges (excluding software development cost). The low end of the cost range

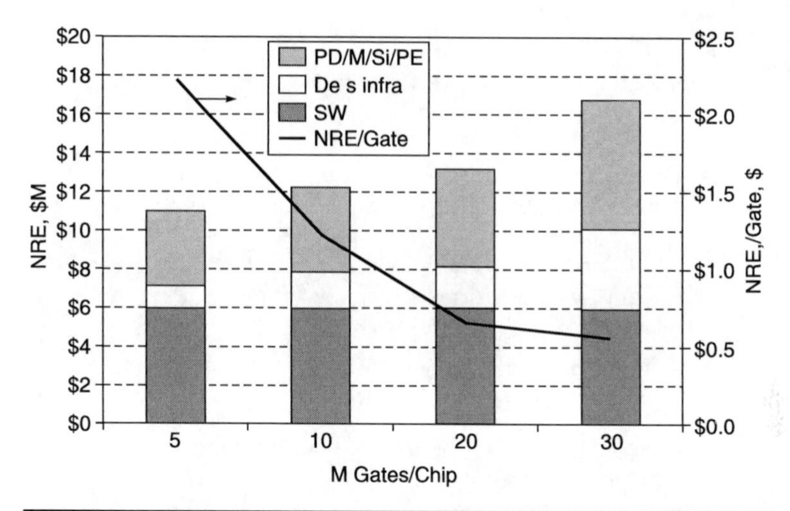

FIGURE 7.17 Simplified estimates of design NRE and NRE/gate for 65 nm node in 2007. Design costs decrease as gate complexity is decreased. However an increase in the NRE cost/gate is observed for less complex ICs.

assumes a mature chip architecture and performance goals that are well within the mainstream capability of the technology and the IP:

- 180 nm $1–5M
- 130 nm $2–8M
- 90 nm $3–15M
- 65 nm $5–20M

Another possible approach is to do an initial design in an older technology, such as 130 nm and then migrating it to a newer process node, or a half-node. There can be some unique problems with such transitions, but they are certainly worth considering. Costs for such implementations could be at or below the low end of the ranges provided.

There are a variety of business models for procuring IP. Libraries are commonly available from the foundries at no up-front charge. The foundries usually include a charge in the wafer price. Some memory blocks and commonly used IPs are also made available by the foundries. Specialty blocks can be designed by the fabless company or can be designed by IP and design services providers. Such services are usually on an NRE basis. Commonly used embedded microprocessors such as the ARM and MIPS need to be licensed for a fee. The fees are usually a combination of an NRE charge plus a small royalty based on IC unit sales.

It is my hope that from this discussion readers will desire new and creative ways to combat the high cost of design and continue to aggressively implement their ideas in silicon.

7.3.5 Test Costs

Development cost associated with test comes in the following categories:

- Design for Test—this cost is accounted for in the design development cost. Activities include insertion of testability features such as BIST (Built In Test) for memory and logic blocks, ATPG test vector generations, and simulations to verify functionality and performance against the functional and scan vectors.
- ATE test program creation and validation.
- Prototype testing.
- Characterization of the design operation in process, voltage and temperature "corner" limits.
- Production program development. Activities including generation of "guard band" limits that allow each part to be tested at nominal conditions. With the proper limits in place the part would meet operating conditions in the "corners."
- Screening methodology to weed out marginal and weak parts. Tests in this category are a leakage test called "I_{DDQ}" and accelerated voltage testing to screen out infant mortality fails;
- Hardware cost for load boards, sockets and automatic handlers.

Estimation of development cost has three major components—tester time, manpower effort and tooling hardware cost.

ATE tester cost has been relatively flat over the last 10+ years. A high-end tester with the maximum number of pins, the highest operating frequency and the maximum number of allowed test vectors costs $3–5M. There are also low cost and specialized testers available below $0.5M.

SATS as well as independent test houses usually quote a cost of test time per hour. The actual number is a function of the tester used, the manpower and overhead cost, the location of the test facility and market/demand fluctuations. The capital cost of the tester is amortized using an appropriate depreciation schedule. Hourly rates can vary between $32 and $800, with an average of $110 and a median of $120. The data is taken from GSA's Q4 2006 report [7.3]. At $100/hour, the cost of test is 3¢ per second of test time.

The manpower cost for development tests can vary significantly depending on the type and complexity of the design. Complex digital, analog, and RF designs can consume much engineering time.

Support hardware costs for the load board, sockets, fixtures, etc. are usually in the $10–25K range.

7.3.6 Package Development Cost

Package development costs can also vary significantly depending on the type of package and the number of connections (pins, leads or balls). Mature or mainstream packages can have very low NRE cost associated with their development. Here are some examples of such packages:

- Mainstream:
    - less than 300 PBGA with ball pitch greater than or equal to 0.5 mm;
    - leadframe packages such as QFN with less than 65 leads.
- Mature:
    - quad and dual leadframe packages.

NRE for mainstream packages could be in the $10–50K. The cost includes the design and fabrication of a custom leadframe, a custom routing (BGA), and electrical modeling to assess performance match to the application. The lead time for the design and prototyping of a new leadframe could be long, so plan on getting this started early in the development cycle. NRE for mature packages could be less than $10K.

Leading edge packages are ones that push the limits on parameters such as pin count, lead pitch, wirebond pitch. There are also continual enhancements of package types that reduce form factor—board footprint and height. Examples are the evolution of PBGA ball pitch from 1.27 to 1 to 0.8 to 0.5 to 0.4 mm. Package height has been reduced from a few mm to under one mm. Applications such as slim-line cellular phones and portal multimedia players are driving rapid evolution of packaging technology. One must budget for customization of the package for their application as well as additional engineering effort to characterize and qualify the assembly process and the new package. In addition there could be electrical, thermal and mechanical modeling required to customize the package design for the application environment. The range of development cost of leading edge packages could be $30K to $100K, or higher, depending on the complexity.

7.3.7 Qualification Cost

Qualification costs include the development of the life-test boards, other hardware and the cost of performing the lifetests. A discussion of the tests and cost is included in Chap. 8.

7.4 Development and Operations Costs

The development and operations skill-sets required at a fabless IC company were discussed in Chap. 6. Table 7.5 illustrates of an "order of magnitude" estimate of the number of resources required. There is a comparison of resources required to support an ASIC or a COT approach. Words of caution are that the numbers shown could vary significantly based on the particular situation at the fabless company. Factors that can affect the numbers are the maturity of the technology node, design complexity, design features, design classification (leading edge, mainstream, or mature). These numbers assume that the front end, the back end design and the analog block design are implemented in house. The numbers are applicable to a 10–20 M gate design in the 65 or 90 nm process nodes. Another assumption is that there is one design started every year at the fabless company. The infrastructure, once established at the fabless company, can be scaled up to implement multiple design starts. Note that the scale factor is not linear—to implement two designs/year, the resources required will be less than twofold. Even when using an ASIC supplier, it is assumed that the front end design is implemented in house. Fractional resources in the model represent either shared or contracted resources.

	ASIC supplier	COT In-house design 1 Design/Yr @10M Units/Yr
DESIGN ENGINEERING:		
Front end		
Architecture	1	1
RTL	4	4
Verification	6	6
Back end		
Physical design – FP, P&R, Verification		5
Flow, Lib, Mem, IP, Tapeout		3
Analog		3
Design engineering sub-total:	11	22
SUPPLY CHAIN INTEGRATION		
Program & SC management	0.5	1
Silicon		1
Foundry relationship management		
Process e-test, models, yield		
Acceptance criteria, test, disposition		

	ASIC supplier	COT In-house design 1 Design/Yr @10M Units/Yr
Packaging / assembly		0.5
Assembly relationship management		
Packaging expertise		
Bond diagram, build instructions, etc.		
Product/test engineering		2
Test relationship management		
Test program & hardware development		
Design validation/characterization		
Debug/failure analysis/FIB repair		
Yield engineering		
Product certification		
Qual plan, reliability assessment		1
SC integration engineering sub-total:	0.5	5.5
MANUFACTURING ENGINEERING:		
Silicon	0.5	0.5
Packaging/assembly		0.5
Product/test engineering		0.5
Quality & reliability	0.5	0.5
Manufacturing engineering sub-total:	1	2
BUSINESS PROCESSES		
Legal	0.2	0.2
PP&C	1	1
Financial	0.3	0.3
Customer support	1	1
Manufacturing engineering sub-total:	2.5	2.5

TABLE 7.5 Manpower Estimates When Using an ASIC Supplier or a COT Approach

7.5 Overall Development Cost Estimation

An example of a real design implementation is shown in Table 7.6. This was a mainstream design, implemented in the 180 nm process node in 2006–07.

7.6 Some Cost Tradeoffs

There are indeed some management decisions that can help curb the escalating development costs. I will discuss three examples here. For the first two, remember the 80–20 rule, which says that 80% of the effort, cost, and schedule is spent on the last 20% of a project schedule.

Process node (nm)	**180**
Chip size (mm)	**4 x 4**
Gates (K)	50
Analog block	custom
SRAM	1T 64Kx8
ROM	32K
OTP	1K
Package	44 QFN
FE Design [RTL to NL]	
Effort: 2MY	500
Physical Design [NL to gds]	
Outsourced	100
Manufacturing	
Masks	160
1st Si Lot	20
Package design	20
Test dev	70
Qualification	70
Total ($K)	**940**

Table 7.6 An Example of Development Cost for an IC Implemented in 2006–7

- In the physical design phase, the amount of effort needed to minimize die size could increase significantly. Keeping in mind the 80–20 rule, a judicious choice needs to be made to "cut and run." Engineers will usually do the right thing and try to minimize the last little bit of die area. The executive has to make the right trade-off between a slightly higher unit cost/lower NRE/shorter time to market and the lower unit cost/higher NRE/delayed market entry associated with continued optimization and refinement.
- Similar trade-offs can be made in the DFM verification for susceptibility to variations and the DFT context. When to cut off these optimization efforts becomes a challenge because one is trading off unit cost, development cost, time to market, and quality.
- There is a discussion at many companies about whether or not to do a functional test of the die in wafer form ("wafer sort"). The principle of finding and rejecting defective parts as early as possible is the right one. However, it may not be necessary for the fabless start-up to include wafer sort for the first implementation of a new design. If the sort yield is high and relatively inexpensive package is being used, it may never be necessary. This could save cost and effort.

Making trade-offs like these can be crucial in the success of a fabless company. Keeping the customer and market impact in mind is crucial to making the right trade-offs. It requires experienced individuals, maybe some tools, and some guidance to help make knowledgeable decisions.

7.7 Key Points

- Optimizing CPF is critical in selecting the package density and technology targets.
- Although cost per unit area of silicon goes up in each new technology node, CPF is reduced with proper selection of the IC targets. This has been a key element driving the semiconductor industry.
- Design development costs on leading edge technology nodes have escalated rapidly. These high cost make it difficult for emerging companies to pursue leading edge designs and is stratifying the industry. Alternative approaches have been outlined.
- Emerging companies must differentiate their product based on features other than the use of leading edge process technology. One possible approach is to demonstrate the product concept in a mainstream technology. Once the product is a hit with the customer, there will be implementation options in newer technologies.

CHAPTER 8
Managing Quality

Many start-up fabless companies pay little attention to quality management in the early stages of the company. It is believed that quality is something to worry about when they get into high volume. An opposite view is proposed here. If the fabless company intends to become a world-class supplier of integrated circuits, it must focus on quality in all phases of the lifecycle. This chapter provides a holistic view to building in quality in order to get an early start. Figure 8.1 shows a conceptual view of major quality and reliability activities during the lifecycle of the company. This conceptual framework is used to describe the various quality activities in this chapter.

The quality levels required are determined by the market for, and the customers of the company's products. The automotive market, for instance, has much more stringent quality requirements than the consumer marketplace. Some customers require ISO (International Organization of Standardization, http://www.iso.org) certifications. Customers in Japan have traditionally demanded high quality levels.

There is a recent horror story about LiteOn, a Taiwanese manufacturer of disk drives that had to stop selling their name brand drives [2.6]. In order to be price competitive at a very large retail store, they started cutting corners on IC reliability and quality metrics. Too many customer returns caused the retailer to disqualify them as a supplier.

Before I go much further, let me emphasize that the *fabless company* owns the quality and reliability performance of the ICs they ship with their company logo. This is in spite of the fact that a distributed supply chain was used to design and fabricate the ICs, and that parts may be shipped directly from a distributed supply chain supplier/partner to the customer. This is depicted in Fig. 8.2. As is discussed later, the fabless company's burden is reduced if they certify, and audit the quality suppliers being used for the design IP, the silicon process, the package and assembly, and test. The fabless company's burden can also be reduced by fitting their product within the 'envelope' of certification and qualification of the supplier's offering. For instance, the foundry usually maintains quality and reliability levels of their

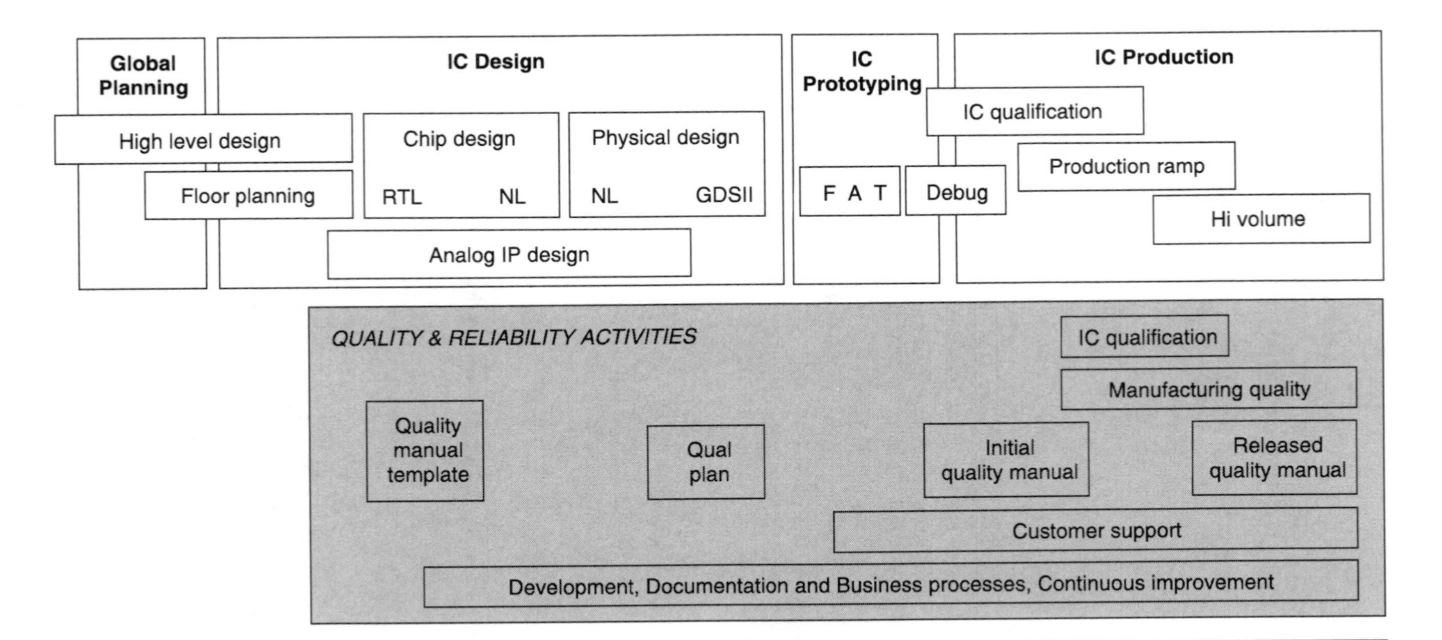

FIGURE 8.1 Summary of quality activities relative to company's IC development phases.

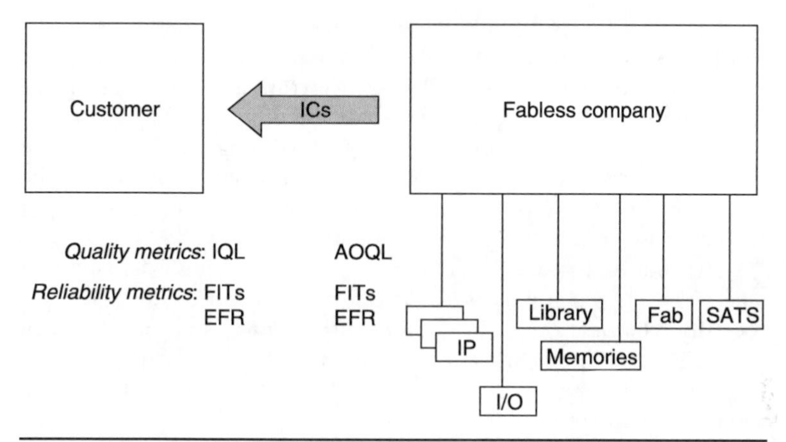

FIGURE 8.2 Simplified landscape showing the various entities in the IC value chain, as well as typical quality and reliability metrics.

baseline process. By using the standard baseline process one will not need to re-qualify the process. Conversely, by requiring the fab to make adjustments to the baseline process, the fabless company may have to perform additional qualification tests. This adds cost, and also risk, because there is generally much more focus on the baseline process. If done right, the quality and reliability levels of the fabless IC are a superposition of the quality and reliability of the individual operations. A case in point—the fabless company's "product qual" leverages information from, and is in addition to the foundry's "process qual."

A fabless semiconductor startup company has no credibility base other than the team's previous track record. It is very important for the startup to work diligently towards establishing their credibility—demonstrating achievable goals, meeting commitments, demonstrating supplier support, and commitments to quality, being a few examples. A good execution plan and demonstration of execution milestones can go a long way towards establishing credibility. Even with an excellent technical solution, the startup company must overcome many peripheral risk factors, e.g., sourcing and lack of supply chain leverage that may keep them from getting the design win, especially at large customers.

Sam Lee, Ph.D.
Semiconductor industry veteran

Let me also clarify the difference between quality and reliability. The term quality refers to the defectivity level of the IC as it is shipped to, or received at the customer. The fabless company needs to ship ICs that meet a certain AOQL (Acceptable Outgoing Quality Level). The customer usually will monitor the IQL (Incoming Quality Level) at their site. The term reliability refers to the longevity of the ICs in the customer's (or their customer's) system. Quality refers to defectivity at time zero, whereas reliability refers to defectivity beyond time zero.

So, what is included in managing quality? Quality is usually a statement of a company's philosophy. There are many example 'Quality Manuals' available online. Reviewing some of them can provide considerable insight. Good examples are ones from Intel, Infineon, Xilinx and Analog Devices [8.1–8.4]. In simple terms, the top three quality considerations are:

- customer focus and satisfaction;
- internal documentation, business processes, procedures, and discipline for continuous improvement;
- quality and reliability metrics.

8.1 Quality Manual

Early in the lifecycle of the fabless company, it is recommended that the company creates a template for a quality manual. Here is a sampling of the elements of a quality manual:

- Quality policy—usually a statement from the CEO.
- Company's quality management system:
  - description;
  - company values, mission;
  - company organization and management responsibilities;
  - documentation system:
    - document structure
    - document control processes, and procedures;
  - business processes;
  - continuous improvement processes.
- Product development:
  - goals;
  - methodologies;
  - revision control;
  - product qualification.

- Design and manufacturing supply chain:
  - partners;
  - business processes for selection, contracts, interface, and coordination.
- Product manufacturing:
  - Fab, assembly and test supplier selection, certification and audit processes;
  - quality metrics and monitoring processes for work in process and finished deliverables;
  - change notification procedures;
  - AOQL (acceptable outgoing quality level) targets and monitoring processes;
  - reliability FIT (failures in time) rates and EFR (early failure rates);
  - ORM (ongoing reliability monitoring) process and information;
  - corrective action process.
- Customer satisfaction support:
  - delivery performance;
  - customer visits;
  - customer scorecards;
  - customer hotline;
  - customer complaints;
  - RMA (return material authorization) process;
  - CAR (corrective action request);
  - customer feedback processes.
- Hiring and employee training.

The quality manual template is envisioned to be a "shell" which provides a high level view of the desired goals and activities towards becoming a world-class supplier of ICs. The first set of activities after the launch of the IC design activities are:

1. a revision control system to manage the various versions of the chip descriptions;
2. documentation of the target chip description, leading up to the specification;
3. a set of execution assumptions including the implementation methodology, the marketing requirements, chip size, gate

count, pin count, package type, preliminary supply-chain partner list, an IP list, etc.

It is recommended that the quality manual (QM) be implemented in phases. A three step approach is indicated in Fig. 8.1. Establishing a QM template is the first step. This can be followed by an initial QM and then a released QM.

The intent is to incorporate an execution discipline and always to have a baseline set of assumptions consistent with the company's goals and the "end in mind," one of Steven Covey's seven habits [8.5]. One typical execution issue is that development engineers seldom like to take the time to document even the overall design targets, methods, flows, and procedures. This can lead to confusion, duplication of efforts, and re-dos, especially in a growing organization. However, the documentation being suggested here should not get in the way of the development engineering, which is the main focus at the start-up. The intent is to keep it simple and not to make this a burdensome methodology. A hierarchical approach starting at the top level can help keep the documentation simple. The organization has a choice to decide on the level of documentation hierarchy worth their investment.

Referring to Fig. 8.1, the next major quality activity is the definition of a qualification (or "qual") plan, details of which will be discussed later. This is followed by the execution of an "early-look" qualification on a limited number of parts from the first pass silicon. The purpose of this "early look" qual is to weed out any major quality and reliability issues in the first implementation of the design. This is followed by the full qualification, manufacturing quality, and customer support activities. These are discussed in more detail later.

ISO 9000 certification of suppliers has become a pervasive requirement during the last 10–15 years. Most of the silicon foundries and SATS are ISO certified. Some of the fabless company's customers will require ISO 9000 and possibly other certifications. The activities being recommended here will form the basis for an easier transition to becoming ISO certified, when the need arises. In the author's opinion, look for a compelling reason to get ISO certified. Going through the process is quite burdensome and time consuming.

8.2 Documentation System

A hierarchical system with a numbering system that is easy to track is the norm. The following levels of hierarchy are recommended as a minimum:

- Company level principles.
- Operational policies, methodologies, and guidelines:
    - requirements to define roles and responsibilities for each functional group;

  - top level description of the IC, the sourcing methodologies, the design flow and design methodologies could all be examples in this category of specifications.
- Development/operational procedures:
  - systematic instructions, details of execution, and supporting information for a specific operation or function;
  - most technical specifications for the design, the process, package, assembly, and test will be part of this hierarchy level.

In addition there needs to be a system for document control and engineering change control.

The company document control activity could also be expanded to include a central repository of information to be shared within the company. Web links facilitating such communication can be of great value.

8.3 Quality in the Development Phase

As discussed in previous chapters, the major activities in the development phase are IC design, package design, operations planning, and supplier selection/contracting.

Quality in the design phase is focused on the following methods [8.1]:

- Correct by construction. These methods ensure that reliability design rules and guidelines are met during the design phase.
  - Example tasks are:
    - power-grid planning for electromigration;
    - clock routing for joule heating [8.6];
    - verification of cell libraries and other IP.
- Design verification.

> Design verification and quality are the most important issues in implementing IC designs these days. The following is a brief checklist of must-do verification activities:
> - make a list of of every item or feature that is to be tested;
> - make a list of tests which will cover the above list;
> - develop the test suite such that all tests are self-checking;
> - write scripts to confirm that all the required tests have been run and passed.
>
> Laury Flora
> Vice President
> Octera Corporation

- Design for reliability. Best known methods and techniques can be used to assure that the product has robust reliability performance for known issues such as the following:
  - ESD (electro static discharge);
  - latch up;
  - gate dielectric degradation;
  - hot carrier injection;
  - threshold voltage stability.
- Reliability verification. Automatic reliability verification checks incorporated in the design flow should check for compliance of the design database to reliability design rules. Design rule checking for the following effects is routine at world class suppliers:
  - electromigration/joule heating;
  - ESD;
  - latch up;
  - IR drop and signal integrity;
  - hot carrier effects;
  - NBTI (negative bias temperature instability);
  - antenna effects caused by in-process charging.
- Design for Test. These methods ensure that the product is testable and has adequate fault coverage (more about test escapes in the next section). Some of these methods are:
  - structured memory and logic test (for example MBIST);
  - quiescent current measurements (IDDQ);
  - guard-banding rules and techniques;
  - techniques for isolating timing faults;
  - techniques for accelerated silicon debug, fault isolation, and fault tolerance.
- Design for Manufacturability. These methods are used to provide rules and guidelines that ensure manufacturability through the fab, assembly and test operations. There is considerable focus on DFM techniques as discussed in previous chapters.

8.4 IC Quality and Reliability Qualification

This section is devoted to a discussion of the IC's initial quality as well as its long-term reliability.

8.4.1 Methodology

The reliability goals of integrated circuits are generally discussed in conjunction with the traditional, "bathtub" shaped curve shown in Fig. 8.3. The curve shows the failure rate (FR) of ICs as a function of time. There is usually a region of rapid decrease in FR, known as "infant mortality." Defects in this regime are characterized by an EFR and is expressed as percentage fails, or in ppm (parts per million). A typical EFR could be 0.1%, or 1000 ppm. Units that fail in this period are associated with latent, silicon or assembly process related defects and are considered weak units.

During the "useful" life time period, the FR is constant until the "wearout" period. Failures in the useful life are usually random failures, with very few occurrences. ICs that function past the infant mortality period have a high probability of surviving the normal and long operating life of the end product. FRs in this period are measured in FITs in a billion device operating hours) or in ppm/1000 operating hours. A typical number is 100 FITs, which implies 100 fails in a billion device hours of operation. As an example, assume there were a million parts shipped. After constant operation for one year (10,000 hours, which is the normal "rounded up" number of operating hours in a year), one would expect 1000 of these parts to have possible failures. Note that in this example, there are two ways to get the resulting number. First, note that a million devices operating for 10 k hours is a total of 10 billion device operating hours. Therefore, the number of expected failures will be 10 times the FIT rate of 100. The second way is to start with the 100 ppm /1 k hours failure rate. Because the devices operate for 10 k hours, the expected failure rate would be 1000 ppm, and hence 1000 parts will be expected to fail.

As is shown later, the calculated FIT rate is inversely proportional to the number of device operating hours. This means that the more the number of parts subjected to operating life tests, or the longer the test time, the smaller the FIT rate that can be resolved. Large semiconductor manufacturers ship very large quantities of parts, sometimes

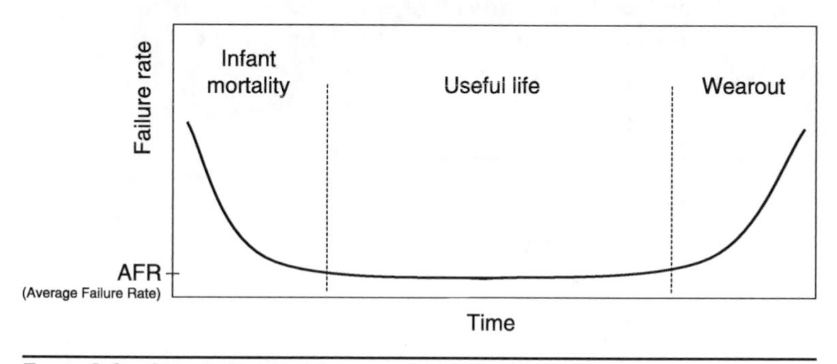

FIGURE 8.3 Classic "bathtub" curve of IC failure rate.

aggregated over many part numbers and therefore can report FIT rates at or below 10 FITs. For an emerging fabless IC company there are only a few designs and limited quantities of parts shipped. Also, in order to limit costs, there is usually a limited number of parts that are subjected to life testing. As a result, a 100–200 FIT failure rate is considered reasonable for most commercial and consumer applications. High reliability, medical, automotive, and other applications may require lower FIT rates. Note that if a larger FIT rate is the only reason the fabless company is having trouble winning a design at the customer, there are some degrees of freedom that a knowledgeable expert can help you with.

During the wearout period the FR rises rapidly. Wearout failures are generally associated with failure mechanisms such as oxide wearout, metal migration, hot electron effects, wire bond inter-metallics, and thermal fatigue. The wafer-fab and assembly staff usually characterizes these prior to certification and qualification of the process. The fabless company should review characterization information available from the wafer fab. There will likely be no additional requirement to investigate these, unless the design is pushing some limit—this should be flagged during the reliability verification of the design.

8.4.2 Acceptable Outgoing Quality Level (AOQL)

When the fabless company begins shipments of parts to customers, it is important to track the quality levels of the parts. A common industry term is AOQL. The customer may track the IQL. The initial quality levels are affected by two major categories of problems—visual/mechanical rejects and electrical rejects. The fabless company and the customer usually set goals for AOQL. A typical number may be 0.1%, or 1000 ppm. It is common practice for the fabless company to accept target AOQL levels based on data from the wafer fab and the SATS. The fabless company then needs to follow up with a monitoring plan in partnership with the customer, and finally, establish a corrective action plan if required.

- **Visual and mechanical rejects**—These rejects are usually found via visual inspections by the incoming quality assurance personnel at the customer site:
  - package pin co-planarity;
  - missing package pins or solder balls;
  - molding defects;
  - marking defects;
  - other mechanical damage.

The incidence of package defects is usually very low provided the SATS has implemented either a 100% package scan or an acceptable sampling plan. It will be important to verify the package co-planarity and mechanical specifications against customer requirements at the SATS. In addition, post scan defects can sometimes be introduced by improper banding of trays prior to transit, improper rotation of the parts, or by mishandling at the customer location.

- **Electrical rejects**—This category represents combined failures due to:
  - improper handling;
  - ESD, latch-up;
  - EOS (electrical over stress);
  - test escapes;
  - infant mortality (early life) failures.

ESD and other handling defects are less likely if the customer adheres to proper handling procedures.

The theoretical test-escape rate for gross failures (opens, shorts, functional fails) can be modeled as a function of the sort yield and the amount of fault coverage achieved during DFT. Details of the formula are discussed in Appendix B.3. As shown in Table 8.1, in order to achieve a 1000 ppm reject rate when the sort yield is 90%, the fault coverage needs to be greater than 99%. In order to achieve such fault coverage, the use of logic BIST techniques is recommended, since SCAN coverage using functional vectors usually falls short of such targets.

- **Infant mortality**—Infant mortality (early life) failures represent the first part of the bathtub curve discussed in the previous section. The industry has used stress tests to screen the weak parts, and thereby achieve acceptable EFRs. There are two types that are generally used:
  - Temperature and voltage stress, typically called "burn-in" and performed on packaged parts. These tests are performed

Y=	90%	
Fault coverage	**Escape rate**	**Escape rate (ppm)**
97.0%	0.32%	3,156
98.0%	0.21%	2,105
99.0%	0.11%	1,053
99.5%	0.05%	527
99.9%	0.01%	105

TABLE 8.1 Escape Rates as a Function of Fault Coverage at 90% Yield

after final test and require special ovens that subject parts to high temperatures, typically 125–150° C, while in operation. Typical voltage stress is 20–40% over the nominal power supply voltage.

- Dynamic voltage stress. These tests are typically implemented at wafer sort. Typical voltage stress is 20–40% over the nominal power supply voltage.

The use of these screens can be expensive. They should be used only if the fabless supplier is unable to meet the customer's EFR requirements. Most of the time the qualification testing described later in this chapter gives an early indication of weak defects. A high priority must be placed in determining the root cause of the weak spots, in ensuring proper fixes at the fab and/or the SATS. In the early days many of the EFR fails were caused by poor assembly and handling practices. As the industry has controlled these processes, EFR defects are now associated with defectivity levels in the silicon process.

Sometimes it is prudent for the IC supplier to use burn-in screens on initial shipments, until the desired EFR has been achieved. The number of parts to be burned in is a function of the target goal and the number of fails observed. For example, if there is one fail in 10,000 parts burned in, one can resolve a 0.01% EFR.

8.4.3 Average Failure Rate (AFR) Calculation

The industry has developed standard stress tests to accelerate failures in order to allow estimation of the average failure rate. There has also been much work in characterizing possible failure mechanisms that allows modeling of the acceleration factors. In estimating the AFR for a given IC in a customer's application, the possible failure mechanisms must be postulated based on experience and based on best known information from the suppliers. There are two common sources of the standard documents:

- JEDEC, Joint Electron Device and Engineering Council (http://jedec.org/).
- AEC, Automotive Electronics Council (http://www.aecouncil.com/AECDocuments.html).

A listing of the appropriate documents in the context of the qualification tests is provided later in this chapter. Accelerated stress testing of ICs is commonly referred to as "life testing."

The methodology and the parameters to be used in the calculation of the AFR require knowledge, experience and information from the suppliers. Stress acceleration is provided in two forms—voltage acceleration and temperature acceleration. Therefore there are two different acceleration factors that are calculated.

These depend on the actual voltage used to stress the ICs, and the temperature during the life tests, respectively. The calculation also assumes a distribution for the failures. The most common distribution used is the "chi-square" distribution. Knowing these parameters the calculation provides estimated AFRs depending on the number of failures observed during life testing. There is also a confidence level (CL) associated with the calculations. Most commonly used confidence levels are 60% and 90%. A 90% CL means that there is a 90% probability that the AFR will be equal to or less than the number stated.

Figures 8.4 and 8.5 show representative results from a calculation of AFR at 60% and 90% confidence levels, respectively. The life test assumes 77 total parts to start with and the calculations show AFR if there were 0, 1, or 2 fails.

As if there weren't enough terms already, there is one the fabless company must learn! While semiconductor suppliers use the FIT to describe part reliability, their customers usually refer to the term MTTF (mean time to failure). This is the inverse of the AFR. Thus 100 FITs implies an MTTF of 10^7, or 10 million hours. Learning to speak the customer's language is essential for good communication.

8.4.4 Customer Requirements

Before embarking on setting a qualification plan for its ICs, it would be preferable if the fabless company has inputs from the customer regarding the normal operating conditions for the IC as well as their target requirements for the quality and reliability. The problem is that this information is generally not easy to come by, especially for the new, start-up company. A mature fabless company that has provided a number of ICs to the same customer should have access to such information.

In the absence of specific customer requirements, it is recommended to set targets yourself based on experience and generally

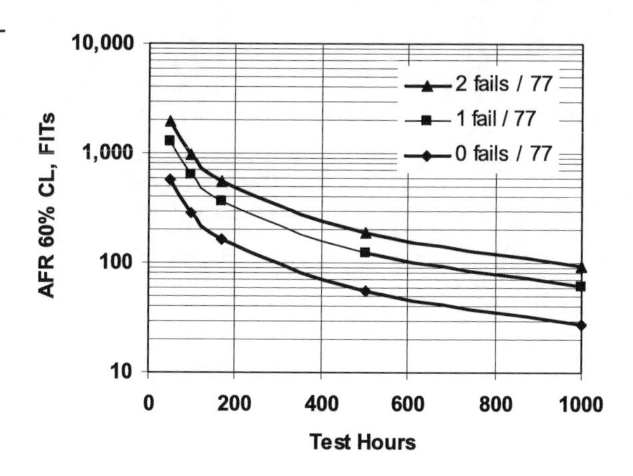

FIGURE 8.4 AFR at 60% CL versus test hours for various number of fails (0, 1, or 2) on a life test with 77 parts.

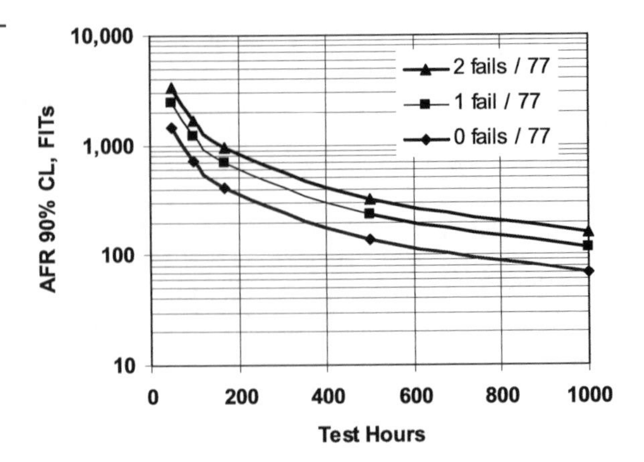

FIGURE 8.5 AFR at 90% CL versus test hours for various number of fails (0, 1, or 2) on a life test with 77 parts.

accepted numbers. It is good to have such a baseline because it helps set the qual plan for the fabless company and also helps start the negotiation of the final spec with the customer. Table 8.2 lists a good "starter set" of requirements for a mainstream IC. Some of the definitions and numbers will make more sense after a review of the qual plan and the details in Appendix B.3.

Note that, for ESD targets, the HBM (human body model) and MM (machine model) limits have become a de-facto standard. This is not the case for the CDM (charge device model). Its use is becoming increasingly popular. However, there are many issues with a uniform implementation method.

Note that there are many different requirements emerging for "green" compliance. Lead free was one of the first requirements that

	Target specifications
Initial quality level (including mechanical & electrical rejects)	< 0.1% or 1000 ppm
Average failure rate (λ)	< 100 FITs or <100 ppm/1k hrs
ESD immunity	± 2 kV HBM ± 0.2 kV MM ± 0.5 kV CDM
Latch up immunity	JESD 78 compliant
Minimum MSL	Level 3
"Green" compliance	Pb-free

Notes: HBM: human body model; MM: machine model; CDM: charge device model; MSL: moisture sensitivity level; Pb: lead.

TABLE 8.2 Target Specifications for IC Quality and Reliability

has been implemented. Halogen free is another example. There are different requirements based on the global region where your customer is located. It is best to check with the customer and the SATS for the latest requirements that need to be met.

8.4.5 Qualification Plan

A good qualification ("qual") plan comprehends all dimensions of the quality and reliability of the IC:

- design, process, packaging, assembly, test;
- its usage conditions—electrical, mechanical, thermal and environmental;
- lifetime requirements.

Key elements of the qual plan are:

- statement of customer requirements;
- statement of usage condition assumptions;
- reliability testing;
- procedures for determining AOQL and IQL;
- verification that the manufacturing silicon and assembly processes are qualified and in control;
- verification of the package characterization and qualification;
- verification that the design and IP are compliant to the process design rules, models, and reliability rules and guidelines;
- verification that the test characterization has been completed, both in an engineering and a manufacturing environment;
- verification that the IC meets its device functionality across operating conditions in the product specification.

The intensity of the due diligence can vary depending on the class of IC under consideration. For example, referring to the definitions in Chap. 4, a leading edge IC may require testing of additional parts, for additional hours and a more thorough analysis of information from the suppliers. As a fabless company it is essential to leverage as much manufacturing quality and reliability information from the suppliers as possible. The more mainstream the product and the more mature the silicon process, the assembly process and the package, the less stringent the requirements on the life testing and the qual conducted by the fabless company.

8.4.6 Qual Tests

Typical accelerated life tests to determine the EFR and AFR are listed in Table 8.3. Some of these tests are focused on silicon and some on package of reliability. Also shown in this table are typical defects that are expected to be identified during the life tests and the acceleration stress. Details of the test conditions are shown in Appendix B.3.

In addition, the following is a typical list of tests conducted to qualify a package.

- Package/assembly integrity related tests.
  - wire bond shear;
  - wire bond pull;
  - die shear;
  - solderability;

Test	Defect categories	Stress
HTOL (Hi Temp Op Life)	Latent defects Oxide defects Si defects Contamination Electromigration Contact/via defects Photo defects	Temp voltage
HTS (Hi Temp Storage)	Ionic contamination Metal void propagation C4 joint degradation Via & contacts stress (Cu processes)	Temp
HAST (Highly Accelerated Stress Test)	Metal corrosion Contamination induced V_T shifts	Temp moisture
TC (Temp Cycling)	Assembly defects Solder joint fatigue Package cracking Intermetal dielectric cracking	Δ Temp
ESD (Electro Static Discharge)	Electrostatic sensitivity	Voltage
Latch up	Sensitivity to parasitic bipolar action	Voltage current temp

Table 8.3 Accelerated Life Tests, Possible Defects They Are Targeted to Identify, and the Stress Factors

- solvent resistance;
- physical dimensions;
- solder ball shear.

- Accelerated environmental stress tests. These test the mechanical integrity of the package under various environmental conditions. The intent is to identify effects such as cracking, de-lamination, moisture sensitivity and other mechanical damage:
  - preconditioning;
  - biased HAST;
  - unbiased HAST, or "autoclave";
  - temperature cycling;
  - temperature shock;
  - power temperature cycling;
  - high temperature storage.

8.4.7 Qual Costs

Qualification costs are dominated by the cost of the high temperature sockets used and the specific sample size selected. The pin count and the type of the package used is the key factor in determining the socket type used. The set of charges are from a real example; it is provided to illustrate an "order of magnitude" estimate for a "starter set." The numbers could be much higher depending on the extensiveness of the testing and analysis.

Board and socket	$ 25 K
Readout and test	$ 5 K
Test hardware	$ 5 K
ESD and Latchup tests	$ 5 K
Package tests	$ 15 K
Manpower	$ 15 K
Total	$ 70 K

8.4.8 Reliability Maintenance

As the fabless company begins shipping high volumes of ICs, it must invest in a reliability maintenance program, sometimes called an ORM (ongoing reliability monitoring). An example is a quarterly life test of a small set of ICs in addition to supplier audits in order to maintain the certification and qualification status.

8.4.9 Parts Designation

As the IC design goes from being an engineering prototype to being "qualifiable" to being qualified, there has to be a way to mark the parts to distinguish their quality level. Many companies have their own designations. One possible set of definitions for "E," "PQ," and "P" parts is as shown in Table 8.4.

Note also that it is usually possible to track the part history from a date code marked on the package. Some companies choose to not use PQ and P markings, and only use the date code to track the parts.

8.5 Manufacturing Quality

Manufacturing quality of ICs supplied by fabless IC companies is a reflection of the quality at the various manufacturing partners and their design. The goal should be to establish communications and a relationship with the suppliers such that this becomes a low maintenance, complete and auditable monitoring system. As the fabless company grows, it is appropriate to formalize the processes for supplier selection, certification, qualification, process control, and continuous improvement. Some of the important activities are:

- Supplier selection—Criteria such as technology, quality record, availability, cost, service, risks must be considered.
- Certification and qualification—This activity includes a review and audit of considerations such as process stability, process capability, high volume capability, process qual data at the supplier.
- Process control—This activity includes statistical process control (SPC) review and audit, yield improvement history review, processes for corrective actions in the event there is a process excursion.

Marking	Classification	Purpose	ATE Test	Qual
E	Engineering	Samples	None/ partial	None
PQ	Potentially qualifiable	Customer qual	Full	500 hours
P (or none)	Qualified / production	Manufacturing	Meet spec	Complete

Table 8.4 Parts Designation Matrix

- Auditing the processes used by suppliers to manage their suppliers:
    - continuous improvement methodologies are a helpful tool.
- Improvement plan.
    - supplier scorecard, to document quarterly ratings of activities required at the supplier in order to improve quality.
    - quarterly business reviews—these are very useful in recapping activities in the previous quarter and to set goals for the next quarter. Allows management involvement and participation;
    - corrective action requests—a central repository of open and completed corrective action can be used to track these.

In addition, the fabless supplier must have processes to track the following quality assurance activities internally:

- yield tracking;
- delivery and cycle time tracking;
- AOQL tracking and reporting;
- customer-related QA, as discussed next.

8.6 Customer Support

The primary goals for customer support are the following:

- Understand customer's quality and reliability requirements and expectations.
- Product change management—Changes can occur at the customer's end, at the fabless company and also at the suppliers. Processes must be in place to evaluate, communicate, implement and manage the changes at the appropriate value chain elements. The fabless company must take responsibility for communicating and getting approval for process changes made at the suppliers. There needs to be a formalized process for such communication and authorization.
- Order fulfillment quality—This refers to the quality of the order placement and logistics processes from start to finish. Problems can occur if processes and procedures are absent, incorrect or are not followed.
- Technical quality—A part that is non-operational in the customer's manufacturing or test operation is included

in this category. There could be many reasons for such failures—for example, the IC could have been damaged at the customer site or enroute, it could be a reliability failure or it could be a test escape. Processes must be defined for root cause analyses, determination of the corrective action and closure of the issue.

- Returns management—Whether you like it or not, there will be returns. Therefore there must be defined processes for dealing with them. There may be slightly different processes for in-warranty and out-of-warranty returns. Processes must be defined for root cause analyses, determination of the corrective action and closure of the issue.
- Measuring customer satisfaction—A well managed program for soliciting customer feedback and making demonstrated improvements can go a long way in establishing the fabless company as a quality supplier. Continuous measurement and improvement can reinforce the company's customer focus and commitment to quality.

8.7 Key Points

- Build in quality from the start.
- Early investments in a QM qualification, documentation processes, and customer support processes are essential elements.
- Lower qualification cost by leveraging data available from suppliers. It realyy helps of your design fits within the envelope of supplier offerings.

CHAPTER 9

Managing the Implementation Program

9.1 Management at the Vertically Integrated Company

In this chapter there is discussion of management practices appropriate for fabless IC companies. The major difference is the mostly external, inter-company focus versus the all-internal, intra-company management, and coordination at the vertically integrated semiconductor company. Some of the concepts have been discussed and reported previously by the author [9.1].

Figures 1.4 and 4.6 illustrate the major IC development activities at vertically integrated companies. These companies are usually organized as a functional organization. There are multiple product groups and some central groups responsible for chip design, technology development, and operations, as shown in Fig. 9.1.

The product groups are generally responsible for the definition, development, and delivery of the specific system or IC product. These groups generally assign a project leader or project manager who makes arrangements for support from other functional organizations—chip design, technology development, and operations. Some organizations form teams of representatives from each of the functional organizations. The project leader is usually a proven technical contributor and a leader. Other companies may use a matrix management approach to assign resources to each product.

All activities are internal to one company. Negotiations for setting of priority for the project, resources and capacity allocation, schedule, and delivery commitments are all within the company's management chain. At many companies the "buddy system" and the charisma of the project manager are keys to securing focus on the project. Any problem or conflict resolution is internal to the company. External relationships are either with customers or suppliers of

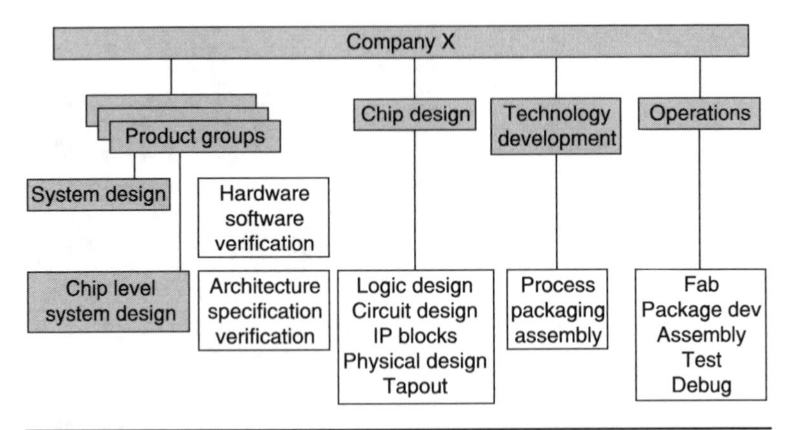

Figure 9.1 Typical organization chart at a vertically integrated company.

raw materials/equipment/tools used in the design and fabrication facilities. Supplier relationships have traditionally been managed by procurement specialists.

Financial management of capital, R&D (research and development), engineering and operations costs is usually a challenge. This makes it difficult to benchmark items such as true wafer cost, product cost and engineering manpower cost.

9.2 Management of the Distributed Supply Chain

Figure 9.2 is an illustration of the "horizontal" supply chain for the design and implementation of an IC using the COT sourcing model, as discussed in Chap. 4. During the design phase the fabless IC company usually develops the IC specification and does the front end design in house. At that point there is a synthesized, gate level netlist available. By then a standard cell library, various memory configurations,

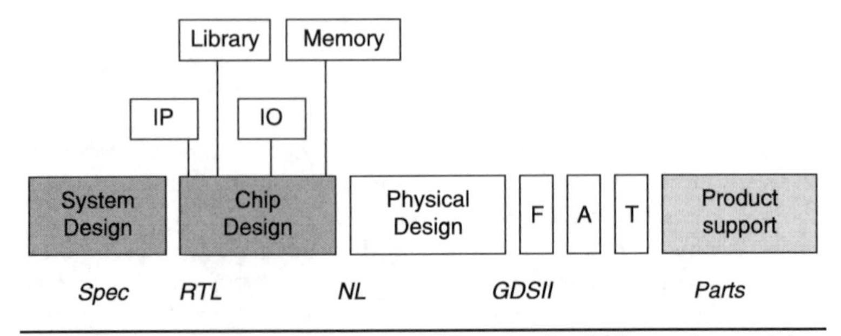

Figure 9.2 A typical supply chain for ASICs using the COT sourcing model.

and I/O cells from one or more suppliers have been integrated into the design process. The fabless IC company could then engage with a design service provider, or complete the design in house. Special analog blocks could be developed either by the fabless IC company or a third party from a specification.

Upon completion of the physical design, the design information is transferred to the wafer fab as a GDSII tape for mask-making and silicon fabrication. Completed silicon wafers are then transferred to an assembly house that performs the assembly and packaging of the silicon chips. The SATS (semiconductor assembly and test supplier) then performs the final test of the IC that gets shipped either to the fabless IC company or directly to their customer. Product support is generally provided by the fabless IC company.

For this supply chain, the foundry fabs and SATS are located around the world with a heavy concentration in the far east countries of ROC (Taiwan), Korea, China, the Philippines, and Malaysia. Design services and IP providers are available primarily in the US, Europe, ROC, India, and China. The entire supply chain is spread around the globe.

Each entity in this distributed supply chain is focused on their core competence. This model has been called "MICRO" optimization [9.1]. Each entity is positioned to provide the best value to their customer via the best features, quality, service and deliverables at a competitive price. Web based inventory management, ERP (engineering resource planning), and MRP (manufacturing resource planning) tools have become pervasive and facilitate smooth operation of the supply chain most of the time. The infrastructure exists to move material between suppliers, and to and from customers.

9.2.1 "Macro" Optimization

While each of the supply chain partners does the best they can, "gaps" still remain and must be filled for the successful delivery of finished IC products. The characteristics of some of these gaps are related to the complexity of the IC being implemented. Other gaps can be related to the logistics and hand-off issues between the many supply chain partners. A "virtual re-integration" of the supply chain is required to deal with cross-functional and inter-active issues that could affect more than one supplier. Vertically integrated companies recognized the need for the resolution of such issues and invested in the required infrastructure. The filling of these gaps has been referred to as "MACRO" optimization [9.1] and will be illustrated with a few simple examples as follows.

Example 1

In the implementation of an industrial control chip, the customer and the assembly house decided to wire bond the Vss pad to the back side of the die and the package paddle in order to have a more solid electrical ground connection. In doing so the entire die was shorted because that is the way the IP provider and the foundry designed the embedded block.

Example 2
In the ramp up phase of an MPEG4 DSP ASIC, a significant yield loss occurred as the foundry engineer decided to center a few process parameters in order to run a process in better statistical process control. It was the right thing for the foundry engineer to do. However the impact was very significant for the ASIC product yield. And determining the root cause took a lot of failure analysis and debug effort and resulted in a schedule delay.

Example 3
Performance of an RF chip for a wireless application was critically dependent on the length of the wire bonds to the ground plane—this was communicated to the assembly partner. What did not get communicated was that a more critical parameter was the equality of the wire bond lengths on the east and west sides of the die. The part performed well once the additional constraint was implemented.

The important point here is that communication and coordination across the many entities in the supply chain in a timely fashion is critical. This is especially true for IC designs that push the limits in one or more areas—performance, complexity, or chip size.

9.2.2 Inter-Company Coordination

Unlike the vertically integrated IC company, coordination of the global, a distributed supply chain for fabless ICs requires management across legally separated entities in many different parts of the world. Such inter-company coordination requires special management skills in addition to the normal program management skill-sets. Some of the important areas that must have adequate focus are discussed next.

9.2.2.1 Contracts

Contracts and/or purchase orders are required to initiate and manage product prototyping and builds. Formal documentation of projects is a must during the definition, execution and completion phases. Documentation internal to the fabless company needs to be formalized to a greater degree than is necessary within a vertically integrated company. Venue, governing law, and procedures for the resolution of legal conflicts must be understood by the fabless company using a global supply chain. The good news is that there are many companies executing successfully using this model. Political and geophysical stability have also been concerns that require some attention.

9.2.2.2 Forecasts

Most suppliers require customers to provide formal IC forecasts for a "rolling" six to 12 month period and a longer term outlook reaching out 12–18 months. These demand forecasts must be updated monthly. Especially in times of tight capacity constraints, the last one to two months of the forecast become a commitment to purchase services or goods and materials.

9.2.2.3 Schedule and Delivery

Each supply chain partner makes commitments of schedule and delivery, usually via web-based systems. Variances do occur and usually have chain reaction impacts on other suppliers. Managing such variances requires careful negotiations with a special focus on human-relations management.

9.2.2.4 Prioritization

The largest customers spending the most money with the suppliers usually get the most attention and the highest priority. However, most of the major suppliers also have programs to support emerging companies, especially those with high growth potential. It is important for the emerging fabless company to establish appropriate relationships with executive management at the suppliers.

9.2.2.5 Relationships and Communication

Establishing good working relationships with the interface personnel as well as the management chain at each of the suppliers is important. Effective communication across many time zones is critical especially with language and cultural differences that must be understood and dealt with. When problems do occur (and they will!) what will the escalation process be? Relationships are also important in getting on the supplier's "radar screen" if you are a small, emerging fabless company.

9.2.3 Managing Cash Flow

Transactions across the legal entities require real money transfer. Negotiating business terms for material acceptance and payment become an important activity. The partners selected and the methodologies used affect the total budget and the cash flow. Some examples are:

- Selection of the process technology and any special options affect wafer price.
- Selection of the foundry affects the wafer pricing.
- Use of shuttles for prototyping lowers prototyping cost.
- Negotiating credit and payment terms affects cash flow.
- Negotiating a payment schedule with IP providers linking the final payment to working silicon maintains leverage with the supplier in addition to delaying some cash flow.

The fabless company must also be aware of currency exchange rates, the price of precious metals and their possible impact on the company's IC product.

9.2.4 Technology Optimization

Compared to internally developed process and design technologies, the IC designer usually has to contend with technologies that are not customized for their application needs. The onus is on the IC designer at the fabless company to distinguish and differentiate their product based on its architectural and design features.

9.3 Comparison of Management Processes

Table 9.1 shows a comparison of the major features of management processes when managing IC development in a vertically integrated environment versus the distributed supply chain.

The fundamental differences are that an internal focus gives way to an interaction with many, legally separated entities around the globe. More formal processes are required to manage the many supplier interfaces in the fabless/distributed supply-chain business model.

Supply chain	Integrated	Distributed
Project coordination	Intra-organization	Inter-organization
Corporate entities	One	Many, legally separate
Contracts	None	Many
Commitments	Internal	Contractual
Forecast process	Informal	Formal
Prioritization	Internal	Business negotiation
Resource and capacity	Internal	Business negotiation
Schedule commits	Internal	Formal
Delivery commits	Internal	Business related
Problem resolution	Internal	External/business related
Cash flow	Internal, if any	External
Relationships	Broad-based, Internal Personal	Few contact points/ external Partnerships
Project management skill set	Technical	Technical Business Relationship focus

TABLE 9.1 Comparison of Integrated and Distributed Supply Chains

9.4 Relationship and Partnership Management

Another major difference for a fabless company in managing the entire value chain, including the distributed supply chain, is in the need to develop and manage relationships and partnerships. This was discussed in Chap. 2. There are a minimum of two relationships that must be established with downstream supply chain partners—one with the foundry and one with the SATS. The number goes up from there as one adds a physical design services provider, one or more IP providers, failure analysis, debug service providers, and possibly others. Then there is at least one customer relationship upstream in the value chain.

There are three levels of interfaces and relationships that could be established with each major partner:

- Technical—usually at a working level as well as a management level.
- Business—usually a program manager or a procurement manager.
- Executive—usually a C-level or a VP level.

While there are no standard best practices, a successful way to manage the relationships is as follows. The door opener of the relationship is the C-level executive, or another executive that has previous relationships with the partner. Assignment of a focal point within the fabless company that can follow up the executive discussions with action plans is very important. Since initial interactions usually are technical in nature, a technical lead is a good example. However, soon thereafter the procurement and contracts people need to get involved. Assignment of the right program manager is one proven way for success.

Maintaining visibility at the suppliers and customers is also extremely important, especially for the emerging company dealing with large players. It is easy to get removed from the radar screen and thereby lose priorities. Even if there are delays in the program at the fabless company, it is important to nurture and maintain the relationships.

9.5 Program Management

It is recommended that the assignment of strong, cross-functional program management be used as the "glue" in pulling together relationships in, and the management of, the value chain. As the company grows the program management team will need to be supplemented with support from the functional organization. It should be noted that in recent years, the role of the traditional program manager has been expanded.

- Traditional program management roles:
  - budgeting;
  - scheduling;
  - monitoring;
  - controlling.
- Recommended additional program management roles:
  - participating in product definition and be involved in the entire product lifecycle in order to better manage risks;
  - coordinating supplier and customer relationships;
  - coordinating business activities.

The "new" program manger must have technical knowledge and experience, must have knowledge of business issues and must be adept at managing people and relationships. It is important for the program manager to be involved in the definition of the IC product from innovation to specification. This is one way to manage and avoid "over-specification" discussed in Chap. 4. It is also a way to maintain reasonableness in projected schedules, especially if the program manager is intimately familiar with the partner capabilities. It is envisioned that the following example considerations will provide a flavor of items to be incorporated in the program plan:

- Benchmarking of the IC complexity relative to industry capability (transistors/chip, transistors/sq mm, click speed, power dissipation, pin count, etc.).
- Identify and maintain risk areas.
- Unit cost estimates consistent with options being considered by the technical team.
- Timely identification of and engagement with supply chain partners.

In a start-up fabless company such activities are usually not formally addressed, or are the responsibility of the VP of engineering. With the overall burden of other responsibilities, some of these "disciplined" activities get unintentionally de-prioritized.

Establishment of partnerships with the key participants of this global supply chain are strongly recommended. This will be especially important as the fabless company grows.

Strong execution is a must if the startup fabless IC company is to be successful. Strong execution means making aggressive and yet achievable goals, establishing a program plan, the relationships with partners, monitoring progress and making required trade-offs as things change. While MS Project is a good tool to set up a project schedule, managing the program is a much broader activity in which the MS Project schedule is a necessary, though not sufficient, element.

In the execution of the XBox chip set, we assigned a strong Program Manager (PM) embedded at the supplier facilities. This PM became part of the supplier execution team and could mitigate issues in real-time based on product priorities. Had we not had such a complete alignment with our value chain suppliers, it would have resulted in significantly reduced volume at our product launch at best, or completely missing the holiday launch window at worst.

Bill Adamec
Senior Director
Semiconductor Technology Group
Microsoft XBox

9.6 Risk Management

The two major components of risk management are risk avoidance and mitigation [9.2].

Risk avoidance entails attempts to reduce the probability of failure. In the complex IC design and implementation process this is usually accomplished through the use of automation and execution processes that are well defined.

Risk mitigation involves the reduction of failure cost. This involves creative ways to make adjustments to the action plan to minimize slippage of schedule and increase of project cost. Good program and project managers are adept at planning in contingencies during the development process, and taking swift actions when problems do occur. This was discussed in a sidebar in Chap. 6.

The two key strategies for reducing mistakes (risk avoidance) in any project are *process* and *testing*. For chip development, strict adherence to process is mandatory. The design tool chain allows the designer to capture his/her intent in code or schematics, have the digital portion compiled into gates, lay the blocks out and assemble the design, generate masks, fabricate wafers, and produce packaged parts. These steps are performed with tools and automation to minimize human involvement; each step generates the input for the next step. To ensure the integrity of each step, tests are used to verify functionality and check for common errors—simulation and design verification testing; timing analysis; design rule checks; electrical rule checks; layout-versus-schematics verification; wafer process monitors; and finally lab testing of the final device.

However, in any complex development, mistakes are nearly inevitable; for the project manager the question isn't "if" or "when", but "how do we recover." Project managers look for ways to mitigate risk and allow for contingencies in the development process. The first place to look for reducing risk is in the product definition phase. If the design is expected to push the envelope in many directions at the same time, the chances of missing the project goals increase substantially. Toning down the marketing requirements can enhance the probability of success. However, reducing the feature set has its own risks of introducing a product that is inadequate or non-competitive. A careful balance and good market knowledge are required at this point.

When a design problem does occur, the first action should consider the market impacts—how severe is the bug; is there an acceptable external work-around; can a feature be eliminated that allows the chip to otherwise operate; is the market better off having the chip sooner without the feature or is the feature required even at the cost of a delayed introduction; can the chip be used to provide workable samples while the fix is implemented.

If a fix is required, solutions that minimize both schedule and development costs will be desirable. The costs of design defects in a chip development project can be great—new mask sets can costs hundreds of thousands of dollars; and, finding a bug, designing a fix, and implementing the fix can take three to six months. However, if you have provided design contingencies, these costs can be substantially reduced.

Kurt Stoll
Director, ASIC and Hardware Development
Echelon Corporation

9.7 Design Productivity

Unfortunately there are not very many tools to predict design productivity, schedules, and risks. Most schedules are "guesstimates" based on past experiences of the team. This makes it very difficult to not be one of the 85% of design teams that missed schedules [4.1]. Another ugly fact is that most new designs need a metal fix, and over one-half of the designs require a second spin of silicon. This information usually does not get reported. The EDA industry has used this phenomenon to drive many of their tools offerings and to justify the business case for tools investments [9.3].

Recently there have been publications that discuss modeling of designer productivity and relate it to design complexity, size of the design team and similar parameters [9.4]. Two charts of interest are reported here. Figure 9.3 shows that productivity, expressed as output per person-week, declines as project team size increases—however, overall rate of team output increases because of the higher staffing level. Such observed decline can be attributed to many factors such as increased communications required between additional players, resulting possibilities of miscommunications, consensus building, organizational battles between competing groups, etc. Also noteworthy is that IC project managers often create project plans that implicitly assume a productivity value far above what is typically realistic. Notice the discrepancy between the manager's anticipated, or assumed, productivity (triangle in the figure) for a project undergoing planning and actual historical performance of similar projects compiled by Numetrics. The assumed productivity puts the project at high risk in terms of meeting the target schedule, because in order to meet the schedule target, the team must achieve a productivity level far above norm.

Figure 9.4 shows an estimate of resources required in the front end and the physical design phases as modeled. Compare the similarities in the conceptual chart shown in Fig. 6.10 with Fig. 9.4.

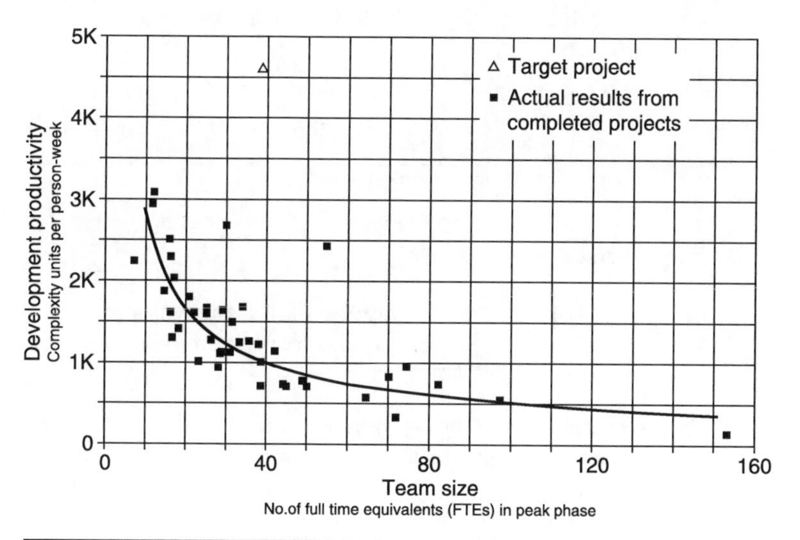

FIGURE 9.3 IC development productivity declines as the team size grows. Planned vs. historical project data. (*Source: Numetrics [9.4].*)

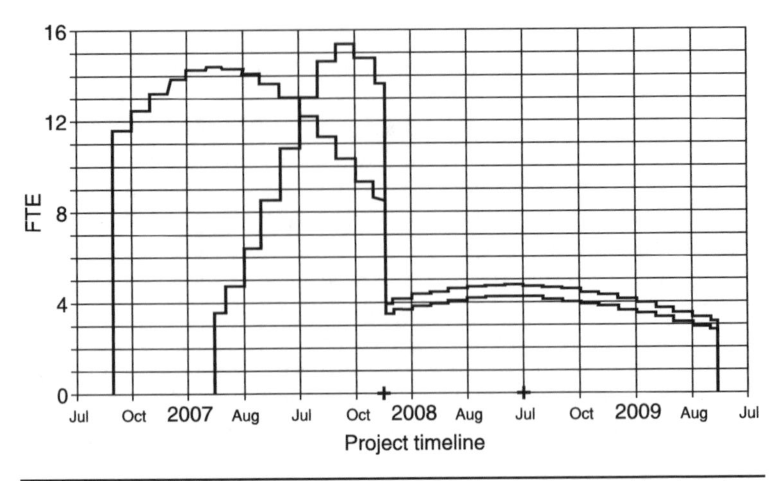

FIGURE 9.4 Example design resource planning for the front- and back-end phases. (*Source: Numetrics [9.4].*)

> The estimate is generated by Numetrics' project plan synthesis engine, which is in production use by many of the top semiconductor companies in the industry. The estimate is based on the development team's estimated productivity combined with a calculation of the target design's complexity.
>
> Such information can be useful in estimating the staffing level needed to meet schedule goals, benchmarking the underlying assumptions of a project plan to determine schedule risk and then tracking and monitoring the schedule risk of an IC development program
>
> Ron Collett
> President and CEO, Numetrics

9.8 Key Points

- Use of a distributed supply chain requires implementation of management process focused on inter-company coordination.
- Management of relationships, partnerships, the program, design productivity, and risks are essential elements for success in fabless semiconductor implementation. Success is defined as an on-time, on-budget delivery of product that meets functionality, performance, and cost targets.

CHAPTER 10

Future Trends

The nearly 40 year association I have had with the semiconductor industry has been exciting. This is in spite of the many industry ups and downs. Being associated with leading edge IC technologies and the resulting products has been fulfilling. I have had the opportunity to be involved in many facets of the industry as discussed in this book. In this chapter I would like to recap some of the major directions and trends from the fabless perspective.

10.1 Industry Stratification and Opportunities

As discussed in Chap. 1, the overall semiconductor industry growth rate appears to be slowing. It has been observed that the growth rate of revenue per wafer is slowing down [1.21]. ASPs are trending downward as the overall IC volumes increase faster than revenue growth. ICs with increased functionality and reduced ASPs have fueled the electronics industry. How many times have you experienced a drop in the price of the new digital camera or DVD player right after you bought it! The average price of a DVD recorder dropped by 12% a year between 2002 and 2007 [2.6]. All this leads to increased competition and pricing pressures on IC suppliers. Couple this with the increasing cost of designing ICs at the leading edge process nodes, and you have a fairly gloomy outlook. The outlook gets a little worse if you realize that high volume applications are mostly in the consumer markets, where low unit cost is a strong driving force.

So, I will have to agree with industry perception that things are gloomy if you consider business as usual. However, if you realize that the people involved in this industry are innovators, there is much hope, especially if we combine technical innovation with business acumen to find creative solutions. A significant positive impact in growing the ubiquity of ICs and the semiconductor industry comes from recent discussion, of the emergence of the second wave of the digital consumer revolution, the digital society [10.1] and surface, tangible and graspable computing [10.2].

However, in the near term, the fabless Company has to overcome harsh realities. In an industry where designing at the very leading edge process node is a must, it gets tough for the $10M fabless IC start-up to even think of playing in that ballpark. The leading edge ballpark has become almost exclusive to IC companies that have the design resources to introduce products at the leading edge technologies and to support high volume products. Trying to design a product that competes with the "big boys" will be a critical mistake for the start-up. What one needs is an innovative IC product with differentiation. It is prudent for the start-up to avoid differentiating their product based only on the use of leading edge process technology node. The differentiation should not be based on process technology alone. The IC must have a new feature or capability that helps solve a real customer problem. The new feature or capability could come from a hardware solution, a software solution or a combination of the two.

One approach for a fabless startup is to embrace a two-stage strategy. This can be done by demonstrating the new idea in a mainstream technology and then licensing the idea to one of the "big boys," or partnering with them. Utilize the big boys' advanced technology design infrastructure and shuttle capacity to demonstrate and potentially launch the idea in the leading edge technology. This approach could also lead to a merger/acquisition exit strategy.

In Chap. 7 there were suggestions of how it is indeed possible to design in the leading edge process nodes. By judiciously choosing the start of a new design about one to one and a half years after the first announcement of a new process node, it is possible to reduce design cost to under $10M. The nay sayers will argue that this means your product will be non-competitive. Obviously, this is true if your product has no other differentiation and you are competing for the same design win as the big boys. A key message is that your product must have differentiating features other than process technology!

Another approach is to play in a ballpark associated with mature technologies ($\geq$ 250 nm). The design costs will be lower, and so will the wafer costs because these technologies are fabricated in fabs with already depreciated capital infrastructure. The CPF will not be the lowest possible, but this could be variable approach if you can meet product cost and margin goals. In Chap.5 data was presented showing that approximately 70–80% of the top four foundries' revenue came from mainstream and mature technologies in 2006. There is additional information available from a VLSI Research report showing that 50% of worldwide wafer starts are on process nodes 1.5 μm or larger [1.21]. What this means is that there is a market to play in even if you are not at the very leading edge process technology with your product.

Hardware related product differentiation comes from factors such as functionality, features, and performance. Differentiation could come in different ways—circuit level enhancement, process combinations and features, packaging creativity and the like. You must also recognize that there can be much competition in this mature technology space from countries around the globe.

Yet another dimension is to play with disruptive ideas using unique technologies, or playing in new emerging areas. One example is MEMS technology. This technology has already been applied extensively in the development and shipment of accelerometers. Many other opportunities exist. An Interferometric Modulator Display (IMOD) is one example among electro-optical applications [10.3, 10.4]. Another area is the application of nanotechnology, in its various forms, to bio-medical applications.

It is clear to me that business considerations will continue to stratify the industry. The top tier will be the users of leading edge technologies that will play in the very high volume markets. The low tier will have players designing and shipping products in mature technologies. The middle tier is where most of the fabless start-ups could play, through creative technical ideas and business approaches. Assume that things will be different and it will not be business as usual. The good news is that opportunities are out there!

Here are a few guidelines, as a reminder of the rules for selecting technologies:

- Design in the **oldest** process technology node that meets the functionality and performance goals.
- Design in the **newest** technology node you can **afford**.
- If the resulting IC is not competitive, you have to re-think the company strategy!
- You also have a serious issue if the technology you can afford will not meet the functionality and performance goals!
- Remember also that your idea and your IC must solve a customer problem! Especially one they are to willing to pay for.

10.2 Funding

Venture capital funding was a popular way to get a new fabless IC company launched in the second half of the 1990s. While this is still a viable alternative, getting funded has become more challenging since the collapse of the dot-com days. Remember, the investors are looking to maximize their return on investment at the lowest risk factor possible. So, if you have a really good idea, a great team with a proven record, a solid marketing, execution business plan, I am sure there are investors that will be interested in assisting the launch of a new company.

It is clear to me that the time is also ripe for new investment models to emerge. Some larger IC companies, e.g., Intel, TSMC, Qualcomm among others, have set up their own venture funds. I believe that such venture funds will need to expand their business models and efforts to encourage and foster innovation in the industry at large. Private equity growth funds are also becoming interested in making investments in fabless companies [10.5].

Through these venture funds, many leading companies look for disruptive ideas and solutions [10.6, 10.7] from the start-up companies they invest in. One of the great assets of a start-up is the ability to "think out of the box" and create value. If the fabless IC company offers such potential, there is hope. Now, it would be ideal if these companies opened up their design infrastructure and/or silicon shuttles to the startup. In this way, it may be possible for the creative start-up to get a jump on a leading edge process node. If done right, this approach could allow the start-up to cut the lag time in half, from one to two years to maybe only six to twelve months behind the first use of the technology. That would be awesome! Such a relationship also establishes a link to the OEM value chain and their product development schedule. In order for this to happen there has to be an open and trusting relationship, maybe a partnership, between the key players at both companies. I believe this whole arena is ripe with opportunities to create innovation, and to foster disruptive ideas. Implementation will take creative solutions from "out of the box" thinking both at the startup as well as the "mother" companies.

10.3 Alternatives to the IDM Model

10.3.1 "Fab-Lite"

While there has been much discussion recently about how many companies can really afford the expensive leading edge wafer fabs, one must recognize that this is not a new problem. Ever since the late 1980s the vertically integrated system companies—Unisys, Digital Equipment, Honeywell in particular—have struggled with the financial justification of an internal wafer fab operation. Unisys shut down its internal wafer fab around 1991 as competitive IC process technology became available from its partners. Unisys then focused on internal IC design and packaging capabilities critical for providing differentiation in its products. As competitive design and packaging capabilities also became available, internal capabilities have generally focused on architecture, software and services.

It has been said recently that only a few IDM companies will be able to afford new wafer fabs [1.22]. Let us discuss a rationale for this. Figure 10.1 shows the 2006 revenue of the top 10 semiconductor companies [10.8]. TSMC's revenue has been included, although it was not included in the Gartner information. It is postulated that only companies with annual revenue more than $10B will be able to afford their own wafer-fab [1.22]. The cost of a 300 mm wafer-fab is typically $2.5–3 and a 450 mm fab is projected to be over $5B. If the company is looking for a 10 fold return from this investment in five years, the annual revenue stream required will be $6–10B. A more realistic goal may be a 5 fold return in 10 years. This implies required annual revenue of $1.5–2.5B from *that* fab. This kind of a revenue

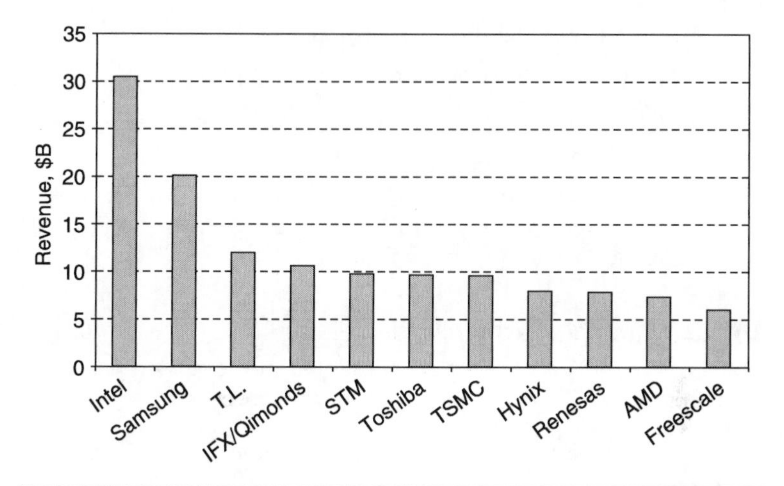

FIGURE 10.1 2006 revenue for the top semiconductor companies. (*Source: Gartner [April 2007].*)

stream also recoups the original investment in approximately two years. A more detailed financial modeling of wafer-fabs and their justification is beyond the scope of this publication.

However, such investment analyses are causing the IDMs to make bold moves towards going "fab-lite," or "asset-lite," and forming strategic partnerships. During the last few years there have been announcements from large IDMs (Texas Instruments [10.9], Freescale, others) about leveraging the commercial foundries for some of their IC manufacturing. In this "fab-lite" or "asset-lite" model, the IDMs can select a variety of operating modes. In one extreme they may request the foundry to create a special version of the process technology dedicated to them. In the other extreme they may use the standard foundry process, just like any other fabless company. There are also many variants in between these two extremes. A likely model these days establishes a cooperative relationship where the IDM and the foundry work together to define next generation process technologies.

Such relationships can be a win for the foundry as well as the IDM. The foundry gets increased manufacturing volume from the IDMs. The IDM gets access to new process technologies while avoiding enormous capital and process development investments.

The overall trend is towards a world where only a few vertically integrated and IDM companies will invest in building new wafer fabs. For some years the IDMs will likely leverage their existing fabs for specialty technologies that are not readily available from the foundries. The semiconductor ecosystem will consist of the following:

- A few big IDM companies.
- A few big foundries.

- A number of "fab-lite" IDM companies.
- Many fabless IC companies.
- SATS.
- Many IP and Design Services providers.
- Many companies providing support infrastructure—qualification and test services, failure analysis services, yield diagnostics, tester companies etc.

10.3.2 Strategic Partnerships

Collaborative efforts can help share the cost and competencies across members of an alliance. Examples of such alliances are:

- Fujitsu, NEC, Renesas, Sony, and Toshiba.
- IBM, Samsung, and Chartered Semiconductor, who have created a common technology platform. In this case the companies pool their resources in research and development. The customer then has a choice of fabricating designs at any of the partners, one of which is a commercial foundry. The cooperative arrangement has been expanded to include Infineon and Freescale.
- In 2000, there was an alliance set up at Crolles in France between ST Microelectronics and Philips. The alliance was later joined by Freescale and TSMC. NXP (formerly Philips) announced their withdrawal from the alliance in early 2007 [10.10] causing much uncertainty.

Collaborative innovation begins when companies come together to solve problems and/or develop customer-centric solutions that are beyond the scope, scale, or capabilities of the individual companies. Then they are collectively able to solve problems, develop new products or implement winning new business models as a result—an innovative process that was not possible when they acted within the "box" of their own individual space.

Chartered participates in two significant collaborative innovation alliances at present. The first one is with semiconductor technology leader IBM, which also now includes Samsung, Infineon, STMicro, and Freescale for the development of advanced semiconductor process technology. This collaboration has allowed us to close the technology gap with the leading companies in the semiconductor industry.

Additionally, Chartered has partnered with IBM and Samsung on a new customer-centric initiative called the Common Platform where we are collaborating in developing an ecosystem of partners to offer customers an open and flexible new model for manufacturing services. This collaborative go-to-market initiative brings our collective manufacturing and support capabilities together to help customers solve complex design problems and gives them new choices and flexibility.

Chartered has derived both direct financial and intellectual benefits from our collaborative alliances.

Kevin Meyer
Vice President, Chartered
Semiconductor

10.4 Virtual "Re-Integration"/IFM

While the disaggregated silicon supply chain works well for the fabless IC community, it is interesting to note that there are some trends that point to the need for a virtual "re-integration" of the supply chain. As good as each of the supply chain partners is in their core competency area, the fabless company has to provide the "glue" in bridging some of the "gaps" in the distributed manufacturing model. In this chapter I will discuss the need for this re-integration from three perspectives—engineering, operations and competitiveness.

10.4.1 Engineering considerations

Chapters 5 and 6 discussed the need for co-optimization of process and design considerations, especially on the latest nanometer process nodes. Examples given related to the management of leakage currents, process variability, and restricted design rules among others. In order to enhance manufacturability of advanced ICs, adjustments are required in the design phase. This creates a need for interactive efforts between designers and process folk, including the sharing of process information and models that have traditionally not been available outside the foundry. By the way, such information was readily available to the designer within an IDM company—a distinct advantage. In recent years the foundries and the EDA companies have stepped up to the challenge and now make available PDKs for use by designers. The PDKs include process related information that becomes readily available in the EDA tools. While these moves are good steps in the right direction, there needs to be close cooperation between the foundry and the design houses on the leading edge process nodes. This requires a partnership arrangement between the fabless IC company and the foundry. This is important for the fabless company to proactively consider process and manufacturability issues in the design phase.

As shown in Fig. 10.2, a new integrated model has been proposed for the interactions between a fabless company and the foundry [10.11].

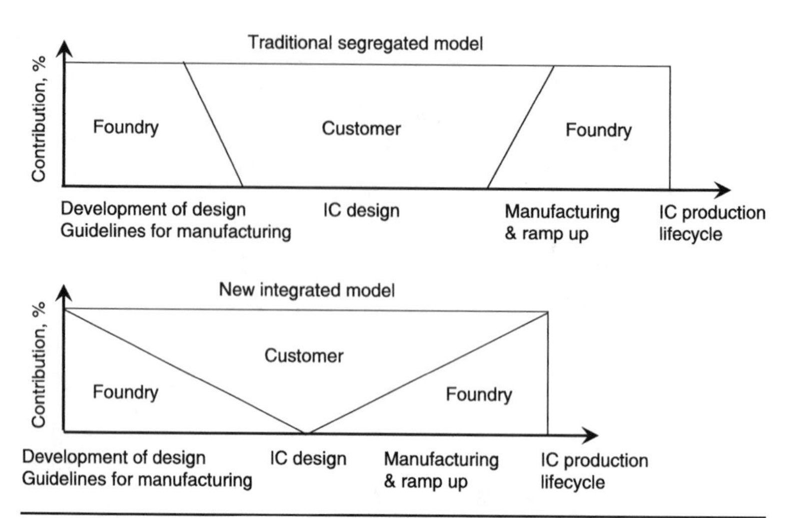

FIGURE 10.2 Comparison of a new foundry engagement model with the traditional model. (*Source: TSMC [10.11].*)

In the traditional model, the foundry provides a set of design rules and models, and may make available a library. Then it is up to the design house to complete the design, which gets prototyped and manufactured by the foundry. The new model shows a cooperative engagement throughout the design and manufacturing phases.

A similar concept, called IFM (Integrated Fabless Manufacturing) has been proposed by the largest fabless IC company [10.12]. The proposal is to align the technology chain through collaboration, not ownership. The methodology is proposed as a way to derive the best of all worlds—the business advantages and the flexibility of the fabless world along with the technical alignment and integration of the IDM world. A representation of the IFM approach compared to the vertical system company is shown in Fig. 10.3.

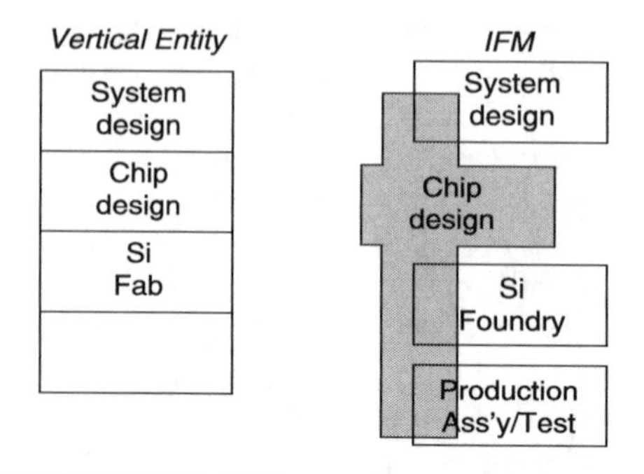

FIGURE 10.3 Comparison of the IFM model with the vertically integrated approach.

> The IFM business model derives the best of the long term learning at captive semiconductor companies (IDMs). It realizes benefits of the integrated solutions without requiring asset investments. Some keys to the successful execution of the IFM model are:
>
> 1. We invest selectively in areas where we can derive benefits such as time to market, lower cost and feature enhancements. Some of these investment areas are special libraries, circuit design, models, analog, and R.F. design, packaging, and test.
> 2. We leverage investments made by supply chain partners in their own core competency areas.
> 3. We keep our focus on differentiating our offerings.
>
> Behrooz Abdi
> President & CEO, Raza Microelectronics
> Former SVP & GM, Qualcomm CDMA Technologies

10.4.2 Competitiveness considerations

One of the perceived disadvantages of the fabless IC model is the delay in getting ICs to market on leading edge technologies, relative to the IDMs. This has been true, although the gap is shrinking. I would like to illustrate this through the following figures.

Figure 10.4 illustrates a maturity curve for silicon process development. Also illustrated are the library development maturity curve and the typical phases in the development cycle. The figure is not to scale. For instance, from the start to completion of qual could take two to three years. Library development starts later in the process development cycle because the design rules and the models have to firm up before library design can start. Completion of process and a product qual are usually a signal for the beginning of the production ramp.

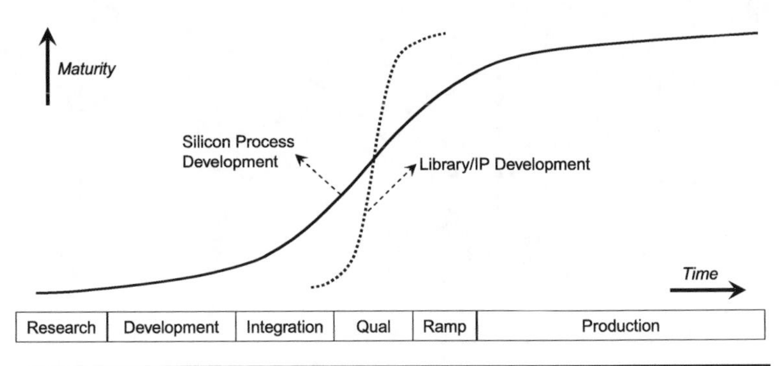

FIGURE 10.4 Silicon process and library development maturity curves. Also shown are typical phases in the silicon development lifecycle.

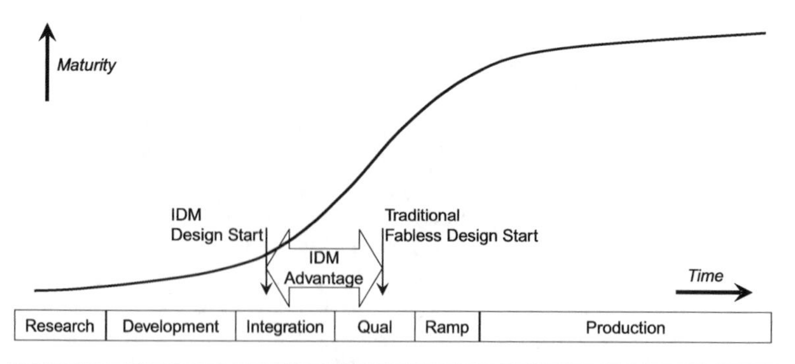

FIGURE 10.5 IDMs can start design sooner than a fabless IC company in the traditional model.

Traditionally, this is when any fabless company can get design rules and models from the foundry and begin IC design. This benchmark is shown by the right vertical arrow in Fig. 10.5. The arrow shows an aggressive schedule for the start, being before the actual completion of the qual. Within the IDM, library and product development could start early, even when the design rules and models are not quite mature. The assumption here is that the process maturity curve is identical at the foundry and the IDM. Yet the IDM has a significant advantage in starting the design relative to a fabless IC company. An IDM with a more aggressive process maturity curve can add to this advantage in design lead time.

Now let us consider the collaborative, IFM approach. If done right, the foundry works with a few key fabless partners in the process definition, development, and integration phases as if they were an internal design house, as illustrated in Fig. 10.6. While its execution is a challenge, such an approach can reduce the IDM advantage significantly.

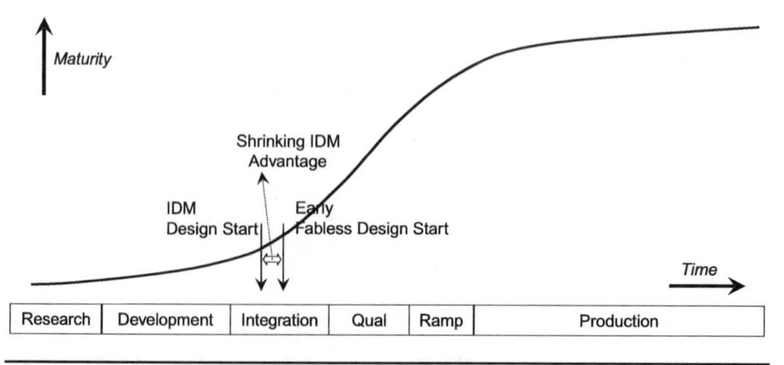

FIGURE 10.6 Shrinking IDM schedule advantage for design starts through the use of a collaborative model such as IFM.

Of course the risks are higher for the fabless company as it starts designing while the process is maturing. The foundries have been very aggressive in introducing new processes and making available design ecosystem to facilitate their technologies to be competitive with the IDMs.

10.4.3 Operations considerations

An important reason for the virtual re-integration of the supply chain became evident as the industry started to ship SiP parts with multiple die assembled inside. Procuring KGD from multiple suppliers is usually a challenge in itself. To take responsibility for yield and any interactive issues is an even bigger challenge. It is very difficult to find a supplier that will take responsibility for the yield and the interactive issues involved in the assembly of the SiP. Realistically, the fabless IC company is the only entity that can take on this responsibility.

Another important reason for the fabless IC company to manage the supply chain itself has to do with margin stacking. When using a turn-key supplier, you'll have to pay the margin of each of sub-contracting suppliers, plus an additional charge to the turn-key supplier. This can be an important limitation for cost sensitive ICs shipping in very high volume.

> The fabless company's ability to scale its operations is crucial for success. Initially the start-up company has the challenge of finding a foundry and other suppliers that have the right technology and will accept their business. As the business grows their challenge becomes one of ensuring adequate capacity to meet their demand. As they become a top tier customer they have additional challenges related to managing the suppliers to reserve upside capacity. Partnering and maintaining proper relations with the suppliers is a must.
>
> Another aspect of the scaling issue is the fact that small percentage changes in wafer and part availability can cause huge swings in the company's revenue stream.
>
> The entire supply chain must be managed to operate in alignment "on all cylinders." For example, if the test yield has not ramped up per plan it could affect huge swings in the wafer and assembly capacity demand. If the production test time has not been cut down as planned, the product cost could escalate and affect margins.
>
> Jim Clifford
> Sr. V.P. and General Manager
> Qualcomm CDMA Technologies

10.5 Order Entry Methodology

The normal process for a fabless IC company to place orders on the foundry is as follows:

- provide a 12-month rolling forecast;
- place orders a month in advance;
- calculate number of wafers required based on unit demand and the foundry's projected yield.

Typical lead time for delivery of production wafers is 8–14 weeks, depending on the process node and the foundry. During the fabrication time there is much "juggling" of priorities, especially if the fab is operating at or over capacity. Recently a streamlined methodology has been proposed [10.13]. In the new "outs" driven methodology, the fabless IC company will place orders for a certain number of die or wafers on a required date. The foundry would then manage their factory loading to fulfill the order. Similar methodologies have been proposed at the SATS.

> While the planning tools were not available initially, the "outs based" methodology has worked well for us as well as our suppliers. We get our parts when we expect them. The suppliers get to manage the start dates and quantities based on their internal projections of demand, capacity, and yield expectations. This approach is a win for the customers and the suppliers.
>
> Jim Clifford
> Sr. V.P. and General Manager
> Qualcomm CDMA Technologies
> Managing Innovation

10.6 Managing Innovation

In the last 15 years, fabless companies have flourished by incorporating product innovations into ICs that are manufactured by best-in-class suppliers. As the fabless segment grows beyond 20% of the semiconductor business, it is appropriate to think about the source of new technology innovations. This problem gets worse as IDMs cut back their investments in capital and process development. Technology challenges related to low power, device leakage, increased performance and memory incorporation will dominate the need for solutions over the next few years. Management of these challenges will require increased cooperation between the disciplines of IC process, design, EDA tools, packaging, test, architecture, and software. Business challenges related to the high cost of design, short time to market, short product life times, and low unit cost will require judicious choices at fabless IC companies. All this will requrie technical and business innovation. Issues related to managing innovation have been addressed by Geoffrey Moore [10.14]. There is a looming innovation crisis as more energy is spent in the deployment and optimization phases rather than in innovating. This is a significant challenge facing the industry as we move forward.

10.7 The Role of Research Organizations

In light of questions about how the semiconductor industry will maintain the "Moore" and the "more than Moore" treadmills, it is important to recognize the value and contributions of research organizations. There are three major research organizations worth identifying. While fabless companies generally assume that the technology will be there when they need it, large fabless companies interested in reaching for leading edge technology must be aware of the issues and opportunities in this space. These organizations are proving to be crucial in fostering pre-competitive research where major industry partners cooperate. Such cooperation helps defray the cost of capital investments. New ideas can be discussed and evaluated at the consortia. The partners pick up the idea after feasibility demonstration and integrate it into their individual processes. These organizations are also fertile ground for new developments in innovative, 3D packaging and other such technologies in support of "more than Moore" concepts.

- **SRC** (Semiconductor Research Corporation, http://www.src.org)
  - Established in 1982 as a non-profit organization.
  - Goal is to nurture and grow university capability in performing pre-competitive research.
  - Contributes to the continuous flow of research results and technical talent.
  - SRC funds 300+ projects at over 100 universities worldwide.
  - Over 5500 PhDs produced through SRC funding.
  - Nearly $1B distributed to universities since 1982.
  - Members review and analyze technology trends, and set strategic direction of research investments.
  - Research areas—Materials/Process, Design, CAD, Packaging, NanoManufacturing.
  - Research focused on ITRS driven issues, 3–12 year focus.
  - Members—AMD, Freescale, IBM, Intel, LSI, Spansion, TI, HP, Applied Materials, Novellus, Cadence, Mentor Graphics, Tokyo Electron, Axcelis, Rohm, and Haas.

- Members can assign mentors/liaisons that provide research direction and bring back early research info back to their company.
- Strong support infrastructure maintained through member driven advisory boards—Intel, TI, FSL, IBM, and AMD. Very strong proponents.

> SRC is a unique consortium with a mission to manage a range of university research programs worldwide that provide a competitive advantage to its members. It achieves this by funding and guiding graduate university research in areas of interest to its members that are pre-competitive in nature. This model has worked for 25 years where now the research funded by SRC has become a key component of the long-term business plans of numerous major semiconductor companies.
>
> Steven Hillenius, Ph.D.
> Vice President, SRC

- Members do an assessment of all projects annually, identify compelling reasons.
- Excellent web-based research engine.

- **SEMATECH** (Semiconductor Manufacturing Technology, http://www.sematech.org)
  - Formed in 1987 to re-invigorate the US semiconductor industry.
  - Member companies cooperate pre-competitively in key areas of semiconductor technology.
  - Key areas of effort are:
    - lithography;
    - interconnect;
    - FE processes;
    - manufacturing productivity;
    - environment, safety, and health;
    - coordination of global standards;
    - enhancing relations between semi companies and equipment and materials suppliers.
  - Consortia members make up 50% of the WW chip market.

> SEMATECH is a consortium for helping our members turn semiconductor technology innovations into manufacturing solutions. We work with 14 member companies and numerous partners and associates to address critical challenges in advanced technology and manufacturing, finding ways to speed development, reduce costs, share risks, and increase productivity.
>
> SEMATECH is highly effective because:
>
> - We assemble a critical mass of people and ideas that drive real change in chip manufacturing.
> - We provide flexible opportunities for participation in our breakthrough technical and manufacturing programs.
> - We return more than $5 in R&D value for each $1 our members invest in SEMATECH, an achievement that equates to more than $2 billion over the past five years.
> - We are the only consortium that provides and drives technology strategy for the entire industry.
>
> Dynamic and responsive, SEMATECH has evolved from a bold experiment in industry/government cooperation with a US focus to an international collaboration of leading manufacturers who represent more than 50% of global semiconductor revenues. At each stage of our evolution, we have shown our customers that SEMATECH brings measurable benefits to their R&D and economic development goals.
>
> Michael Polcari, Ph.D.
> President and CEO, SEMATECH

- Members—AMD, HP, IBM, Infineon, Intel, Micron, NEC, NXP, Panasonic, Qimonda, Renesas, Samsung, Spansion, TSMC, Texas Instruments.

- **IMEC** (Interuniversity Microelectronics Center, http://www.imec.be)
  - Located in Leuven, Belgium.
  - Established in 1984 by Belgian Government as a non-profit organization.
  - Currently ~1500 people, including 400 industrial residents.
  - €240M revenue (> 80% from Industry) in 2007.
  - Strategic partners: Infineon, Qimonda, Intel, Micron, NXP, Panasonic, Samsung, ST Microelectronics, Texas Instruments, TSMC, Hynix, and Elpida.
  - Research areas bridging the gaps between fundamental research at universities and technology development in the industry:

- silicon process and device technology;
- design technology;
- packaging;
- embedded wireless communication systems;
- develops processes and modules for heterogeneous integration of MEMS, sensors, etc. with ICs;
- wireless autonomous transducer solutions;
- photovoltaics;
- biomedical electronics;
- Organic electronics.

- Offers forum and opportunities for collaboration between industry and academics.
- Key to managing escalating research costs for major breakthroughs in new materials, new lithography, interconnects, and transistor concepts.

IMEC is a world-leading independent research center in nanoelectronics and nanotechnology with multi-disciplinary research ranging from IC process technology, system design and packaging to photovoltaics, biomedical, and organic electronics. Our research bridges the gap between fundamental research at universities and technology development in industry.

We offer the semiconductor industry a broad spectrum of research programs to tackle their ever growing challenges. Our research collaboration model is recognized worldwide as one of the most successful international partnership models for joint development of next-generation technologies. The concept is based on a sharing of cost, risk, talent, and IP. As an example, IMEC has set up the world's largest collaboration partnership on CMOS scaling uniting the world's leading IDMs, memory suppliers, and foundries together with leading equipment and material suppliers.

IMEC is positioned as a key research partner for shaping technologies for future systems with our unique balance of processing and system know-how, intellectual property portfolio, state-of-the-art infrastructure, and a strong network of companies, universities and research institutes worldwide

Gilbert Declerck, Ph. D
President and CEO, IMEC

10.8 Key Points

- The high cost of fabrication and process development is causing companies to collaborate and parter in order to reduce R&D costs. Research organizations are also playing a key role in implementing technology advancements.
- An important rule for your IC—"use the oldest process techonology node that allows meeting the product functionality, performance, and cost targets and the newest technology that you can afford."
- Differentiate you IC based on features other that the use of leading edg process technology.
- Your IC must solve a real customer problem.
- By making investments in a few key areas (e.g., IC package design and test), leading edge fabless IC companies are able to leverage the available ecosystem in order to combat some of the traditional advantages of being an IDM.
- A virtual re-integration of the supply chain is essential to bridge gaps that result from the use of a distributed supply chain.

Fabless Semiconductor Implementation is REAL. GO FOR IT!

Appendices

A.1 Business Plan Example

Business Plan Table of Contents

- Executive Summary
  - Objectives
  - Mission
  - Key to Success.
- Company Summary
  - Startup Summary
  - Management Team
  - Technical Team
  - Company Locations and Facilities.
- Market Analysis
  - Industry Overview
  - Market Size
  - Market Opportunities
  - Competitions.
- Product Summary
  - Product Description
  - Sourcing and Technologies
  - Product Development Schedules
  - Competitive Analysis
  - Product Advantages
  - Product Roadmaps.
- Marketing and Sales Strategy
  - Targeted Markets

- Customers
- Strategic Alliances
- Advertising and Promotion
- Selling Tactics.

- Manufacturing and Operations Plan
  - Wafer Sourcing
  - Backend Manufacturing Plan.
- Organization and Personnel Plan
  - Organization
  - Personnel Plan.
- Financial Plan.

A.2 Term Sheet Outline

Summary of Principal Terms

- Issuer:
- Closing Date:
- Form: Series C Preferred Stock.
- Price:
- Automatic Conversion: The Series C Preferred will be automatically converted into Common Shares upon ....
- Board of Directors:
- Voting and Protective Provisions:
- Restrictions on Stock Transfers:
- Rights of First Refusal:
- Pre-emptive Rights:
- Liquidation Preference:
- Antidilution Provisions:
- Dividends:
- Redemption:
- CoSale Rights
- Registration Rights:
- Standoff Provision:
- Information Rights:
- Expenses.

Rank	Company	Revenue CY 2006, $M	Stock Symbol
1	Qualcomm (QCT)	4,331	QCOM
2	Broadcom	3,668	BRCM
3	SanDisk	3,258	SNDK
4	NVIDIA	3,069	NVDA
5	Marvell	2,238	MRVL
6	LSI Logic	1,928	LSI
7	Xilinx	1,872	XLNX
8	MediaTek	1,624	2454 (TSEC)
9	Avago	1,576	Private
10	Altera	1,286	ALTR

TABLE A.1 Top 10 Fabless IC Companies. (*Source: FSA [1.23, 1.24].*)

	Revenue CY 2006, $M	Mkt. Share %
ASE	3,089	16%
Amkor	2,729	14%
SPIL	1,733	9%
STATS ChipPAC	1,617	9%
UTAC	638	3%
Others	9,263	49%
Total	19,069	100%

TABLE A.2 Top SATS (Semiconductor Assembly and Test Suppliers). (*Source: Gartner [http://www.gartner.com/it/page.jsp?id=501415].*)

B.1 Transistor Scaling

B.1.1 Tailoring Transistor Performance [5.10]

The following is a listing of transistor parameters and the associated areas of transitor device engineering.

μ_{eff}	:	Strained Silicon, High mobility channel
ε_{ox}	:	Hi-K gate dielectric
V_T	:	Gate workfunction (metal gate)
$\sum C$	:	Intrinsic and parasitic capacitance
L_g, V_{dd}, t_{ox}	:	Traditional scaling

$$Perf = \frac{1}{\tau} = \frac{Ids}{C * Vdd} \approx \frac{\mu_{eff}\varepsilon_{ox}}{\sum C} \frac{1}{L_g t_{ox}} \frac{(V_{dd} - V_T)^{1\sim 2}}{V_{dd}}$$

B.1.2 Process Roadmap

<table>
<tr><th>Technology Generation, nm</th><th>180</th><th>130</th><th>90</th><th>65</th><th>45</th></tr>
<tr><td>Start Risk Production</td><td>1999</td><td>2001</td><td>2003</td><td>2006</td><td>2008</td></tr>
<tr><td>V_{dd} Core/External, Volts</td><td>1.5 / 1.8, 2.5, 3.3</td><td>1.2 / 1.8, 2.5, 3.3</td><td colspan="2">1.2 /1.8, 2.5, 3.3</td><td>1.1 / 1.8</td></tr>
<tr><td>SRAM - 6T HD, sq μm</td><td>5</td><td>2.5</td><td>1.2</td><td>0.5</td><td>0.3</td></tr>
<tr><td colspan="6">PROCESS Technology Modules</td></tr>
<tr><td>Well</td><td>Retrograde</td><td colspan="4">Super Steep Retrograde</td></tr>
<tr><td>Isolation</td><td colspan="5">Shallow Trench</td></tr>
<tr><td>Gate Material</td><td colspan="5">Salicided-Poly-Si/SiO_2(nitrided)</td></tr>
<tr><td>Mobility</td><td></td><td></td><td colspan="2">Strained Silicon</td><td>Enhanced</td></tr>
<tr><td>Lithography (critical layers)</td><td>248 nm OPC PSM</td><td colspan="3">193 nm</td><td>193 nm Immersion</td></tr>
<tr><td>Silicide</td><td colspan="3">$CoSi_x$</td><td colspan="2">$NiSi_x$</td></tr>
<tr><td>Metal Layers/Material</td><td>Aluminum</td><td colspan="4">Copper</td></tr>
<tr><td>IMD</td><td>Std. (K=3.6)</td><td>FSG (K=2.9)</td><td>Low K (2.6)</td><td>Low K / ELK</td><td>ELK</td></tr>
</table>

FIGURE B.1 Process roadmap showing major changes in nodes from 180 nm to 45 nm.

B.1.3 Immersion Lithography Basics [5.10]

Rayleigh equation:

$$Minimum\ Feature\ Size = k_1 * \frac{\lambda}{NA}$$

where, λ is wavelength,

248 nm for KrF light source
193 nm for ArF light source
13.5 nm for EUV source

NA is numerical aperture with practical limits of 0.93 for Air and ~1.3 for water

k_1 is a constant representing system capability. The range of its values are:

State of the art	0.3–0.4
Practical limit	0.26–0.3
Physical limit	0.25

B.1.4 OPC Basics

The effects of sub-wavelength lithography were discussed in Chap. 5. The need for OPC arises because rectangular design images get distorted when printed using the lithographic process. The distortions are caused by optical diffraction and can be affected by process effects, reticle quality, proximity effects, and the like. The resulting patterns can cause effects such as narrow lines and shrinking or bulging at the corners. Complex algorithms have been developed that make adjustments to the design database to compensate for such process effects.

OPC operations are generally part of the mask making process at the mask making entity.

Another adjustment for extending the use of 193 nm optical lithography to 65 nm and 45 nm nodes is the use of "phase shift masks" (PSM) [7.4]. The mask making process adjusts the phase of the light on either side of a critical feature. This allows printing of tight features that would not be possible otherwise.

B.2 Yield Models

B.2.1 Poisson

$$Y = e^{-AD}$$

where A is area and D is the defect density per unit area

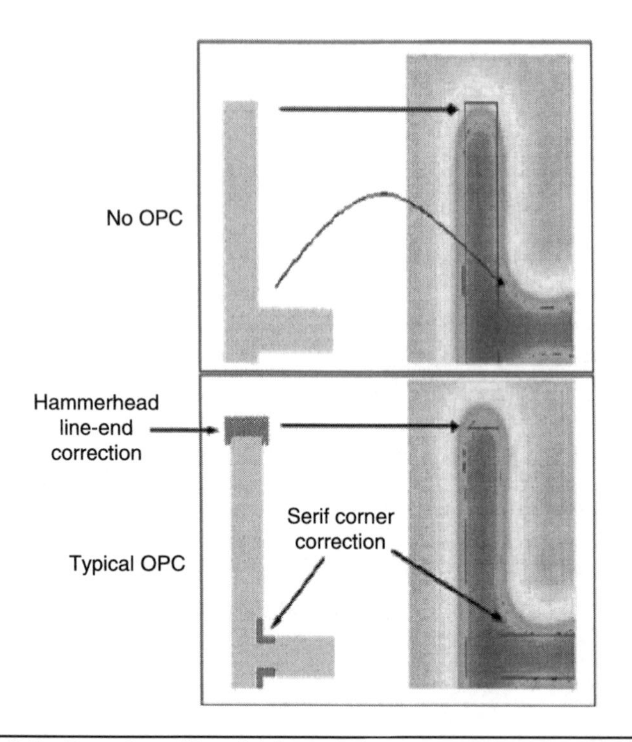

Figure B.2 Examples of line shrinkage and corner rounding without OPC correction, a typical hammerhead line-end correction, and a serif-corner correction.

B.2.2 Murphy

$$Y = \left[\frac{(1-e^{-AD})}{AD}\right]^2$$

where A is area and D is the defect density per unit area

B.2.3 Bose-Einstein

$$Y = [1 + AD_o]^{-n}$$

where A is critical area, D_o is the defect density per unit area per critical layer and n is a complexity factor representing the number of critical layers and $D_o = D / n$

B.3 Quality

B.3.1 Test Escapes

The theoretical test-escape rate for gross failures (opens, shorts, functional fails) can be modeled as:

$$\textit{Escape Rate} = 1-\left[Y^{(1-FC)}\right]$$

where FC is the true Wafer Sort defect coverage.

Table 8.1 shows escape rate as a function of probe yield and fault coverage:

The most common functional test metric, single stuck-at fault coverage, needs to be high in order to obtain an acceptable defectivity rate. ASIC providers that supply to consumer product manufacturers typically set a goal at 0.1% (1000 ppm). The escape rate expression, the target fault coverage must be better than 99% (with 90% probe yield), and closer to 99.5% (with 85% probe yield).

B.3.2 Reference Documents

JEDEC, AEC, and other References
http://www.jedec.org
http://www.aec.org

- J-STD-020, Moisture-Induced Stress Sensitivity for Plastic Surface Mount Devices.
- JESD22-A104, Temperature Cycling.
- JESD85 Methods for Calculating Failure Rates in Units of FITs.
- JESD22-A101, Steady State Temperature Humidity Bias.
- JESD22-A108, Temperature, Bias Operating Life.
- JESD22-A110, Highly Accelerated Stress Test (HAST).
- JESD22-A113, Preconditioning of SMT Devices prior to Reliability Testing.
- JESD22-A114, ESD Sensitivity Testing HBM.
- JESD47B, Stress test Driven Qualification of ICs.
- JESD78, IC Latch-up Test.
- JESD85, Methods for Calculating Failure Rates in Units of FITs.
- JEP122B, Failure Mechanisms and Models for Semiconductor Devices.
- ANSI ESD STM5.1 ESD—Human Body Model.
- ANSI ESD STM5.2 ESD—Machine Model.

- ANSI ESD STM5.3.1 ESD—Charged Device Model.
- AEC 002—Statistical Yield Analysis.
- AEC 003—Characterization.
- AEC 100-002D—HBM.
- AEC 100-003E—MM
- AEC 100-007—Fault Simulation and IDDQ.
- AEC 100-009—ELFR.
- AEC 100-011B—CDM.

B 3.3 Life Test Applicable Documents and Typical Test Conditions

HTOL (Hi Temp Op Life) JESD22-A108
125°C (Grade 1) 1000 hours 77 parts from 3 Lots, fails allowed (77 × 3/0)
Parts tested at room temperature, hot and cold upon completion of life test

HTS (Hi Temp Storage) JESD22-A103B
150°C (Grade 1 Plastic); 200°C (Ceramic) 1000h; 72h 45 × 1 Lot/0

HAST (Highly Accelerated Stress Test) JESD22-A102, ESD22-A118
121°C/100% relative humidity/15psig or 130°C/85% 96h 77 × 3 Lots/0
Pressure cooker tests. Parts tested at room temperature upon completion of life test

TC (Temp Cycling) JESD22-A104B
–65°C /150°C (Grade 1) 500 cycles 77 × 3 Lots/0
Pressure cooker tests. Parts tested at room temperature and hot conditions upon completion of life test

ESD (Electro Static Discharge) AEC-Q100-002, AEC-Q100-003
2kV HBM or 1kV CDM or 200V MM 6 × 1 Lot/0
Parts tested at room temperature and hot conditions upon completion of life test

Latch Up AEC-Q100-004
6 × 1 Lot/0
Parts tested at room temperature and hot conditions upon completion of life test

Glossary

3D IC	Three dimensional integrated circuits fabricated by stacking two or more die.
A/d	Mostly analog IC with a small amount of digital content.
AEC	Automotive electronics council.
AFR	Average failure rate, bottom of bathtub curve, Fig. 8.3.
AOQL	Acceptable outgoing quality level.
ASIC	Application specific integrated circuit.
ASP	Average selling price.
Assembly	Process used to mechanically attach and electrically connect a die into a package.
ASSP	Application specific standard product.
ATE	Automatic test equipment.
ATPG	Automatic test pattern generation.
Ball	Refers to solder balls attached either to the package substrate to fabricate a BGA family of packages, or to the die for flip chip assembly.
BGA	Ball grid array package. If using a laminate substrate it is called a plastic BGA (PBGA). If using a ceramic substrate it is called a ceramic BGA(CBGA).
BiCMOS	Combination of bipolar and CMOS devices on the same IC.
Bipolar	Transistors formed by the combination of two p-n semiconductor junctions, either n-p-n or p-n-p, using both electrons and holes as carriers.
BIST	Built in self test. Methodology for improving fault coverage and reducing test time.
Blind build	Refers to the assembly and packaging of parts without prior functional testing in wafer form.
C4	Controlled collapse chip connect. Technique to reflow solder balls for making connections to the package, originally used at IBM.
CAD	Computer aided design.
CAGR	Compound annual growth rate.

Captive	A vertically integrated system or semiconductor company performing all aspects of IC design and fabrication in house.
CAR	Corrective action request, usually in response to an RMA or another customer request.
CDM	Charge device model, a configuration used to test for ESD.
CDMA	Code division multiple access.
Chip	An IC in silicon form, prior to assembly in a package. Used interchangeably with "die."
Chip complexity	The number of transistors or gates or functions on an IC chip.
CMOS	Complementary metal oxide semiconductor, process that incorporates both n- and p-channel transistors.
Corners	Refers to the envelope of limits of the process, voltage or temperature in which the design must be operational.
COT	Customer owned (mask) tooling, refers to approach where the customer has designs fabricated by a foundry.
CPF	Cost per function.
CPU	Central processing unit.
CSP	Chip scale package, similar to a BGA but with a smaller footprint; package is only slightly bigger than the die.
D/a	Mostly digital IC with a small amount of analog content.
DD or Do	Defect density or defectivity per unit area.
DFM	Design for manufacturability.
DFT	Design for test.
DRC	Design rule checking, is a process that verifies compliance of the design database to the foundry's design rules.
Die	An IC in silicon form, prior to assembly in a package. Used interchangeably with "chip."
DIY	Do it yourself, approach where a fabless company implements the COT approach themselves.
DRAM	Dynamic random access memory.
ECO	Engineering change order.
EDA	Electronic design automation.
EFR	Early failure rate.
ELK	Extreme low K, dielectric constant of the insulting material used in the inter-metal layers, K values less than 2.5.
EOS	Electrical over stress.
ESD	Electrostatic discharge.

e-test	Electrical test performed on a wafer upon completion of the silicon processing.
Extraction	Calculation of parasitic parameters from a design data base.
Fab	Fabrication facility, usually a silicon wafer fab.
fab-lite	Business model where an IDM outsources some or all of their wafer fabrication to a foundry, thereby reducing their capital asset investments.
FBGA, FPBGA	Fine-pitch (plastic) ball grid array package.
FIT	Failures in time, failures per billion hours of device operation.
Flash	A type of non-volatile memory.
Flip Chip	A technology using solder balls for connecting a "flipped" die, face down to the package.
Floorplanning	A design process where the designer lays out the blocks of an IC for initial planning, estimates of connectivity, die area, etc.
FT	Unity gain cut-off frequency.
Full Mask	IC fabrication using all mask layers in the selected process technology.
GaAs	Gallium arsenide is an alternative material to silicon. It is used for making high performance discrete semiconductors and low complexity ICs compared to CMOS.
Gate	A grouping of transistors that is used to perform a logic function, e.g., a 2-input NAND gate.
Gate array	An array of logic gates. Traditionally used as a vehicle for the designing ASICs by customizing the metal interconnect layers only.
GDP	Gross domestic product.
GDPW	Gross die per wafer, the total number of possible die on a wafer for any given design.
GDSII	Graphical data system. A de-facto standard for the exchange of design database to the wafer fab.
GSM	Global system for mobile communications.
HAST	Highly accelerated stress test.
HBM	Human body model, a configuration used to test for ESD.
HCI	Hot carrier injection, refers to injection and trapping of hot carriers at the gate oxide-silicon interface.
HDMI	High definition multimedia interface.
HiK gate	Transistor gate structure fabricated using a high K dielectric gate oxide.
HTOL	High temperature operating life test, typically run at 125°C.
HTS	High temperature storage test, typically run at 150°C for plastic packages.

I/O	Input output of the IC.
IC	Integrated circuit.
IDM	Integrated device manufacturer, refers to a vertically integrated company performing all aspects of IC design and fabrication in house.
IFM	Integrated fabless manufacturer, refers to an evolving business model where a fabless company gets IDM-like benefits without large capital investments.
IP (blocks)	Pre-configured intellectual property blocks performing desired functions.
IQL	Incoming quality level.
ITRS	International technology roadmap for semiconductors.
JEDEC	Joint electron device engineering council.
JTAG	Joint test action group.
Leading edge	Refers to ICs using any or all the following—leading edge process node, packing density, package pin count, performance, etc.
LF	Lead frame, usually made of copper, used in a package to make connections between the die and the external PCB.
Library	A library of functional elements and/or standard that cells is used as building blocks to implement IC designs.
Lithography	The photolithographic process used to transfer the IC design graphical elements onto the silicon wafer.
Logic Synthesis	Conversion of a software description of circuit behavior (usually in RTL code) into logic gates.
Lot	Refers to a batch of wafers or packages during fab and assembly respectively.
Low K	Refers to the dielectric constant of the insulating material used in the inter-metal layers, K values less than 2.9.
LSI	Large scale integration, refers to ICs with up to about 10K gates.
LVS	Layout versus schematic refers to a process for checking if the design's physical database matches the original circuit schematics.
M&A	Mergers and acquisitions, a process used to assess and implement a merger between two companies or an acquisition of one company by another.
Mainstream	Refers to ICs using any or all the following—mainstream process node, packing density, package pin count, performance, etc.
Masks	A set of glass or quartz plates with a metallic coating that has been personalized with particular layers of an IC design.

Mature	Refers to ICs using any or all the following—mature process node, packing density, package pin count, performance, etc.
MCM	Multi chip module, refers to a package with more than one chip.
MEBES	Manufacturing electron beam exposure system. Now generally refers to a design database format that drives the e-beam mask making equipment.
MEMS	Micro electro mechanical systems, or micro machines fabricated using modified semiconductor IC processes.
Micrometer	A unit of length equal to one millionth of a meter.
MiM cap	Metal-insulator-metal capacitor.
Minimum feature	The smallest dimension, either line width or space fabricated on an IC.
Mixed Signal	The presence of both digital and analog signals on an IC.
MM	Machine model, a configuration used to test for ESD.
Moore's law	Gordon Moore's projection of a 2 fold increase in transistor complexity every year (1975); currently every two years.
More than Moore	A methodology for maintaining, and perhaps exceeding, transistor growth rate using alternative technologies and packaging.
MOSFET	Metal oxide semiconductor field effect transistor, where source to drain current flow is controlled by gate voltage.
MPEG4	Moving picture experts group standards for video compression.
MPW	Multi project wafer. A technique for simultaneous fabrication of multiple designs by incorporating multiple design databases on the same glass mask or reticle.
MSI	Medium scale integration, refers to ICs with up to about 100 gates.
MSL	Moisture sensitivity level, specifies how long parts can be exposed to the environment before they will need to be baked prior to board assembly.
MTTF	Mean time to failure, in hours. This is the inverse of the FIT rate, e.g., 1000 FITs equates to a 1 million hour MTTF.
NAND	Logic gate with functional equivalence of "not and."
Nanometer	A unit of length equal to one billionth of a meter.
NBTI	Negative bias temperature instability, refers to a negative shift in PMOS threshold voltage.
NDPW	Net die per wafer, the number of functional die on a wafer. This is the product of the GDPW and the sort yield.

NMOS	N channel MOSFET, where electrons are the majority carriers.
Node	A process generation of technology, usually defined by the minimum feature size allowed in the technology, e.g., 45 nm.
NRE	Non-recurring engineering. Charge for engineering services, usually paid up front, and not connected to the product's unit cost.
NVM	Non-volatile memory. Memory elements that retain their information when power is turned off and then back on.
OEM	Original equipment manufacturer.
OPC	Optical proximity correction. Methodology used to make images printed on the wafer replicate the designed images.
ORM	Ongoing reliability monitor. Procedures to monitor IC reliability after it has been released to production. Based on a sampling of parts.
OTP/MTP	One (multi) time programmable, elements that can be programmed after fabrication is completed.
Packing density	The number of transistors or gates or functions on an IC chip per unit area; sometimes refers to the total number instead.
PCB	Printed circuit board. Used to assemble and interconnect electronics part in a system.
PDK	Process design kit, includes process related information important for the designer, e.g., design rules, models, DRC, etc.
PMOS	P channel MOSFET, where holes are the majority carriers.
PoP	Package on package is a configuration where one or more packaged ICs are stacked on top of another packaged IC.
Prototyping	The process used to fabricate a handful (usually 10–40) of ICs for sampling a new design.
PSM	Phase shift masks, are fabricated such that phase of the light on either side of critical features is adjusted.
PVT	Process, voltage and temperature, corner limits to ensure IC performance in manufacturing.
QFP	Quad flat pack, a leadframe package with leads on all four sides.
Qual	Qualification validates that the IC will meet the reliability requirements as set forth in the Quality manual.
RC	Resistor capacitor, usually refers to signal propagation delay due to line resistance and capacitance.
RDL	Redistribution layer, is a metal layer on top of the die that connect die bonding pads to pad where solder bumps are located.

RF	Radio frequency
Risk production	Production wafer starts in anticipation of qual completion, design acceptance and customer orders to allow a rapid volume ramp.
RMA	Return Materials Authorization, a process for authorizing, channeling and follow up of customer returns.
ROA	Return on assets.
ROE	Return on equity.
RTL	Register transfer level, is a software description of a circuit's behavior in terms of signal flow between hardware registers, and the logical operations performed thereon.
SATS	Semiconductor assembly and test suppliers.
Scaling	Refers to the scaling of devices from one process generation to the next. Typical scale factor between nodes is 1.4x, with feature reduction to 0.7x.
Scan	A DFT methodology that connects all flip-flops in the design into a shift register chain to verify functionality.
SD	Stacked die, refers to the assembly of one or more die on top of another in the same package.
Shrink	Refers to the reduction in feature size applied to a finished design through logical and optical operations.
Shuttle	A periodic lot run by foundries that incorporates MPWs for design validation of multiple designs.
Silicon	Semiconductor material used as the substrate in IC fabrication.
SiGe	Silicon germanium refers to a process used for fabrication of silicon bipolar transistors incorporating germanium in the base.
SiO2, SiON	Silicon dioxide (or oxynitride) is a dielectric used in silicon fabrication.
SiP	System in package, is a loosely used term to describe packaging of multiple die in the same package, e.g., a logic die plus a memory die.
SMT	Surface mount technology, allows attachment of electronics parts on a PCB with connections on the surface only (no holes).
SoC	System on a chip, refers to ICs at the leading edge, incorporating 10M or more gates/chip and/or multiple system blocks on the same chip.
SOI	Silicon on insulator, is an alternative starting substrate material that offers lower junction capacitance and improved performance compared to bulk CMOS.
Solder bump	Solder balls that are attached to pads on the die. The die is then flip-chip attached in the package or directly on the board.

SP	Stacked package, same as PoP.
SPC	Statistical process control, refers to the methodology for managing and controlling process parameter distributions in manufacturing.
SRAM	Static random access memory. A memory element that stores the data until new data is re-written.
SSI	Small scale integration, refers to ICs with up to about 10 gates.
STA	Static timing analysis, checks for setup, hold and transition time violations in PVT corners to validate that the design performance,
SSTA	Statistical STA, refers to the deterministic timing of gates and interconnects with probability distributions; result is a distribution of possible circuit outcomes.
Standard cell	Pre-designed logic elements that are used as basic building blocks in standard cell based ASICs.
STI	Shallow trench isolation, is a process used to etch partial trenches in the silicon which are used to form isolation areas between adjacent devices in the die.
Structured array	ICs that have pre-defined functional blocks along with areas that have customizable standard cell or gate array areas.
Substrate	Refers to the wafer (usually silicon) in wafer fab, the package body in assembly.
Subthreshold	Refers to current that flows between the source and drain of a MOSFET at gate voltages below the gate threshold voltage.
SY	Sort yield, represents the fraction of die that are functionally good.
TC	Temperature cycling, refers to a life test that checks for mechanical integrity of the packaged IC. Temperatures are usually cycled from –65°C and 150°C.
TH	Through hole, refers to PCBs with holes. Packaged ICs with leads are mounted on the PCB with the leads protruding through the holes.
Timing analysis	Processes used to verify that the physical design process will result in an IC that meets its performance goals.
Time units	pS (pico-second, 1e-12), nS (nano-second, 1e-9), µS (micro-second, 1e-6), mS (milli second, 1e-3).
TT$	Time to revenue $, refers to lead time until the fabless company begins generating revenue.
Virtual reintegration	In the IFM business model, virtual reintegration of the distributed supply chain refers to the replication of behaviors and activities like at an IDM.

VLSI	Very large scale integration, refers to ICs with up to about 100K gates. In the absence of another common term, VLSI is the ubiquitous term referring to complex ICs.
V_T	Gate threshold voltage of the MOSFET.
Wafer	A round, highly polished, single crystal semiconductor substrate (usually Silicon) used in IC fabrication. Diameters up to 300 mm are currently used.
Wafer fab	Silicon wafer fabrication facility.
Wafer lot	A batch of wafers in wafer fab. An engineering lot usually has 12 wafers, a production lot commonly has 24 wafers.
Wire bond	Refers to the connection of a wire to the bonding pad on the die or in the package.
WLCSP	Wafer level CSP, refers to attachment of flip-chip die directly to the PCB.
WS	Wafer sort, refers to functional testing of the die in wafer form prior to assembly and test.
Yield	At any step, yield refers to the fraction of parts (or die, or wafers) that are acceptable.

Acronyms

3D	3 Dimensional
AFR	Average Failure Rats
AMD	Advanced Micro Devices
ADC	Analog Digital Converter
AOQL	Acceptable Outgoing Quality Level
ASICs	Application Specific ICs
ASP	Average Selling Price
ASSPs	Application Specific Standard Products
ATE	Automated Test Equipment
ATPG	Automatic Test Pattern Generation
BE	Back End
BEOL	Back End of Line
BGA	Ball Grid Array
BIST	Built in Self Test
C4	Controlled Collapse Chip Connect
CAA	Critical Area Analysis
CAD	Computer Aided Design
CAGR	Compound Annual Growth Rate
CAM	Content Addressable Memories
CD	Critical Dimension
CEO	Chief Executive Officer
CMP	Chemical Mechanical Polishing
COGS	Cost of Goods Sold
CDMA	Code Division Multiple Access
COT	Customer Owned Tooling
CP	Circuit Probed
CPF	Cost per Function
CPU	Central Processing Unit
CS	Customer Services
CSPs	Chip Scale Packages
CUP	Circuit under Pad
DAC	Data Convertors
DD	Defect Density
DFM	Design for Manufacturability

DFQ Design for Quality
DFT Design for Test
DFY Design for Yield
DRAMs Dynamic Random Access Memories
DRC Design Rule Check
ECL Emitter Coupled Logic
ECOs Engineering Charge Orders
EDA Electronic Design Automation
ELK Extremely Low K
EOS Electrical Over Stress
ERP Engineering Resource Planning
ES Engineering Sample
ESL Electronic System Level
FBGAs Fine Pitch BGA
FE Front End
FF Flip Flop, or Fast Fast
FIB Focused Ion Beam
FITS Failures in Time
FPGA Field Programmable Gate Array
FR Failure Rate
FSA Fabless Semiconductor Association
G Generic
GDPW Gross die on the Wafer
GDS II Graphic Data System II
GP General Purpose
GSM Global System for Mobile Communication
HCI Hot Carrier Injection
HP High Performance
IC Integrated Circuit
IDM Integrated Device Manufacturers, or Manufacturing
IFM Integrated Fabless Manufacturing
IMMOD Interferometric Modulator Display
IP Intellectual Property
IPO Initial Public Offering
IQL Incoming Quality Level
ITRS International Technology Roadmap for Semiconductors
JTAG Joint Test Action Group
JVS Joint Ventures
KGD Known Good Die
LCD Liquid Crystal Display
LER Line Edge Roughness
LF Lead Frame
LOI Letter of Intent
LP Low Power
LPC Layout Parameter Check
LSI Large Scale Integration
LVS Layout versus Schematic

LWR	Line Width Roughness
M&A	Mergers and Acquisitons
MCM	Multi-Chip Modules
MEMS	Micro Electro Mechanical Systems
MF	Mainframe
MiM	Metal insulator Metal
MLF	Micro Lead Frame
MM	Multi Media
MoM	Metal Oxide Metal
MOS	Metal Oxide Semiconductor
MOSFET	Metal Oxide Semiconductor Field Effect Transistor
MOU	Memorandum of Understanding
MPW	Multi-project Wafers
MRP	Material Resource Planning/Manufacturing Resource Planning
MSI	Medium Scale Integration
MTP	Multi-time Programmable
MTTF	Mean Time to Failure
NA	Numerical Aperture
NBTI	Negative Bias Temperature Instability
NDAs	Non-Disclosure Agreements
NDPW	Net Die per Wafer
NL	Net List
NRE	Non-Recurring Engineering
NVM	Non-Volatile Memory
OCV	On-chip Variation
OEM	Original Equipment Manufacturers
OLED	Organic Light Emitting Diode
CMOS	Complementary Metal Oxide Semiconductor
OPC	Optical Proximity Correction
ORM	Ongoing Reliability Model
OTP	One-time Programmable
PC	Personal Computer
PCBs	Printed Circuit Boards
PDKs	Process Design Kits
PGA	Pin Grid Array
PiP	Package in Package
PLLs	Phased Lock Loops
PMOS	P-Channel MOS
PO	Purchase Order
PoP	Package on Package
PP&C	Production Planning and Control
PSM	Phase Shift Mask
PVT	Process Voltage and temperatures
QA	Quality Assurance
QFN	Quad Flat No-lead
QM	Quality Manual
QS	Qualifiable Samples

RC	Resistance Capacitance
RDL	Redistribution Layer
RFID	Radio Frequency Identification
RMA	Return Material Authorization
ROA	Return on Asset
ROE	Return on Equity
RTL	Register Transfer Level
SAM	Served Available Market
SATS	Semiconductor Assembly and Test Suppliers
SAW	Surface Acoustic Ware
SD	Stacked Die
SEM	Scanning Electron Microscope
SHL	Super Hot Lot
SIA	Semiconductor Industry Association
SiP	System in Packages
SM	Surface Mount
SOC	System on a Chip
SOI	Silicon on Insulator
SP	Stacked Packages
SPC	Statistical Process Control
SRAM	Static Random Access Memory
SS	Slow Slow
SSI	Small Scale Integration
SSTA	Statistical Static Timing Analysis
STA	Static Timing Analysis
STI	Shallow Trench Isolation
SxS	Side by Side
SY	Sort Yield
TAM	Total Available Market
TH	Through Hole
TSMC	Taiwan Semiconductor Manufacturing Company
TSS	Through Silicon Stacking
TSV	Through Silicon Via
TT$	Time to $
TTM	Time to Market
ULSI	Ultra Large Scale Integration
VLSI	Very Large Scale Integration
VP	Vice-President
WIP	Work in Progress
WLCSPs	Wafer Level CSP
WS	Wafer Sort Cost

Bibliography

1.1 W. F. Brinkman et al., "A history of the invention of the transistor and where it will lead us," IEEE Journal of Solid State Circuits, Vol. 32, December 1997, pp. 1858–1865.
1.2 Texas Instruments, Bell Labs and Sony History, http://www.pbs.org/transistor/
1.3 Fairchild History, http://www.fairchildsemi.com/mediaKit/history_1950.html
1.4 Intel History, http://www.intel.com/museum/corporatetimeline/index.htm
1.5 Motorola History, http://www.motorola.com/content.jsp?globalObjectId=7710
1.6 IBM History, http://www-03.ibm.com/ibm/history/history/history_intro.html
1.7 AMD History, http://www.amd.com/us-en/Weblets/0,,7832_10554_10555,00.html
1.8 National Semi History, http://www.national.com/company/pressroom/history.html
1.9 IEEE SSCS Newsletter, Vol. 12, No. 2, Spring 2007, pp. 16–54
1.10 G. E. Moore, "Cramming more components onto integrated circuits," Electronics, Volume 38, Number 8, April 19, 1965, pp. 114ff.
1.11 G.E. Moore, "Progress in Digital Electronics," 1975 IEEE IEDM, pp. 11–13.
1.12 G.E.Moore, "No Exponential is Forever; but 'Forever' can be delayed," ISSCC 2003, Paper 1.1
1.13 M. Bohr, "A 30 Year Retrospective on Dennard's MOSFET Scaling Paper," IEEE SSCS Newsletter, Vol. 12, No. 1, Winter 2007, p. 13.
1.14 Semiconductor Industry Association. The International Technology Roadmap for Semiconductors, 2005 edition. SEMATECH: Austin, TX, http://public.itrs.net
1.15 IC Knowledge, http://www.icknowledge.com
1.16 R. H. Dennard et al., "Design of Ion-Implanted MOSFET's with Very Small Physical Dimensions," IEEE Journal of Solid State Circuits, Vol. SC-9, October 1974, pp. 256–268.

1.17 Mark Bohr, "The new era of scaling for energy efficient processors," Microprocessor Forum, May 2007, http://download.intel.com/technology/silicon/neweva.pdf/
1.18 R. Wawrzyniak, "SOC Trends in the year 2000," Electronic News, January 3, 2000, http://www.edn.com/index.asp?layout=articlePrint&articleID=CA48636
1.19 R. Tsai, Corporate Overview, TSMC Technology Symposium, April 9, 2007.
1.20 T. C. Chen, "Where CMOS is going: Trendy Hype vs. Real Technology," ISSCC 2006, Paper 1.1.
1.21 E. T. Heyen, "IDM's and the Fabless," FSA Conference, Munich, May 9, 2006.
1.22 O. H. Kwon, "Perspective of the Future Semiconductor Industry," DAC 2007 Keynote, June 5, 2007.
1.23 FSA, "Global Fabless Fundings and Financials Report – CYQ4/2006/Year-end 2006," http://www.fsa.org/publications/financials/0604/report.asp
1.24 J. Hurtarte et al., "Understanding Fabless IC Technology," Newnes/Elsevier, 2007.
1.25 N. Sakkran et al., "The Implementation of the 65nm Dual Core 64b Merom Processor," ISSCC 2007, paper 5.6.
1.26 B. Stakhouse et al., "A 65nm 2-Billion-Transistor Quad-Core Itanium Processor," ISSCC 2008, paper 4.6.
2.1 Semiconductor Industry Association, www.sia-online.org/pre_facts.cfm
2.2 Semiconductor Industry Association, www.sia-online.org/pre_release.cfm?ID=426
2.3 R. Rajgopal, "Evolution of Yield Enhancement Methodology for Advanced Technologies," SI China Yield Enhancement Seminar, March 17, 2005.
2.4 Apple Inc., 2003 10-K Annual Report, http://media.corporate-ir.net/media_files/irol/10/107357/121903_10K.pdf
2.5 Bloomberg, http://www.bloomberg.com/apps/news?pid=newsarchive&sid=agKtcy7mt0nM
2.6 Gartner, Inc., "WW Semiconductor Forecast: Boom, Bust, or None of the Above?," Bryan Lewis, December 8, 2006.
2.7 Gartner, Inc., "New Features Drive Growth in Consumer Electronics," Jon Erensen, December 8, 2006.
2.8 C. K. Prahalad and S. L. Hart, "The Fortune at the Bottom of the Pyramid," http://www.cs.berkeley.edu/~brewer/ict46/Fortune-BoP.pdf
2.9 B. Abdi, personal communication.
2.10 Bill Adamec, "Xbox 360: Collaboration in Silicon", Chartered Semiconductor Technology Forum, September 2006.
2.11 J. Andrews and N. Baker, "Xbox 360 System Architecture," IEEE Micro, Vol-26, No. 2, March–April 2006, pp. 25–37.
3.1 H. Chang, R. Harjani, "Can the Analog/RF Designer/Enterpreneur Make Money in a Fabless Startup?," IEEE CICC, Panel Discussion, San Jose, CA, Sept. 12, 2006.

4.1 V. Christopher Moezzi, personal communication.
4.2 W. Krenik, D. Buss, P. Rickert, "Cellular Handset Integration – SiP versus SoC," IEEE Journal of Solid State Circuits, Vol. 40, No. 9, September 2005, pp. 1839–1845.
4.3 Z. Li et al., "A Dual-Band CMOS Transceiver for 3G TD-SCDMA," Intl. Solid State Circuits Conference 2007, February 2007, Paper 19.5.
4.4 Ron Collett, "Benchmarking IC Development Capability," FSA Fabless Forum, March 2004.
4.5 Adam Traidman, "Design for Cost," FSA Forum, September 2005, http://www.gsaglobal.org/publications/forum/article.asp?article=0509/traidman
4.6 Naveed Sherwani, "ASICs aren't merely surviving, they're thriving," Electronic Design, September 1, 2005, http://electronicdesign.com/Articles/ArticleID/10934/10934.html
5.1 B. S. Landman and R. L. Russo, "On a Pin Versus Block Relationship For Partitions of Logic Graphs," IEEE Trans. on Computers, Vol. C-20, pp. 1469–1479, 1971.
5.2 P. Christie and D. Stroobandt, "The Interpretation and Application of Rent's Rule," IEEE Trans. on VLSI Systems, Special Issue on System-Level Interconnect Prediction, Vol. 8, No. 6, 639–648, pp. 2000.
5.3 P. Verplaetse, "Refinements of Rent's rule allowing accurate interconnect complexity modeling," 2001 International Symp. on Quality of Electronic Design, March 2001, pp. 251–252.
5.4 R. Kumar, "The Business of Scaling," IEEE Solid State Circuits Newsletter, Vol. 12, No. 1, Winter 2007, pp. 22–27.
5.5 Mark Bohr, "The New Era of Scaling for Energy Efficient Processors," Microprocessor Forum, May 2007, http://download.intel.com/technology/silicon/newera.pdf
5.6 P. Gelsinger, P. Gargini, G. Parker, A. Yu, "2001: A Microprocessor Odyssey," pubished in "Technology 2001", MIT Press, July 1992, pp. 295–113.
5.7 Mark Bohr et al., "130nm Logic Technology Featuring 60nm Transistors, Low-K Dielectrics, and Cu Interconnects," Intel Technology Journal, Vol. 6, No. 2, May, 16, 2002, http://download.intel.com/technology/itj/2002/volume06issue02/art01_130nmlogic/vol6iss2_art01.pdf
5.8 P. Gelsinger, "Moore's Law – The Genius Lives On," IEEE SSCS Newsletter, September 2006.
5.9 TSMC Technology Symposium, April 2007, p. 79.
5.10 H. Stork, "Structuring Process and Design for Future Communication Devices," DAC Keynote, July, 2006.
5.11 M. Racanelli and P. Kempf, "SiGe BiCMOS Technology for Communication Products," IEEE CICC 2003, http://www.jazzsemi.com/docs/CICC_2003_presentation.pdf
5.12 Foundry Technology Symposia, TSMC 2000–2007, UMC 2001–2002, CSM 2003–2006.
5.13 Anand Srinivasan, personal communication.
5.14 T. C. Chen, "Where CMOS is going: Trendy Hype vs. Real Technology," ISSCC 2006, Paper 1.1.

5.15 E. J. Nowak, "Maintaining the benefits of CMOS scaling when scaling bogs down," IBM Journal of R&D, Vol. 46, No. 2/3, March/May 2002.
5.16 W. Haensch, E. Nowak, R. H. Dennard et al., "Silicon CMOS Devices beyond scaling," IBM J. Res. and Dev., Vol. 50, April/May 2006.
5.17 S. Chou, "Integration and Innovation in the Nanoelectronics Era," ISSCC 2005, Paper 1.3.
5.18 S. Borkar, "Exponential Challenges, Exponential Rewards – The Future of Moore's Law," http://www.nanohub.org/resource_files/2005/11/00299/2004.11.01-borkar.pdf
5.19 C. Poirier et al., "Power and Temperature Control on a 90nm Itanium Family Processor," ISSCC 2005, Paper 16.7, pp. 304–305
5.20 A. deGeus, http://www.synopsys.com/news/pubs/compiler/2006/artlead_3ps-may06.html?NLC-insight&Link=May06_Issue_Art1
5.21 HCI Effects, http://www.dfrsolutions.com/page.asp?id=115&mstr=4
5.22 Reliability Simulation in IC Design, Cadence White Paper, http://www.cadence.com/whitepapers/5082_ReliabilitySim_FNL_WP.pdf
5.23 NBTI Effects in 90 nm PMOS, Xilinx White Paper, November 2005, http://www.xilinx.com/bvdocs/whitepapers/wp224.pdf
5.24 Charlie Kahle, personal communication.
5.25 Merrill Hunt, personal communication.
5.26 B. Chown, "System-level validation increases design productivity and saves errors," http://www.mentor.com/products/es/techpubs/mentorpaper_30123.cfm
5.27 C. Rowen, "Engineering the Complex SOC," Prentice Hall PTR, 2004.
5.28 K. Kundert, H. Chang, "Validating Specifications and keeping them Synchronized," http://www.designers-guide.com/newsletters/newsletter0708. html#specs
5.29 V. Berman, "IP: Service of Product," FSA Forum, Vol. 14, No. 3, September 2007, pp. 28–29.
5.30 N. S. Nandra, "High Performance Connectivity IP: Avoiding Pitfalls When Selecting An IP Vendor," Synopsys White Paper, January 2007, http://www.synopsys.com/products/designware/pdfs/high_perf_conn_ip_wp.pdf
5.31 M. Bohr, "Silicon Technology Leadership and the New Scaling Paradigm," http://download.intel.com/technology/silicon/techleadership.pdf
5.32 E. Gerritsen et al., "Evolution of materials technology for stacked-capacitors in 65 nm embedded-DRAM," Solid State Electronics, Vol. 49 2005 pp. 1767–1775, http://www.cambridgenanotech.com/papers/Evolution%20of%20materials%20technology%20for%20stacked-capacitors.pdf
5.33 G. Imthurn, "The History of Silicon on Sapphire," http://www.psemi.com/articles/History_SOS_73-0020-02.pdf
5.34 G. G. Shahidi, "SOI Technology for the GHz era," IBM J. R&D, Vol. 46, Nos 2/3, March/May 2002, pp. 121–131.
5.35 C. Raynaud, "SOI for Low Power," LETI 8th Annual Review, May 2006, http://www.minatec-crossroads.com/pdf-AR/Raynaud.pdf

5.36 P. Fazan, "Eight SOI Advantages every designer should exploit," Electronic Design Online article 14601, January 2007, http://www.elecdesign.com/Articles/Index.cfm?ArticleID=14601
5.37 T. H. Ning, "Why BiCMOS and SOI BiCMOS." IBM J. R&D, Vol. 46, Nos 2/3, March/May 2002, pp. 181–186.
5.38 Willy M. C. Sansen,"Analog Design Essentials," Springer, 2006, p. 2.
5.39 B. P. Wong et al., http://media.wiley.com/product_data/excerpt/07/04714661/0471466107.pdf
5.40 R. Chau, "Application of High-K Gate Dielectrics and Metal Gate Electrodes to enable Silicon and Non-Silicon Logic Technology," J. of Microelectronic Engineering, Vol. 80, June 2005, pp. 1–6, http://www.intel.com/technology/silicon/INFOS_2005_Micro_Engg.pdf
5.41 C. Chian and J. Kawa, "Design for Manufacturability and Yield for Nano-Scale CMOS," Springer, 2007.
5.42 T. Quan, "Introducing TSMC Active Accuracy Assurance Initiative and Reference Flow 8.0," TSMC Press Presentation, June 4, 2007.
5.43 T. Gregorich, personal communication.
5.44 P. A. Totta and R. P. Sopher, "SLT Device Metallurgy and its Monolithic Extension," IBM J. Res. Development, Vol. 5, May 1969, pp. 226–238.
5.45 L. S. Goldman and P. A. Totta, "Area Array Solder Interconnections for VLSI," Solid State Technology, June 1983.
5.46 W. Krenik, "Case Study of Single-Chip Integration for Wireless," IEEE ISSCC 2006, SE5-4, p. 151.
6.1 R. Kumar, B. Henderson, "Glue-Ware" Essential for Streamlined SOC Execution at Emerging Fabless IC Companies," http://www.fsa.org/resources/whitepapers/tcx_paper_ops_dilemma_Dec_02.pdf
6.2 B. Henderson, R. Kumar, "Addressing the Operations Dilemma at Emerging Fabless Companies," *FSA Fabless Forum*, March 2003.
6.3 H. Chang, K. Kundert, "Verification of Complex Analog and RF IC Designs," *Proc. IEEE*, Vol. 95, No. 3, March 2007, pp. 622–639.
6.4 H. Chang and K. Kundert, www.designers-guide.com/documents.html
6.5 J. Remmers et al., "Hierarchical DFT with Enhancements for AC Scan, Test Scheduling and On-chip Compression – A Case Study," International Test Conference, 2005, Paper 29.3, http://www.mentor.com/products/dft/upload/1ITC2005_Hierarchical_DFT_29_3.pdf
6.6 A. Simon, R. Vogel, "COT Planning Guide," Simon publications, 2002.
6.7 Craig Bousquet, personal communication.
6.8 R. Kumar, "Becoming a Best-in-Class Fabless Company – an Overview of Operations Practices for Emerging IC Suppliers," FSA Forum, December 2005.
6.9 R. Kumar, "Operations Practices for Emerging Fabless Companies," http://www.fsa.org/resources/bestpractices/kumar.pdf
7.1 R. C. Leachman, "Yield Modeling," http://www.ieor.berkeley.edu/~ieor130/yield_models.pdf
7.2 M. Sydow, "Compare Logic-Array To ASIC-Chip Cost per Good Die," *Chip Design Magazine*, February/March 2006.
7.3 FSA, Wafer Fab, Assembly and Test Pricing Report, Q4 2006, http://fsa.org/publications/pricing/index.asp

7.4 C. Chian and J. Kawa, "Design for Manufacturability and Yield for Nano-Scale CMOS," Springer, 2007.
7.5 J.H. Huang, "Fabless challenges and opportunities," FSA Taiwan Suppliers Expo, November 2005, http://fsa.org/resources/presentations/05expotaiwan_huang.pdf
7.6 Morris Chang, "Foundry future – challenges in the 21st century," ISSCC 2007, February 2007, paper 1.1.
7.7 E. Clarke, "FPGAs and Structured ASICs: Low-Risk SoC for the Masses," http://www.us.design-reuse.com/articles/article13080.html
8.1 Intel Quality System Handbook, http://www.intel.com/design/quality/isyh.htm
8.2 Infineon Quality management System, http://www.infineon.com/upload/Document/cmc_upload/documents/083/336/QualityManagement.pdf
8.3 Xilinx Quality Manual, http://www.xilinx.com/products/quality/Quality Manual.pdf
8.4 Analog Devices Reliability Handbook, http://www.analog.com/UploadedFiles/Reliability_Handbook__/519895744307104648156533575258240637 5reliability_handbook.pdf
8.5 S. R. Covey, "Seven Habits of Highly Effective People," Simon and Schuster, 1990.
8.6 D. Abercrombie, K. Chow, M. Basel, "Reliability concerns to the designer," *IEE Electronics Systems and Software*, October/November 2005, pp. 16–21, http://www.mathysco.com/images/portfolios/devlin/IEE-ES&S%20Article.pdf
9.1 R. Kumar, "Evolution of Management Practices for ASIC Development and Implementation," IEEE International Engineering Management Conference, September 2006.
9.2 Kurt Stoll, personal communication.
9.3 Synopsys White Paper, Design Cost http://www.synopsys.com/ps06/pdfs/semic_licycle_study_wp.pdf
9.4 R. Collett, "Benchmarking IC Development Capability," White Papers, Parts 1-4, FSA Fabless Forum, March 2004, July 2004, September 2004, December 2004.
10.1 H. K. Lim, "The 2nd Wave of the Digital Consumer Revolution," IEEE ISSCC 2008, Paper 1.1.
10.2 B. Buxton, "Surface and Tangible Computing, and the 'Small' Matter of People and Design", IEEE ISSCC 2008, Paper 1.2.
10.3 S. Bains, "Micromechanical display uses interferometric modulation," Optical engineering report SPIE, no. 199, July 2000.
10.4 K. Wang et. al., "Micro-optical Components for a MEMS Integrated Display," http://www.ee.washington.edu/research/mems/publications/2003/conferences/iwpsd-kwang-03.pdf
10.5 Ray Bingham, personal communication.
10.6 C. M. Christensen, "The Innovator's Dilemma," Collins Books, 1997.
10.7 C. M. Christensen, M. E. Raynor, "The Innovator's Solution," HBS Press, 2003.

10.8 Gartner, Inc., "Gartner's Final Semiconductor Market Share Results," April 2007 Press Release, http://www.gartner.com/it/page.jsp?id=503221

10.9 M. Lapedus, "TI takes two approaches to IC Manufacturing," EE Times, May 13, 2007, http://www.eetimes.com/showArticle.jhtml?articleID=199500883

10.10 M. Lapedus, J.Yoshida, "Crolles2 teeters as NXP quits," EE Times, January 22, 2007, http://www.eetimes.com/showArticle.jhtml?articleID=196902152

10.11 M. Chang, "Foundry Future: Challenges in the 21st Century," IEEE ISSCC 2007, Paper 1.1.

10.12 B. Abdi, "Trends in Mobile Wireless Semiconductors," IMEC ARRM, October 23, 2006.

10.13 Jim Clifford, personal communication.

10.14 G. A. Moore, "*Dealing with Darwin*," Penguin Books, 2005.

Index

A

B

C

G

H

I

J

N

O

P